MW00814939

Remotely Sensed Albedo

Remotely Sensed Albedo

Editors

Jean-Louis Roujean
Shunlin Liang
Tao He

MDPI • Basel • Beijing • Wuhan • Barcelona • Belgrade • Manchester • Tokyo • Cluj • Tianjin

Editors

Jean-Louis Roujean	Shunlin Liang	Tao He
CESBIO	University of Maryland	Wuhan University
France	USA	China

Editorial Office
MDPI
St. Alban-Anlage 66
4052 Basel, Switzerland

This is a reprint of articles from the Special Issue published online in the open access journal *Remote Sensing* (ISSN 2072-4292) (available at: https://www.mdpi.com/journal/remotesensing/special_issues/Remotely_Sensed_Albedo).

For citation purposes, cite each article independently as indicated on the article page online and as indicated below:

LastName, A.A.; LastName, B.B.; LastName, C.C. Article Title. *Journal Name* **Year**, *Volume Number*, Page Range.

ISBN 978-3-03943-941-6 (Hbk)
ISBN 978-3-03943-942-3 (PDF)

Cover image courtesy of pexels.com.

Contents

About the Editors

Jean-Louis Roujean has received a Ph.D. degree in environmental science with specialty remote sensing from the University Paul Sabatier in Toulouse in 1991. His domain of expertise concerns the use of remote-sensing observations for studies of surface–atmosphere interactions involved in weather forecast and climate modeling. He worked on the development of an optical radiation transfer codes for vegetation and led ground experiments to measure the biophysical parameters during international field campaigns (HAPEX-Sahel, 1992; BOREAS, 1994). Emphasis is placed on the Bi-directional Reflectance Distribution Function (BRDF) and scaling issues and the search for a new strategy of model inversion for the retrieval of biophysical parameters. Among the many applications are satellites techniques and time series analysis methods. He coordinated the short-wave radiation branch of Satellite Application Facilities (SAF) on Land Surface Analysis, a program supported by EUMETSAT to exploit data from MSG and EPS sensor systems responsible for the development and the operational implementation of processing algorithms for albedo and down-welling surface radiation. Other domains of interest are land cover mapping, assimilation of surface parameters, data fusion and aerosol retrieval. He was the PI of the Snow Reflectance Transition Experiment (SNORTEX) project aiming to study the characteristics of the angular and spectral signatures of snow-forest in Finnish Lapland based on ground, airborne and satellite measurements. He was also responsible for the development of surface albedo algorithms for the Copernicus Global Land Service program using PROBA-V (a SPOT-VGT follow-on) observations, with an operational implementation at VITO. He was PI of the Agricultural Health Spectrometry (AHSPECT) experiment consisting of acquiring airborne hyper-spectral measurements over crops and forested areas in southwestern France, and is leading the Centre d'Expertise Scientifique (CES) Albedo of the French data center THEIA, which concerns high-resolution products from Sentinel-2 and Landsat. He is president of the national committee on land surface remote-sensing, under the auspices of CNES, the French space agency. For CNES, he is Principal Investigator (PI) of the Indo-French space mission Thermal Infrared Imaging Satellite for High-Resolution Natural Resources Assessment (THRISHNA) devoted to acquiring high-resolution data in optical and thermal domains from 2025.

Shunlin Liang received a Ph.D. degree in remote-sensing and GIS from Boston University, Boston, MA. He was a Postdoctoral Research Associate with Boston University from 1992 to 1993 and a Validation Scientist with the NOAA/NASA Pathfinder AVHRR Land Project from 1993 to 1994. He is currently a professor. His main research interests focus on estimation of land surface variables from satellite observations, studies on surface energy balance, and assessing the climatic, ecological and hydrological impacts of afforestation in China. He has published about 200 peer-reviewed journal papers, and authored the book *Quantitative Remote Sensing of Land Surfaces* (Wiley, 2004), co-athored the book *Global LAnd Surface Satellite (GLASS) Products: Algorithms, Validation and Analysis* (Springer, 2013)), and edited the book *Advances in Land Remote Sensing: System, Modeling, Inversion and Application* (Springer, 2008), and co-edited the books *Advanced Remote Sensing: Terrestrial Information Extraction and Applications* (Academic Press, 2012) and *Land Surface Observation, Modeling and Data Asssimilation* (World Scientific, 2013). Dr. Liang was a co-chairman of the International Society for Photogrammetry and Remote Sensing Commission VII/I Working Group on Fundamental Physics and Modeling, and an Associate Editor of *IEEE Transactions on Geoscience and Remote Sensing* (2001–2013), as well as a guest editor of several remote sensing journals.

Tao He received a B.E. degree in photogrammetry and remote sensing from Wuhan University, Wuhan, China, in 2006, and a Ph.D. degree in geography from the University of Maryland, College Park, MD, USA, in 2012. He is currently a Professor with the School of Remote Sensing and Information Engineering, Wuhan University. His research interests include surface anisotropy and albedo modeling, data fusion of satellite products, and long-term regional and global surface radiation budget analysis.

Preface to "Remotely Sensed Albedo"

A regular and timely monitoring of surface albedo from local to global scales is vital for determining the radiation exchanges in the continuum soil–vegetation–atmosphere in the context of a changing climate. The surface albedo is a quantity of particular interest that has been identified as a primary essential climate variable. An accurate assessment of surface albedo is relevant for vast domains, such as climate, agriculture, hydrology, meteorology, glaciology, urbanism, and geology.

Land surface albedo has become a standard deliverable of most space missions. Remote sensing measurements have been proven to have a high potential to provide valuable information regarding the mapping of land surface albedo at various spatial and temporal scales. The role of radiation forcing versus atmosphere forcing requires a thorough knowledge of the surface albedo.

With this Special Issue, we will compile state-of-the-art research that addresses various aspects of land surface albedo: mapping from patch, landscape to continental scales, impact of directional sampling, surface radiation modelling, spectral albedo conversion, satellite data merging, environmental monitoring, criteria for quality and uncertainty assessment, link with land cover and land use classification, data assimilation, thematic applications, satellite missions, field campaigns, ground observation networks, and validation. Review contributions are welcome, as well as papers describing new measurement concepts/sensors.

Jean-Louis Roujean, Shunlin Liang, Tao He

Editors

Editorial

Editorial for Special Issue: "Remotely Sensed Albedo"

Jean-Louis Roujean [1,*], Shunlin Liang [2,3] and Tao He [3]

1 CESBIO 18 avenue Edouard Belin, 31401 Toulouse, France
2 Department of Geographical Sciences, University of Maryland, College Park, MD 20742, USA
3 School of Remote Sensing and Information Engineering, Wuhan University, Wuhan 430079, China
* Correspondence: jean-louis.roujean@cesbio.cnes.fr

Received: 24 July 2019; Accepted: 4 August 2019; Published: 20 August 2019

Land surface (bare soil, vegetation, and snow) albedo is an essential climate variable that affects the Earth's radiation budget, and therefore, is of vital interest for a broad number of applications: Thematic (urban, cryosphere, land cover, and bare soil), climate (Long Term Data Record), processing technics (gap filling, data merging), and products validation (cal/val). The temporal and spatial patterns of surface albedo variations can be retrieved from satellite observations after a series of processes, including atmospheric correction to surface spectral Bidirectional Reflectance Factor (BRF), and Bidirectional Reflectance Distribution Function (BRDF) modelling.

The processing chain for deriving surface albedo introduces cumulative errors that can affect the accuracy of the retrieved satellite albedo products (MISR, MODIS, VEGETATION, and Proba-V). A new method is proposed to estimate Directional Hemispherical Reflectance (DHR) and Bi-Hemispherical Reflectance (BHR) from measured variables (downwelling, upwelling, and diffuse shortwave radiation) at 19 tower sites from the FLUXNET network, Surface Radiation Budget Network (SURFRAD), and Baseline Surface Radiation Network (BSRN) networks. The pixel-to-pixel comparison between DHR/BHR retrieved from coarse-resolution satellite observations and upscaled from tower sites from 2012 to 2016 emphasizes the parameters involved (land cover type, heterogeneity level, and instantaneous vs. time composite retrievals) [1].

Global warming effects pose a significant change in the albedo of the boreal forest areas as revealed by observed trends in AVHRR satellite albedo magnitude before and after the snow/ice melt season between 40°N and 80°N from 1982 to 2015. Absolute change is 4.4 albedo percentage units per 34 years. The largest changes in pre-melt-season albedo are concentrated in boreal forest, rather than tundra, and are consistent over large areas. The mean of absolute change of start date of the melt season is 11.2 days per 34 years, 10.6 days for end date of the melt season, and 14.8 days for length of the melt season. The albedo intensity preceding the start of the melt season correlates with climatic parameters (air temperature, precipitation, and wind speed) but is primarily affected by the changes in vegetation [2].

Still, at high latitudes, ice albedo feedback affects the global climate based on LTDR of MODIS and VIIRS product routinely disseminated by NOAA. An angular bin regression method acting as gap-filling supports the simulations of a physically-based sea-ice BRDF representing different types and mixing fractions (snow, ice, and seawater). A comparison of six years of ground measurements at 30 automatic weather stations gathered/derived from the Programme for Monitoring of the Greenland Ice Sheet (PROMICE) and the Greenland Climate Network (GC-NET) shows low bias (~0.03) and root mean squared error (RMSE) about 0.07) [3].

Long-term surface albedo datasets are essential for global climate analysis. A method originally developed for MODIS was applied to AVHRR LTDR reflectance to estimate daily surface albedo, which corrects for directional effects using the instantaneous Normalized Difference Vegetation Index (NDVI) and multiyear MODIS BRDF shapes. To reduce the high noise in the red band caused by atmospheric effects, different approaches were analyzed. It was reported that deriving BRDF parameters from 15+

years of observations reduces the average noise by up to 7% in the Near Infrared (NIR) band and 6% in the NDVI, in comparison to using 3-year windows. By successively estimating the volumetric BRDF parameter (V) and geometric BRDF parameter (R), an extra 8% and 9% noise in the red and NIR bands can be further reduced [4].

LTDR of MODIS surface albedo supports decision-makers for climate mitigation in complement with requirements of a time-evolving high spatial resolution albedo that can be estimated by the 30 m Landsat data merged with a time-averaged MODIS BRDF product. Validation over different land covers (cropland, deciduous broadleaf forest, evergreen needleleaf forest, grassland, and evergreen broadleaf forest) using ground measurements provides a root mean squared error (RMSE) of 0.0085–0.0152 [5].

Since surface albedo is known to be related to land cover type and vegetation structure, the question is: How can it be separated from environmental drivers such as temperature and snow cover? A case study for topographically complex regions in Norway was selected to spectrally unmix MODIS albedo in using high resolution observations. The outcomes are improved constraints on land cover-dependent albedo parameterizations for the purpose of climate and hydrological models. Forecasting surface albedo on a monthly basis is possible from the forest structure, snow cover, and near surface air temperature. New insights are offered between the impact of a changing climate on albedo and anthropogenic land use/land cover change (LULCC) [6].

In urban areas, surface albedo determines the heat storage, depending on landscape alteration, air quality, and human activities. The impact of these factors is studied by a partial derivative method, vegetation index data, and night time light data. Quantitative estimates of the contribution from natural climate change and human activities looked at the Jing-Jin-Ji region of China during its highest population growth, between 2001 and 2011. Albedo trends are equal to 0.0065 and 0.0012 per year, before and after urbanization, respectively, meaning that an increase from 15% to 48.4% infers a decrease in albedo of 0.05 [7].

Gridded satellite albedo refers to a large footprint. A total of 1,820 paired high-resolution Landsat TM and MODIS albedo data from five land cover types were used to evaluate the spatial representativeness of the MODIS albedo product based on semivariograms and coefficients of variations. Landsat TM albedo data was aggregated to 450 m–1800 m using two different methods. Comparison with MODIS albedo indicates that, for evergreen broadleaf forests, deciduous broadleaf forests, open shrub lands, woody savannas, and grasslands, the MODIS 500 m daily albedo product represents a spatial scale of approximately 630 m. For mixed forests and croplands, the representative spatial scale is about 690 m [8].

MODIS 500 m albedo was used to derive spectral and broadband bare soil products over the United States using a soil line approach based on red and green spectral signatures. Compared with 30 m Landsat data, MODIS bare soil albedo indicates a bias of 0.003 and an RMSE of 0.036. Soil moisture from the Advanced Microwave Scanning Radiometer–Earth Observing System (AMSR-E) reveals a reduction of bare soil according an exponent law due the darkening effect of moisture. Land cover type is an indicator for determining the magnitude of bare soil albedos, whereas the soil type is an indicator for determining the slope of soil lines over sparsely vegetated areas, as it describes the soil texture, roughness, and composition [9].

The Harmonized Landsat/Sentinel-2 (HLS) project aims to generate a seamless surface reflectance product by combining observations from USGS/NASA Landsat-8 and ESA Sentinel-2 satellites. Observations are associated with invariant viewing geometry, but still yearly illumination variations. BRDF normalization applied to the HLS product at 30 m spatial resolution relies on MODIS BRDF parameters at 1 km spatial resolution. Unsupervised classification of HLS images is used to disaggregate the BRDF parameters to build a BRDF parameters database at HLS scale. Tested over a desert target and an Amazonian forest, the method reduces the coefficient of variation (CV) of the red and near infrared bands by 4% in forest and keeps a low CV of 3% to 4% for the deserts [10].

Landscape albedo can be estimated using images acquired with a consumer-grade camera on board an unmanned aerial vehicle (UAV). Flight experiments conducted at two sites in Connecticut shows that the UAV estimate of visible-band albedo of an urban playground (0.043 ± 0.077) under clear sky conditions agrees reasonably well with the estimate based on the Landsat image (0.052 ± 0.013). Shortwave albedo estimate, as suited for climate applications, would require the deployment of a camera with a near-infrared waveband [11].

UAV can provide small-scale, mobile remote measurements that fill this resolution gap, as shown for a deciduous northern hardwood forest, a spruce plantation, and a cropped willow field. Estimated albedo from concomitant UAV and fixed tower measurements agrees well and UAV measurements captured site-to-site variations in albedo-like surface heterogeneity related to land use. Clearly, UAV measurements are valuable as a useful tool to stratify the landscape albedo in terms of biomass, phenology, foliar chemistry, and canopy water content [12].

Acknowledgments: We thank the authors who contributed to this special issue on "Remotely Sensed Albedo" and to the reviewers who provided the authors with useful/valuable/insightfulhelpful comments and constructive feedback. This study was partially funded by The National Key Research and Development Program of China (NO.2016YFA0600101) and the National Natural Science Foundation of China (NO. 41771379).

Conflicts of Interest: The authors declare no conflict of interest.

References

1. Song, R.; Muller, J.-P.; Kharbouche, S.; Woodgate, W. Intercomparison of surface albedo retrievals from MISR, MODIS, CGLS using tower and upscaled tower measurements. *Remote Sens.* **2019**, *11*, 644. [CrossRef]
2. Anttila, K.; Manninen, T.; Jääskeläinen, E.; Riihelä, A.; Lahtinen, P. The role of climate and land use in the changes in surface albedo prior to snow melt and the timing of melt season of seasonal snow in northern land areas of 40°N–80°N during 1982–2015. *Remote Sens.* **2018**, *10*, 1619. [CrossRef]
3. Peng, J.; Yu, Y.; Yu, P.; Liang, S. The VIIRS Sea-Ice Albedo Product Generation and Preliminary Validation. *Remote Sens.* **2018**, *10*, 1826. [CrossRef]
4. Villaescusa-Nadal, J.L.; Franch, B.; Vermote, E.; Roger, C. Improving the AVHRR Long Term Data Record BRDF correction. *Remote Sens.* **2019**, *11*, 502. [CrossRef]
5. Zhang, G.; Zhou, H.; Wang, C.; Xue, H.; Wang, J.; Wan, H. Time Series High Resolution Land Surface Albedo Estimation Based on Ensemble Kalman Filter Algorithm. *Remote Sens.* **2019**, *11*, 753. [CrossRef]
6. Bright, R.M.; Astrup, R. Combining MODIS and national land resource products to model land cover-dependent surface albedo for Norway. *Remote Sens.* **2019**, *11*, 871. [CrossRef]
7. Tang, R.; Zhao, X.; Zhou, T.; Jiang, B.; Wu, D.; Tang, B. Assessing the impacts of urbanization on albedo in Jing-Jin-Ji Region of China. *Remote Sens.* **2018**, *10*, 1096. [CrossRef]
8. Zhou, H.; Liang, S.; He, T.; Wang, J.; Bo, Y.; Wang, D. Evaluating the Spatial Representativeness of the MODerate Resolution Image Spectroradiometer Albedo Product (MCD43) at AmeriFlux Sites. *Remote Sens.* **2019**, *11*, 547. [CrossRef]
9. He, T.; Gao, F.; Liang, S.; Peng, Y. Mapping climatological bare soil albedos over the contiguous United States using MODIS data. *Remote Sens.* **2019**, *11*, 666. [CrossRef]
10. Franch, B.; Vermote, E.; Skakun, S.; Roger, J.-C.; Masek, J.; Ju, J.; Villaescusa-Nadal, J.L. A New Method for Landsat and Sentinel 2 (HLS) BRDF Normalization and Surface Albedo. *Remote Sens.* **2019**, *11*, 632. [CrossRef]
11. Cao, C.; Lee, X.; Muhlhausen, J.; Bonneau, L.; Xu, J. Measuring Landscape Albedo Using Unmanned Aerial Vehicles. *Remote Sens.* **2018**, *10*, 1812. [CrossRef]
12. Levy, C.; Burakowski, E.; Richardson, A.D. Novel measurements of fine-scale albedo: Using a commercial quadcopter to measure radiation fluxes. *Remote Sens.* **2018**, *10*, 1303.

remote sensing

Article

Combining MODIS and National Land Resource Products to Model Land Cover-Dependent Surface Albedo for Norway

Ryan M. Bright * and **Rasmus Astrup**

Norwegian Institute of Bioeconomy Research, P.O. Box 115, 1431 Ås, Norway; rasmus.astrup@nibio.no
* Correspondence: ryan.bright@nibio.no; Tel.: +47-9747-7997

Received: 25 January 2019; Accepted: 28 March 2019; Published: 10 April 2019

Abstract: Surface albedo is an important physical attribute of the climate system and satellite retrievals are useful for understanding how it varies in time and space. Surface albedo is sensitive to land cover and structure, which can vary considerably within the area comprising the effective spatial resolution of the satellite-based retrieval. This is particularly true for MODIS products and for topographically complex regions, such as Norway, which makes it difficult to separate the environmental drivers (e.g., temperature and snow) from those related to land cover and vegetation structure. In the present study, we employ high resolution datasets of Norwegian land cover and structure to spectrally unmix MODIS surface albedo retrievals (MCD43A3 v6) to study how surface albedo varies with land cover and structure. Such insights are useful for constraining land cover-dependent albedo parameterizations in models employed for regional climate or hydrological research and for developing new empirical models. At the scale of individual land cover types, we found that the monthly surface albedo can be predicted at a high accuracy when given additional information about forest structure, snow cover, and near surface air temperature. Such predictions can provide useful empirical benchmarks for climate model predictions made at the land cover level, which is critical for instilling greater confidence in the albedo-related climate impacts of anthropogenic land use/land cover change (LULCC).

Keywords: spectral unmixing; empirical modeling; linear endmember; forest cover; forest management; forest structure; BRDF/Albedo; NDSI Snow Cover

1. Introduction

In many regions, strategic land use/land management projects that enhance terrestrial carbon sinks or reduce terrestrial carbon emissions are viewed favorably and analogouslyto mitigating climate change. However, it is increasingly understood that it is important and necessary to include other climate regulating services on land in climate impact assessment studies [1,2]. This includes the surface albedo, which is a biogeophysical property that partly determines Earth's shortwave radiation balance [3]. To exclude the surface albedo in the assessments of land-based mitigation can result in the implementation of policies that are suboptimal or even counterproductive [4,5]. Indeed, recent research has consistently demonstrated the need to value the surface albedo alongside carbon in order to maximize mitigation benefits, particularly for forestry projects [6–10].

However, the credibility of such valuations largely rests on the underlying accuracy and spatial-temporal representativeness of the surface albedo data employed in the research. Although satellite remote sensing analyses of surface albedo have been incredibly useful for constraining the surface albedo by land cover type at a regional or global scale [11–13], the land cover classifications underlying such constraints are still insufficiently broad for subregional applications, as evidenced by the large albedo variations observed across both time and space within individual land cover types [11,12,14]. This is particularly true for forests [15], in which the surface albedo is determined as much by vegetation structure [16–18] and functioning [19,20] as it is by local environmental factors such as snow. Large spatial variations in the surface albedo exist for other land cover types, such as croplands and grasslands, which are heavily influenced by local land management practices [21–24].

Compared to global or regional land cover products, national mapping authorities often provide classifications of land cover and structure at a higher spatial resolution and accuracy. Such classifications often combine multiple information sources, including those obtained from optical satellite remote sensing, aerial LiDAR and photogrammetric remote sensing and local expert judgments. For instance, Wickham et al. [25] recently developed a land cover-dependent albedo dataset for the continental United States by combining the National Land Cover Database together with a MODIS climatology of surface albedo. Given that the mitigation policies of the land-based sectors are implemented and monitored nationally, the use of national land resource maps and national land cover classifications can serve to further improve the accuracy of land-cover dependent albedo estimates based on satellite remote sensing. Furthermore, the use of a national land classification makes pragmatic sense both from a management and reporting perspective.

In the present study, we employ observation-based datasets of Norwegian land cover and structure, near surface air temperature, and MODIS-based snow cover (MOD10A1 v6) to spectrally unmix MODIS surface albedo (MCD43A3 v6) and to study spatial-temporal variations in surface albedo as a function of land cover, forest structure, and the environmental state. Our primary objective is to develop and present a set of simple land cover-dependent empirical models for Norway that facilitate high fidelity predictions of the surface albedo at a monthly resolution. This resolution is deemed appropriate as major intra-annual surface albedo dynamics play out over seasonal timescales. Furthermore, the monthly resolution makes the models amenable to inputs obtained from gridded historical climate observation products or from climate model scenario runs whose outputs are often provided at the monthly resolution. Unlike existing global [12] and national [25] land cover-dependent albedo datasets based on MODIS surface albedo products, our method does not require constraining the analysis to pixels that are homogeneous with respect to single land cover types, thus enabling a more efficient use of MODIS data. Given the relatively low nominal spatial resolution of the MODIS albedo product (i.e., 500 $\times$ 500 m), this is particularly important for regions, such as Norway, where the land cover and structure are relatively heterogeneous at small spatial scales. Furthermore, because spatial-temporal variations in the surface albedo not only depend on variations in land cover and structure but on local environmental conditions affecting the state of vegetation, soils, and snow, we include snow cover and near surface air temperature in our analysis since these factors are known to greatly affect the surface albedo either directlyor indirectly [26–29].

Given their conformity to national land cover products and classifications, such models will be useful in the studies seeking to quantify albedo-related impacts connected to national land use activities, or for constraining land cover-dependent albedo parameterizations in models employed in regional climate and hydrological research making use of the national land cover mapping and classification. In addition, such tools can be applied to create a seamless monthly surface albedo dataset that is land-cover dependent, thus providing a means to benchmark climate model predictions of surface albedo made at the scale of individual land cover or plant functional types—a task that is challenging at present.

We start by detailing our method and datasets in Section 2, which is followed by a presentation of results in Section 3 and a discussion of their merits and uncertainties in Sections 4 and 5.

2. Materials and Methods

The general workflow is divided into two parts: (i) model training and (ii) model validation. Both are limited to the southern portion of mainland Norway (Figure 1) in order to include a larger wintertime sample of good quality MODIS snow cover and surface albedo retrievals (described in Sections 2.5 and 2.6) since these have a low frequency at higher latitudes during winter. Furthermore, the study region contains the full range of land cover and climate variation found in Norway (Figure S6 of Supporting Information).

2.1. Study Region

Forests make up the dominant land cover type within the study domain, which covers a region with a total land surface area of approximately 167,500 km^2 (Figure 1, inset). As such, preserving a similar proportion of forest area between the model training and validation regions was the main criterion when partitioning the domain into the training and validation subsets. Most of the forests within the full domain may be considered part of the boreal forest belt that extends almost continuously around the upper northern hemisphere. Forests are dominated by Norway spruce (*Picea abies* H. Karst.), Scots pine (*Pinus sylvestris* L.) and two birch species (*Betula pendula* Roth and *B. pubescens* Ehrh.), with the understory vegetation typically dominated by ericoid dwarf shrubs (*Vaccinium* spp.) and various herb communities [30].

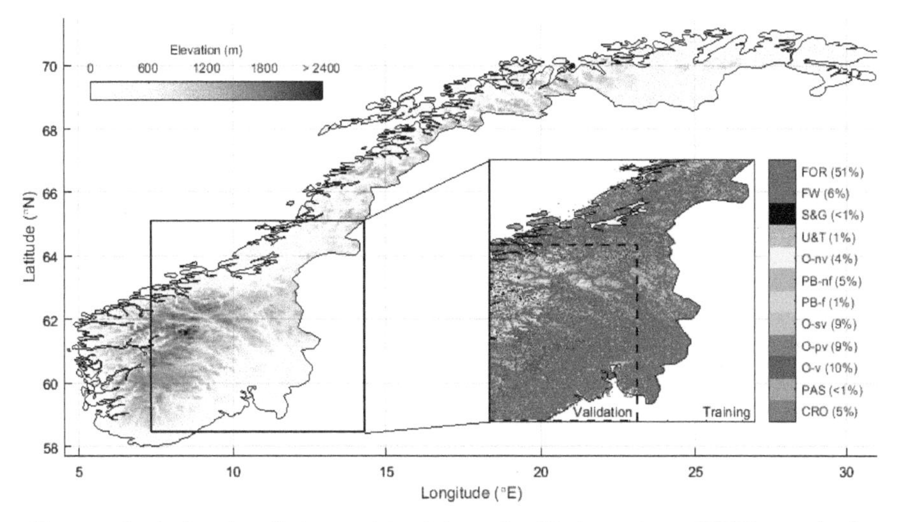

Figure 1. Study domain split into model training and validation regions. "CRO" = croplands; "PAS" = pasture; "O-v" = Open, vegetated; "O-pv" = Open, partly vegetated; "O-sp" = Open, sparsely vegetated; "PB-f" = Peat bog, forested; "PB-nf" = Peat bog, non-forested; "O-nv" = Open, non-vegetated; "U&T" = Urban & transport; "S&G" = Snow & glacier; "FW" = freshwater; and "FOR" = forest.

The eastern parts of the region experience continental climates that are characterized by long cold winters, short mild summers and moderate, seasonally distributed precipitation. Forests in the northwestern coastal regions are more influenced by an oceanic climate, which is characterized by greater amounts of precipitation, warmer temperatures during winter, and cooler temperatures during summer. Snow covers the ground from December through late March/early April in the lowland regions (< 400 m). At higher elevations (> 600 m), permanent snow cover may commence in November and can persist through early May (Norwegian Meteorological Institute, 2013b).

2.2. Spectral Unmixing Regression Analysis

Satellite retrievals of the surface albedo are often provided at a spatial resolution that is too coarse for direct attribution to individual forest stands and other fine-scale features of the landscape. For instance, the nominal spatial footprint of the MODIS albedo product employed in this study (described in Section 2.5) is 25 hectares (250,000 m^2), whereas the footprint of the typical even-aged forest stand in Norway rarely exceeds 1–2 hectares (< 20,000 m^2) [31]. Linear spectral unmixing techniques based on the ordinary least squares regression are increasingly employed to overcome this spatial mismatch challenge (e.g., refs. [32–34]). Unlike conventional spectral unmixing techniques based on linear mixture models [35] in which the endmember spectral signatures are known *a priori* and the goal is to determine endmember fractions within any given pixel, the known endmember fractions in the present study are obtained from the land cover dataset, which are used to estimate their unique spectral signatures (albedos).

Under the premise that the surface albedo (or rather the surface reflectance) signal registered by the satellite spectroradiometer represents a linear combination of the individual albedos (reflectances) of all endmembers (land covers/forest stands) within its footprint, the linear unmixing model may be described [32] as:

$$\alpha + \varepsilon_\alpha = \sum_{i=1}^{n} (ef_i(\alpha_i + \varepsilon_i)) \quad (1)$$

where α is the albedo of the grid cell (described in Section 2.5), ef_i is the fractional coverage of endmember type i within the pixel size (Section 2.3), α_i is the albedo of endmember i, ε_α is the residual error of the pixel, and ε_i is the standard error of the estimator (α_i). In Equation (1), the endmember albedo α_i is essentially the slope that minimizes the sum of squared ε_α.

Endmember albedos α_i are highly sensitive to the presence of snow. Equation (1) is therefore modified following Bright *et al.* [34] where α is described as a weighted combination of the mixed endmember albedos under snow-free and snow-covered conditions, with the weights determined by snow cover:

$$\alpha = SC\sum_{i=1}^{n} ef_i(\alpha_{sc,i}) + [1 - SC]\sum_{i=1}^{n} ef_i\left(\alpha_{sf,i}\right) \quad (2)$$

where α is the albedo of the grid cell (described in Section 2.5), SC is the snow cover of the grid cell (described in Section 2.6), and $\alpha_{sc,i}$ and $\alpha_{sf,i}$ are the albedos for endmember i under snow-covered and snow-free conditions, respectively. This model form is used in some climate models [36] and has been found to perform consistently well over large spatial domains at high latitudes [34,37]. Unlike in Bright et al. [34], however, the model employed here is further modified to capture additional variation in endmember-dependent albedos—$\alpha_{sc,i}$ and $\alpha_{sf,i}$—owed to important local differences in vegetation structure and other environmental factors as described below.

For the vegetated endmembers and in particular forests, both $\alpha_{sc,i}$ and $\alpha_{sf,i}$ are influenced by the structure of the vegetation. In Fennoscandic boreal forests, $\alpha_{sc,i}$ and $\alpha_{sf,i}$ at the stand level have been found to be negatively correlated with canopy cover [38], leaf area index [38,39], aboveground biomass [38,40], volume [41], height [41] and age [27,33]. Because forest canopies are rarely fully buried by snow, ground masking by forest canopies is particularly influential as a control of surface albedo during the snow season [16,18,42], although the snow intercepted and held by forest canopies can be important during the coldest and calmest winter months [43–45].

Following Kuusinen et al. [33], we modeled the forest endmember albedos as functions of stand structure. Although the models of Kuusinen et al. [33] were fit separately for different seasons, it was possible to obtain universal endmember models that were not specific to individual months or seasons by including snow cover as an environmental state predictor. The albedo of an endmember under snow-covered conditions ($\alpha_{sc,i}$) is largely determined by the albedo of snow, which depends on the effective snow grain area and snow water content [46–49]. These are two physical properties that exhibit strong relationships with air temperature [50,51]. Furthermore, given the importance of air temperature as a control over vegetation phenology [19,52,53] and canopy snow dynamics (i.e., snow slippage and melt) [39,44,45,54], we included air temperature as an additional environmental state predictor.

For the forest endmembers, a model function was chosen that gives identical predictions for a zero structure forest—or when the structural predictor (i.e., volume, biomass, age and so on) equals zero. In other words, the forest endmember models have common y-intercepts. For snow-covered conditions, the functional form of the forest endmember model is given as:

$$\alpha_{sc,i}(x_i, T) = (\alpha_{0,sc} + \rho_{0,sc}T) - (\beta_{sc,i} + \rho_{sc,i}T)\left[1 - e^{\lambda_{sc,i}x_i}\right] \quad (3)$$

where T is the air temperature (in °C) of the grid cell (Section 2.7), $\alpha_{0,sc}$ is the y-intercept (albedo) for forests with zero structure and when air temperature equals zero, $\rho_{0,sc}$ is a temperature sensitivity parameter for forests with zero structure, β_i is the difference between $\alpha_{0,sc}$ and the minimum albedo (i.e., the asymptote) for forest endmember i when air temperature equals zero, $\rho_{sc,i}$ is a temperature sensitivity parameter unique to the forest endmember i; $\lambda_{sc,i}$ is a shape parameter unique to the forest endmember i; and x_i is the grid cell mean stand structural attribute for the forest endmember i (Section 2.4). Here, separate fits are performed using either the mean stand volume (x = *Volume*; $m^3\ ha^{-1}$) or mean stand aboveground biomass (x = *Biomass*; $t\ ha^{-1}$) as structural predictors.

The same function represented as Equation (3) was applied to estimate the forest endmember albedos under snow-free conditions:

$$\alpha_{sf,i}(x_i, T) = \left(\alpha_{0,sf} + \rho_{0,sf}T\right) - \left(\beta_{sf,i} + \rho_{sf,i}T\right)\left[1 - e^{\lambda_{sf,i}x_i}\right] \quad (4)$$

where the subscript "*sf*" denotes snow-free conditions.

In Equations (3) and (4), the zero-structure y-intercept parameter α_0 shared by all forest endmembers was based on the mean of the forest endmember-specific zero-structure y-intercept parameters obtained from an initial regression analysis (i.e., $\alpha_0 = \sum_{i=1}^{n} \alpha_{0,i}/n$).

Air temperature modifiers were also included for the non-forested endmembers given its importance as a control over seasonal phenology (i.e., vegetation dynamics) and snow physical processes (i.e., snow albedo):

$$\begin{aligned} \alpha_{sc,i}(T) &= \alpha_{0,sc,i} + \rho_{sc,i}T \\ \alpha_{sf,i}(T) &= \alpha_{0,sf,i} + \rho_{sf,i}T \end{aligned} \quad (5)$$

where the intercepts $\alpha_{0,sc,i}$ and $\alpha_{0,sf,i}$ are the surface albedos for endmember i under snow-covered and snow-free conditions when air temperature equals zero, respectively, while $\rho_{sc,i}$ and $\rho_{sf,i}$ are air temperature modifiers for endmember i (i.e., slopes of the albedo–air temperature relationship).

Separate models were fit for both intrinsic albedos (black-sky/directional hemispherical and white-sky/bidirectional hemispherical) at the local solar noon and for each broad spectral band: visible (VIS; 300–700 nm), near infrared (NIR; 700–5000 nm) and the entire shortwave broadband (SW; 300–5000 nm).

2.3. Land Cover Data

Endmember (land cover) mapping employed in model fitting was based on the high resolution land resource database "AR5", which employs a standardized Norwegian land cover classification system [55]. Although AR5 is a national seamless database, detailed information about land resources is only available for areas below the treeline. As a result, a recent mapping campaign for areas above the treeline was recently undertaken, resulting in a complimentary database "AR-Fjell" [56]. Furthermore, an additional campaign to improve the mapping of agricultural resources was more recently carried out, resulting in the database "FJB-AR5" [57]. These products, which were all produced at a scale of 1:5000, were merged and re-projected to a MODIS sinusoidal grid with a resolution of 16 m × 16 m.

2.4. Forest Cover and Structure Data

The forest species composition and structure employed in model fitting were based on the forest resource map "SR16", which was developed using photogrammetric and LiDAR point cloud data with ground plots from the Norwegian National Forest Inventory (NFI) [58]. The forest cover in SR16 is based on updating the forest cover in AR5 with object-based image analysis methods [59].

Stand attributes, such as tree species, tree height (Lorey's), biomass, and volume are predicted with generalized linear models at a resolution of 16 m × 16 m with an accuracy (normalized RMSE) of 50% [58]. The forests in SR16 are classified as one of four classes: (1) newly clear-cut; (2) spruce; (3) pine; and (4) deciduous broadleaf. Pixels that were classified as newly clear-cut comprised ~1% of all forested pixels in the training dataset and were thus re-classified as spruce since spruce was the most abundant tree species of the model training region.

After fitting the models (Section 2.2), they were applied to reconstruct the monthly mean surface albedos in the validation region (Figure 1). Because the SR16 product used in the fitting did not yet extend to this region of Norway, an alternative product "SAT-SKOG" [60] was instead used in order to provide information about the forest cover and structure. The SAT-SKOG product is based on a *k*-nearest neighbor algorithm, which combines the spectral information from Landsat 5 & 7 with information from permanent NFI plots [60,61]. The structural variables in the SAT-SKOG product are limited to stand volume and height; thus, only the forest models fit using the stand volume as the structural predictor (i.e., x_i) were applied in the validation exercise. The forest area in AR5 and SAT-SKOG are identical so only the species information in SAT-SKOG was used to update the AR5 classification for forests. The SAT-SKOG forest classification differs from SR16 in that it includes a "mixed evergreen needleleaf" and a "mixed needleleaf-broadleaf" classification. For the former, we weighted the pine and spruce endmember predictions evenly while for the latter, we weighted predictions for all three species (spruce, pine, birch) evenly.

2.5. Albedo Data

The albedo data employed in fitting and validation were based on the MODIS Bidirectional Reflectance Distribution Function (BRDF) and Albedo Product (MCD43) algorithm utilizing directional surface reflectance values [62,63] from both the Aqua and Terra satellites [64,65] and a semi-empirical kernel-driven BRDF model (Ross-thick, Li-sparse Reciprocal) [66]. Specifically, we employed the latest version (v6) of the MCD43A3 BRDF/Albedo product with a nominal spatial resolution of 500 m in a sinusoidal projection (tiles v2 h18 & v3 h18), which is now provided at a daily resolution [67]. Although they are now provided at a daily resolution, the v6 MCD43A3 product retains the 16-day observation window where the observations closest to the composite date (9th day) are given a greater weight in determining the appropriate reflectance anisotropy model (i.e., BRDF) [68]. The accuracy of the v6 product is generally higher for snow-free and full inversion retrievals, with RMSEs < 0.02 for most land cover types. For snow-covered conditions, RMSEs are typically < 0.05 for most land cover types [69]. Furthermore, the v6 product obtains more retrievals than the v5 product at higher latitudes from the use of all available observations (as opposed to four observations per day in v5) [69].

The albedo data were downloaded for a temporal extent spanning 1 January 2006–31 December 2010. This temporal extent was chosen because it falls in between the earliest satellite imagery underlying the original AR5 product (2002) and the latest aerial remote sensing data used to produce the SR16 product (2014). The quality flags of the MCD43A2 [70] companion product were used to filter and discard all non-full BRDF inversions, which includes those with solar zenith angles greater than 70°. Data were then averaged into interannual monthly means, and composite dates and locations for the good quality data were stored for subsequent temporal synchronization with the snow cover and temperature data (next sections).

2.6. Snow Cover Data

The snow cover data for the same temporal extent as the albedo data were based on the latest version (v6) of the MOD10A1 Snow Cover product [71], which is also provided at a daily resolution and has a nominal spatial resolution of 500 m in a sinusoidal projection. Unlike the previous product (v5), only the Normalized Difference Snow Index (NDSI) Snow Cover [72] is provided. The NDSI Snow Cover represents the visible fraction of a grid cell covered in snow and rarely exceeds 0.75 in forests. The detailed descriptions of the v6 product, including other important changes from v5, may be found in refs. [73,74]. Recent evaluations suggest notable improvements in the v6 product over v5, which is largely due to refinements in the snow detection algorithm [75,76]. Evaluations at three locations after the conversion of NDSI Snow Cover to Fractional Snow Cover using linear regressions with global parameters suggest an accuracy (RMSE) of 0.2–0.35 [75]. Prior to aggregating these to multi-year monthly means, the temporal signature of the snow cover data was synchronized to match that of the albedo dataset. In other words, the quality flags of the albedo dataset were applied to filter and discard snow cover retrievals for dates that did not correspond to the dates of the retained good quality albedo retrievals. NDSI snow cover (henceforth referred to as simply snow cover; *SC*) values outside the 0–100 range were also discarded prior to the application of the albedo quality filter.

2.7. Temperature Data

The monthly mean temperatures from 2006–2010 were based on 1 × 1 km grids of daily (24-hr) mean air temperatures (2 m) created from three-dimensional spatial interpolation of air temperatures observed at meteorological stations distributed throughout Norway [77]. The accuracy of the gridded daily temperature product (SeNorge 2.0) is ~1 °C based on the "leave-one-out" cross validation score [77]. Temperature data were re-projected and downscaled to the nominal resolution of the MODIS products using a nearest neighbor interpolation method. The same temporal synchronization procedure that was applied to the snow cover data was applied to the temperature data prior to monthly aggregation.

Figure 2 provides an overview of the climate predictor space of the model fitting domain associated with the post-processed dependent variable (albedo) dataset.

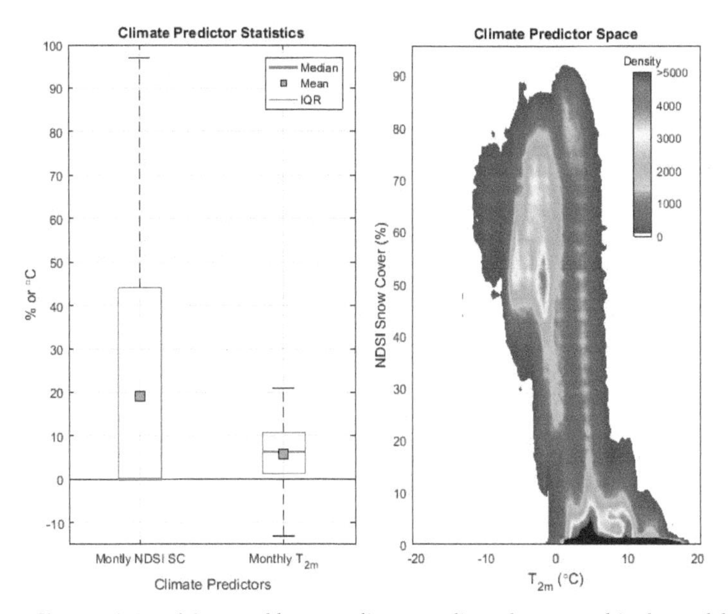

Figure 2. Characteristics of the monthly mean climate predictor dataset used in the model fitting exercise: (**Left**) Statistics; (**Right**) Density scatter. "IQR" = interquartile range; "NDSI" = Normalized Difference Snow Index; and "SC" = Snow cover.

Monthly mean air temperatures rarely fell below −10 °C, with a median close to the mean of ~6 °C (Figure 2, left panel). Over 75% of the monthly mean snow cover (*SC*) retrievals had values less than 44%, with a median of 0% and mean of ~19%. It is important to note that the highest *SC* values were not necessarily found in the coldest months but in months where the mean *T* was around 2–4 °C. These temperatures are characteristic of late winter/early spring when the snowpack is near its deepest (i.e., when a larger amount of short-statured vegetation or other landscape features are buried in snow) [78].

2.8. Endmember Data Processing

Prior to calculating the endmember fractions (ef_i) corresponding to each MODIS product grid cell, the original forest area of the AR5 land cover product was replaced by the updated area of the SR16 product. It is becoming increasingly understood that the effective spatial resolution of the MCD43A BRDF/Albedo product differs from its nominal resolution of ~500 m [15,68,79]. Recently, Campagnolo et al. [80] applied point spread functions to quantify the effective spatial resolution of various MODIS and other optical satellite remote sensing products. For the v6 MCD43A BRDF/Albedo product, they reported a median effective resolution of 833 m along the east–west transect and 618 m along the north–south transect.

Endmember fractions were computed at both the nominal (500 m × 500 m) and the effective (618 m × 833 m) resolutions before model fitting was executed for both resolutions. Similar to Hovi et al. [81], we apply an elliptical point spread function modeled with a Gaussian distribution, which was defined such that 75% of the signal was assumed to originate from within an ellipse having a diameter of 618 m in the Y (north–south) and 833 m in the X (east–west). Forest structural predictors (Section 2.4) were computed as the weighted means within each pixel using the point spread function as weights, which is illustrated conceptually in Figure S5 of the Supporting Information.

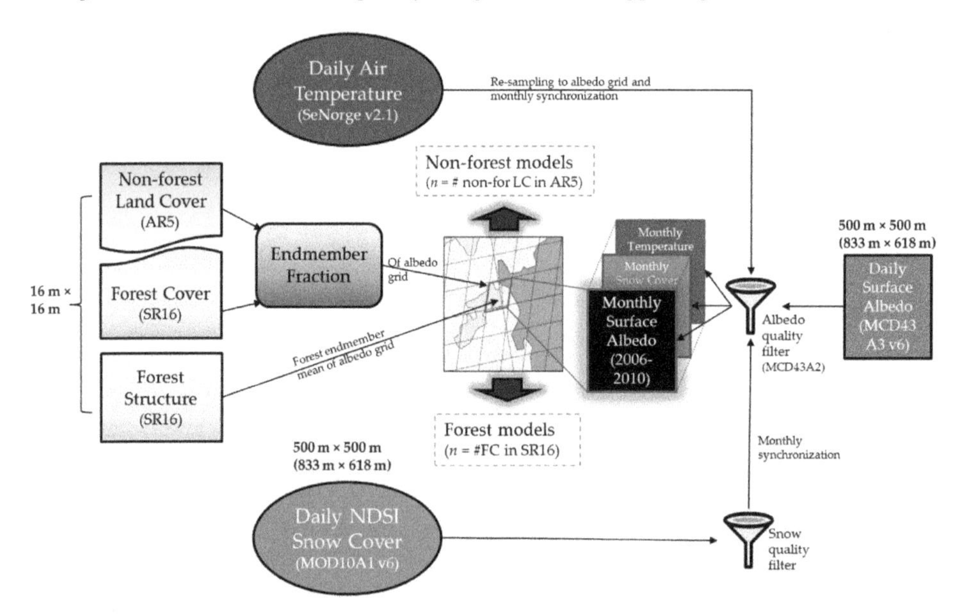

Figure 3. Workflow schematic of the linear unmixing model fitting procedure.

The general workflow of the entire linear unmixing (model fitting) procedure is illustrated in Figure 3. Figure S1 of the Supporting Information provides an overview of the distribution of the quality-filtered and temporally-synchronized observational records employed in model fitting and validation by land cover (endmember) type and month.

2.9. Model Validation

Pixel errors (E) for all broad bands and for both intrinsic albedos were calculated as the difference between the predicted and MCD43A albedo retrievals (both at the local solar noon):

$$E_{p,m} = \overline{\alpha}_{p,m} - \alpha_{p,m} \quad (6)$$

where $\overline{\alpha}_{p,m}$ is the predicted and $\alpha_{p,m}$ is the MCD43A albedo retrieval for month m and pixel p. Because our main interest was to assess the prediction performance of the individual endmember (land cover type) models, we limited the validation sample to a subset of pixels containing homogeneous land cover—defined here as those whose total area contained ≥95% of a single endmember type.

3. Results

3.1. Fit Statistics

The relevance of including both air temperature and forest structure as predictor variables can be appreciated when looking at the coefficients of determination (R^2), which are presented in Table 1. Although snow cover (SC) alone explained 55–81% of the variance (depending on model complexity and albedo broad band), the addition of both air temperature and forest structure (volume or biomass) further increased the percentage of variance explained to 85–88%.

Table 1. Coefficients of determination (R^2) for all candidate models. Number of observations = 4,524,377. "VIS" = Visible broadband (0.3–0.7 μm); "NIR" = Near-infrared broadband (0.7–5 μm); "SW" = Shortwave broadband (0.3–5 μm); "BS" = Black-sky (directional hemispherical); "WS" = White-sky (bidirectional hemispherical); "ef" = Endmember fraction; "SC" = NDSI snow cover fraction; "T" = Air temperature (2 m; °C); "V" = Stand volume (m^3 ha^{-1}); and "B" = Stand aboveground biomass (t ha^{-1}).

Model	R^2					
	SW		NIR		VIS	
	BS	WS	BS	WS	BS	WS
	Nominal resolution					
ef + SC	0.756	0.747	0.593	0.543	0.805	0.804
ef + SC + T	0.791	0.780	0.642	0.598	0.832	0.830
ef +SC +T + V	0.845	0.833	0.707	0.668	0.880	0.876
ef +SC +T + B	0.845	0.833	0.707	0.668	0.880	0.876
	Effective resolution					
ef + SC	0.761	0.752	0.601	0.553	0.808	0.808
ef + SC + T	0.796	0.786	0.650	0.608	0.836	0.834
ef +SC +T + V	0.851	0.839	0.715	0.679	0.884	0.881
ef +SC +T + B	0.851	0.839	0.715	0.679	0.885	0.881

In general, for any given model permutation, the fits for the VIS broad band explained more of the variance than the fits for SW and NIR broad bands. Furthermore, the fits for black-sky albedo explained more of the variance than the fits for the white-sky albedo (Table 1). The models fit at the effective spatial resolution did not lead to as large R^2 improvements as those reported elsewhere [81], although it should be noted that the study of ref. [81] was restricted to forests where there is a much larger variation in vegetation structural controls of the surface albedo.

3.2. Model Parameters

Only the results for the black-sky albedos are presented henceforth. For other results, please refer to the Supporting Information. Starting with non-forest land cover types, the regression parameters for the three broadband albedos (black-sky) at the local solar noon are presented in Table 2. The zero degree albedos (all bands) under snow-covered conditions ($\alpha_{0,sc,i}$) were generally highest for the "Open" and "Peat bog" land cover types.

Table 2. Black-sky albedo model parameters for the non-forested land cover types. "CRO" = croplands; "PAS" = pasture; "O-v" = Open, vegetated; "O-pv" = Open, partly vegetated; "O-sp" = Open, sparsely vegetated; "PB-f" = Peat bog, forested; "PB-nf" = Peat bog, non-forested; "O-nv" = Open, non-vegetated; "U&T" = Urban & transport; "FW" = freshwater; "FOR" = forest. "$\alpha_{0,sc}$" = local noon albedo under snow-covered conditions when air temperature (T) equals 0 °C; "$\alpha_{0,sf}$" = local noon albedo under snow-free conditions when air temperature (T) equals 0 °C; "ρ_{sc}" = air temperature sensitivity parameter for snow-covered conditions; "ρ_{sf}" = air temperature sensitivity parameter for snow-free conditions.

	CRO	PAS	O-v	O-pv	O-sv	O-nv	PB-f	PB-nf	U&T	FW
					Shortwave (SW)					
$\alpha_{0,sc}$	0.570	0.562	0.692	0.643	0.591	0.594	0.755	0.679	0.483	0.562
$\alpha_{0,sf}$	0.126	0.142	0.178	0.142	0.144	0.149	0.198	0.148	0.112	0.059
ρ_{sc}	−0.045	−0.040	−0.027	−0.037	−0.030	−0.012	−0.054	−0.041	−0.033	−0.054
ρ_{sf}	0.002	7.5×10^{-4}	−0.003	−0.002	−0.003	−0.003	−0.004	-6.4×10^{-11}	9.8×10^{-4}	0.001
					Near-infrared (NIR)					
$\alpha_{0,sc}$	0.492	0.457	0.525	0.503	0.470	0.414	0.574	0.523	0.430	0.440
$\alpha_{0,sf}$	0.183	0.241	0.229	0.186	0.177	0.180	0.240	0.224	0.151	0.102
ρ_{sc}	−0.036	−0.027	−0.021	−0.029	−0.022	−0.011	−0.044	−0.033	−0.029	−0.048
ρ_{sf}	0.006	0.001	5.4×10^{-4}	0.001	-1.8×10^{-6}	−0.001	0.001	7.3×10^{-4}	0.003	6.9×10^{-4}
					Visible (VIS)					
$\alpha_{0,sc}$	0.666	0.688	0.855	0.780	0.714	0.753	0.939	0.835	0.564	0.687
$\alpha_{0,sf}$	0.058	0.028	0.104	0.080	0.093	0.134	0.114	0.056	0.066	0.018
ρ_{sc}	−0.057	−0.051	−0.032	−0.049	−0.042	−0.012	−0.067	−0.049	−0.039	−0.063
ρ_{sf}	-4.6×10^{-4}	9.4×10^{-4}	−0.005	−0.003	−0.004	−0.007	−0.006	−0.001	−0.001	1.4×10^{-4}

For shortwave (SW), zero degree albedos under snow-free conditions ($\alpha_{0,sf,i}$) were lowest for the two non-vegetated categories ("U&T" & "FW"). Snow-free albedos for the near-infrared (NIR) broad band tended to increase with increasing vegetation cover. For instance, NIR $\alpha_{0,sf,i}$ was higher for forested peat bogs ("PB-v") than for non-forested peat bogs ("PB-nf"), and higher for vegetated open ("O-v") than for partly or sparsely-vegetated open land cover types ("O-pv", "O-sv"). The influence of vegetation on $\alpha_{0,sf,i}$ was less clear for the visible (VIS) broad band. For non-vegetated open areas ("O-nv") which typically reside at the highest altitudes, $\alpha_{0,sf,i}$ was higher than that of fully vegetated open areas ("O-v") and of vegetated peat bogs ("PB-f").

For all land cover types (endmembers) and for all broad band albedos, the parameter relating the endmember albedo under snow-covered conditions to the monthly air temperature (i.e., $\rho_{sc,i}$) was negative, which is consistent with numerous observations elsewhere [47,50,51,82,83]. For the croplands and pastures occupying lowland regions, $\rho_{sf,i}$ for the SW albedo was positive and appeared to be driven by the positive $\rho_{sf,i}$ for the NIR albedo. For "Open" and "Peat bog" cover types, $\rho_{sf,i}$ for SW was negative and appeared to be driven by a negative $\rho_{sf,i}$ in the VIS band.

For forested endmembers, the parameters for the models that were fit with the stand volume as the structural predictor (x_i in Equations (3) and (4)) are presented in Table 3. Under snow-covered conditions, α_0 was highest for VIS and lowest for NIR broad bands while the opposite was true for snow-free conditions. The magnitudes of the zero structure forest temperature sensitivity parameter ρ_0 follow the same pattern. Under snow-covered conditions, the difference in the albedo between a zero structure and a fully developed forest when air temperature is zero (i.e., β_{sc}) was largest in the VIS band and smallest in the NIR band, whereas the opposite was true for snow-free conditions (β_{sf}). Under snow-covered conditions, the difference in the albedo between a zero structure and a fully developed forest decreased with decreasing air temperature, which provided positive values for ρ_{sc} for all forest endmembers and albedo bands. The lack of foliage during the months with snow at the surface likely explains why the temperature sensitivity parameter ρ_{sc} was larger (in magnitude) for the deciduous endmember ("DBF") than for pine and spruce.

Table 3. Black-sky albedo model parameters fit for forests with stand *volume* as the structural predictor. "DBF" = Deciduous broadleaf forest (*Betula* spp.); "$\alpha_{0,sc}$" = local noon albedo of forests with zero volume under snow-covered conditions when air temperature (T) equals 0 °C; "β_{sc}" = difference between $\alpha_{0,sc}$ and the asymptotic albedo when T equals 0 °C; "$\alpha_{0,sf}$" = albedo (local noon) of forests with zero volume under snow-free conditions when T equals 0 °C; "β_{sf}" = difference between $\alpha_{0,sf}$ and the asymptotic albedo when T equals 0 °C; "λ_{sc}" = an extinction coefficient for snow-covered conditions; "λ_{sf}" = an extinction coefficient for snow-free conditions; "ρ_{sc}" = air temperature sensitivity parameter for snow-covered conditions; "ρ_{sf}" = air temperature sensitivity parameter for snow-free conditions; "$\rho_{0,sc}$" = air temperature sensitivity parameter for snow-covered conditions for zero volume forests; "$\rho_{0,sf}$" = air temperature sensitivity parameter for snow-free conditions for zero volume forests.

			β_{sc}	ρ_{sc}	λ_{sc}	β_{sf}	ρ_{sf}	λ_{sf}
				Shortwave (SW)				
$\alpha_{0,sc}$	0.610	Spruce	0.340	1.2×10^{-3}	−0.025	0.068	-2.5×10^{-4}	−0.023
$\rho_{0,sc}$	−0.020	Pine	0.262	2.5×10^{-3}	−0.022	0.061	-4.4×10^{-4}	−0.025
$\alpha_{0,sf}$	0.151	DBF	0.212	3.0×10^{-3}	−0.007	0.041	6.6×10^{-4}	−0.004
$\rho_{0,sf}$	1.0×10^{-3}							
				Near-infrared (NIR)				
$\alpha_{0,sc}$	0.447	Spruce	0.214	1.7×10^{-3}	−0.023	0.097	-1.1×10^{-4}	−0.021
$\rho_{0,sc}$	−0.014	Pine	0.146	2.1×10^{-3}	−0.021	0.082	-2.6×10^{-4}	−0.019
$\alpha_{0,sf}$	0.242	DBF	0.132	2.7×10^{-3}	−0.004	0.073	-2.2×10^{-4}	−0.002
$\rho_{0,sf}$	1.8×10^{-3}							
				Visible (VIS)				
$\alpha_{0,sc}$	0.784	Spruce	0.470	2.5×10^{-3}	−0.028	0.024	-7.6×10^{-5}	−0.026
$\rho_{0,sc}$	−0.027	Pine	0.391	3.2×10^{-3}	−0.025	0.021	-1.3×10^{-4}	−0.024
$\alpha_{0,sf}$	0.042	DBF	0.309	3.5×10^{-3}	−0.008	0.004	1.1×10^{-3}	−0.007
$\rho_{0,sf}$	7.0×10^{-4}							

As for the snow-free temperature sensitivity parameter ρ_{sf}, the values in all bands were negative for the spruce and pine endmembers. However, we found positive ρ_{sf} values for DBF in the VIS band. A likely explanation is that the leaf area tends to increase with increasing temperature, and for DBF, the VIS albedo of *Betula* spp. bark and branches (especially in the green (560 nm) and red (660 nm) bands [84]) is higher than that for foliage [85,86]. The positive ρ_{sf} in the VIS band appeared to outweigh the negative ρ_{sf} in the NIR band, resulting in a positive ρ_{sf} for the entire SW broadband albedo (Table 3).

The shape parameters λ_{sc} and λ_{sf} in each broad band were similar in sign and magnitude for all the endmembers and were largest in the VIS band. The negative values suggest that the albedos decreased with increasing aboveground volume (or biomass), and that the total surface albedo was driven by the canopy masking of the ground surface and understory. This result can be more fully appreciated when looking at Figure 4.

3.3. Model Behavior in Forests

For forests, the model behavior under both snow-covered and snow-free conditions is only presented for the volume models (Figure 4) although the models behaved similarly for biomass (Supporting Information Figures S3 and S4). To illustrate this behavior, we fixed *SC* at two extremes —0.75 (snow-covered) and 0 (snow-free)—and two temperature extremes at these two *SC* extremes. The shaded area between the solid and dashed curves in Figure 4 illustrates the effect of the temperature sensitivity parameter.

The albedo in all broad bands increased with decreasing temperatures during the snow season. The factors that were likely contributing to the differences between the −12° and 2° curves shown in Figure 4A were the amounts of snow intercepted and held by forest canopies and the physical properties of snow itself. For instance, when the monthly *SC* is 0.75 and monthly *T* is −12 °C, the monthly albedos of a young or newly harvested stand (i.e., when volume = 0) can be as high as 0.86, 0.67 and 0.54 for the VIS, SW and NIR bands, respectively. These values reduce as the forests age and the stand volume (or aboveground biomass) increases. For the pine and spruce endmembers, Figure 4A suggests that the albedo varies only slightly above ~150 m^3 ha^{-1}. However, the asymptotic albedo is not reached within the plotted range for DBF (i.e., 450 m^3 ha^{-1}).

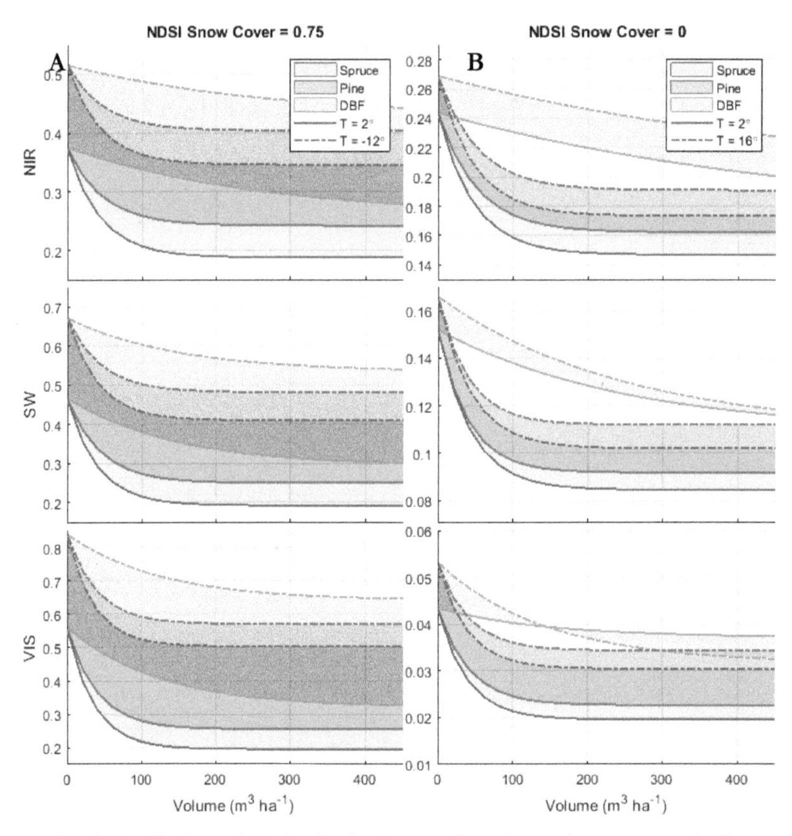

Figure 4. Black-sky albedo model behavior for forest endmembers using mean stand *volume* as the structural predictor: (**A**) with maximum snow cover and (**B**) with zero snow cover. It is important to note the differences in y-axes scaling. "DBF" = deciduous broadleaf forest.

Turning our attention to the forest endmember model behavior under snow-free conditions (Figure 4B), the albedos were highest in the NIR band and lowest in the VIS band. In all broad bands, the albedos were generally highest in DBF and lowest in spruce. The albedo variation with temperature is represented by the shaded areas and is largest for the NIR band. For the spruce and pine endmembers, the highest albedos for any broad band corresponded to the warmest periods. However, the VIS albedo (Figure 4B, solid curve) was highest when the air temperatures were lowest for DBF (i.e., during the shoulder seasons). This may be explained by the lack of foliage that exposed light-colored stems, which are characteristic of the birch species (*Betula* spp.) in our study region. This was in contrast to the relationship between the temperature and NIR albedo where, similar to pine and spruce, the NIR albedos for DBF were highest during the warmest periods (Figure 4B, top panel, dashed green curve). This contrast results in a much narrower range of SW albedo for DBF under snow-free conditions (Figure 4B, middle panel, green shaded area), particularly at high stand volumes.

The range of the albedo variation with temperature is larger for all tree species for the white-sky albedos (Supporting Information Figures S2 and S4).

3.4. Model Benchmarking and Validation

Given their slightly higher accuracy, only the prediction error for the models fit at the effective spatial resolution is presented henceforth. Starting at the pixel level, Figure 5 illustrates the geospatial distribution of the seasonal error (SW broadband, black-sky) within the validation region. The largest errors in all seasons were mostly confined to the higher elevation regions of the north and west. The largest errors were found in winter (DJF) in these regions—regions with monthly T and SC values at the edge of or exceeding the range found in the model training region. In general, the largest positive errors (red values, Figure 5) in winter were found over the largest inland waterbodies ("FW"). Although this error was reduced, the positive error seen over larger freshwater bodies was also evident in spring (MAM) and autumn (SON). On average, the mean SC over these larger freshwater bodies during the colder seasons in 2006–2010 was found to be lower than for the smaller freshwater bodies, which resulted in overestimates of the surface albedo when combined with a low temperature sensitivity parameter (ρ_{SC}, Table 2). For all seasons, the larger negative errors (blue values) were typically found for pixels with larger proportions of the two open area types "O-nv" and "O-sv", which were mostly concentrated at the higher elevation regions of the north and west (Figure 5, third column of sub-panels).

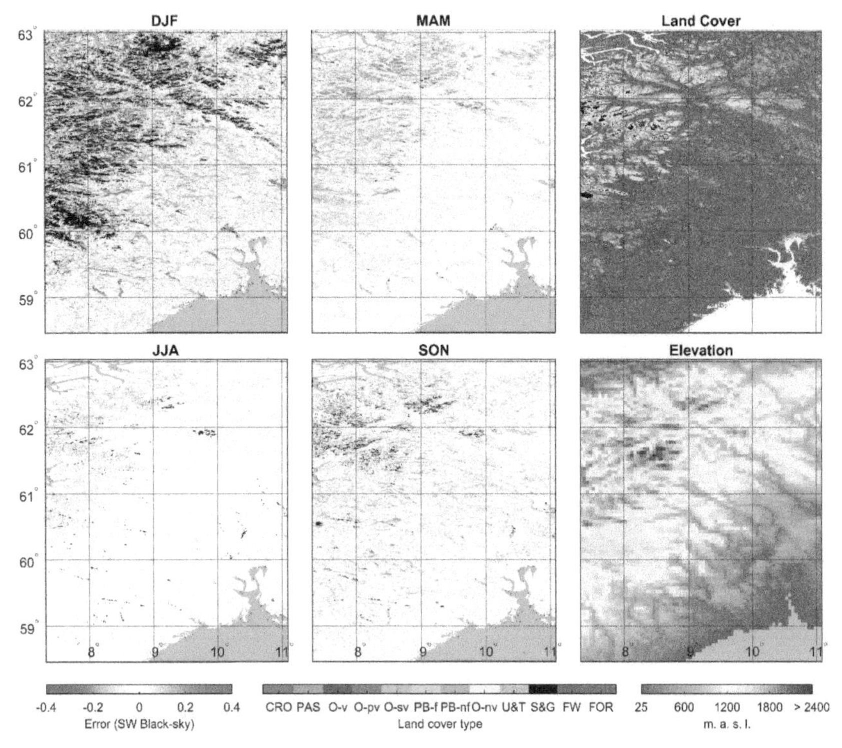

Figure 5. Seasonal prediction error for the shortwave broadband surface albedo (black-sky, at local solar noon) in the validation region. Black values in the error subpanels denote pixels with zero high-quality MCD43A3 or MOD10A1 retrievals during the five-year sample period. "DJF" = December-January-February; "MAM" = March–April–May; "JJA" = June-July-August"; and "SON" = September-October-November.

Moving on to forests, the seasonal prediction errors in the validation domain for the volume-based models are reported in Figure 6 for all bands. The largest errors were found in winter ("DJF"; Figure 6) for DBF where the variations in structure not explained by stand total volume were larger than in snow-free seasons. In winter, the DBF model had a high prediction error, with a median of ~0.05 for all bands. The absolute error for 50% of the predictions (i.e., the interquartile range) for all bands fell within ~0.07–0.17. For other forest types, the median and interquartile errors were smaller. A median positive error of ~0.025 (all bands) was found for spruce. For pine, the median errors were negative for all bands although this error was lower for the NIR band. Combining the parameters for pine and spruce resulted in the lowest median and interquartile error ranges for mixed-conifer forests ("ENF"). Although the median errors were equally low for mixed forests ("MF"), the error interquartile ranges were approximately double that of ENF. For SW, the mean of the median errors for all forest types was 0.01, or ~3% of the mean SW albedo of forests during DJF.

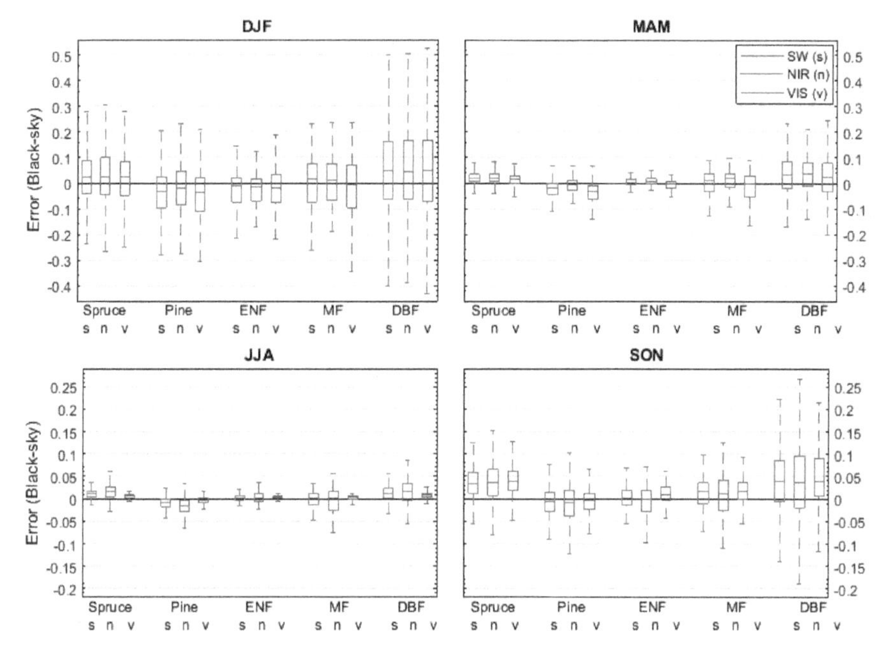

Figure 6. Seasonal error in black-sky albedos (local noon) for pixels with greater than 95% forested area (effective resolution) of one forest type. Predictions are based on the forest models fit with stand volume. "Spruce" = spruce forest (n = 20,260); "Pine" = pine forest (n = 12,167); "ENF" = Evergreen needleleaf forest (n = 100); "MF" = Mixed forest (n = 148); "DBF" = Deciduous broadleaf forest (n = 4175); "SW" = shortwave broadband albedo (250–5000 nm); "NIR" = near infrared broadband albedo (700–5000 nm); and "VIS" = visible broadband albedo (250–700 nm). Horizontal lines represent the medians, boxes represent the interquartile ranges and dashed whiskers represent the extent of the upper and lower quartiles.

Turning our attention to spring (Figure 6; "MAM"), median prediction errors in all forests were found to be similar to those in DJF. However, a major difference was that errors in MAM were much more tightly distributed around the median values. In other words, the error interquartile ranges in MAM were approximately 25–40% of those found for DJF. The improved accuracy in MAM was unsurprising given that a larger share of the high quality MCD43A retrievals representing snow-covered conditions stemmed from MAM, thus influencing the snow-covered model parameters more heavily. Similar to DJF, the largest errors in MAM were found in DBF and MF. For SW, the mean of the median errors for all forest types was 0.01, or 4.5% of the mean SW albedo of forests during MAM.

The lowest errors were found during summer ("JJA"; Figure 6). Whereas median error and error interquartile ranges for the three broad bands were similar in magnitude for winter and spring, the median and interquartile errors in the NIR band in JJA were notably larger than that for the VIS band. For the SW band, the mean of the median error for all forest types was 0.0025, or ~2% of the mean SW albedo of forests during JJA.

As for JJA, the error interquartile ranges for the NIR band were larger than for the VIS band during autumn ("SON"; Figure 5). The spread in error in SON was second largest after DJF although median errors were similar. As for DJF and MAM, the spread in errors (interquartile ranges) was largest for DBF.

For all forested pixels included in the validation exercise, predictions (SW black-sky only) were also compared to predictions from the empirical models of Bright et al. [27] which were based on forest age and T (Figure 7).

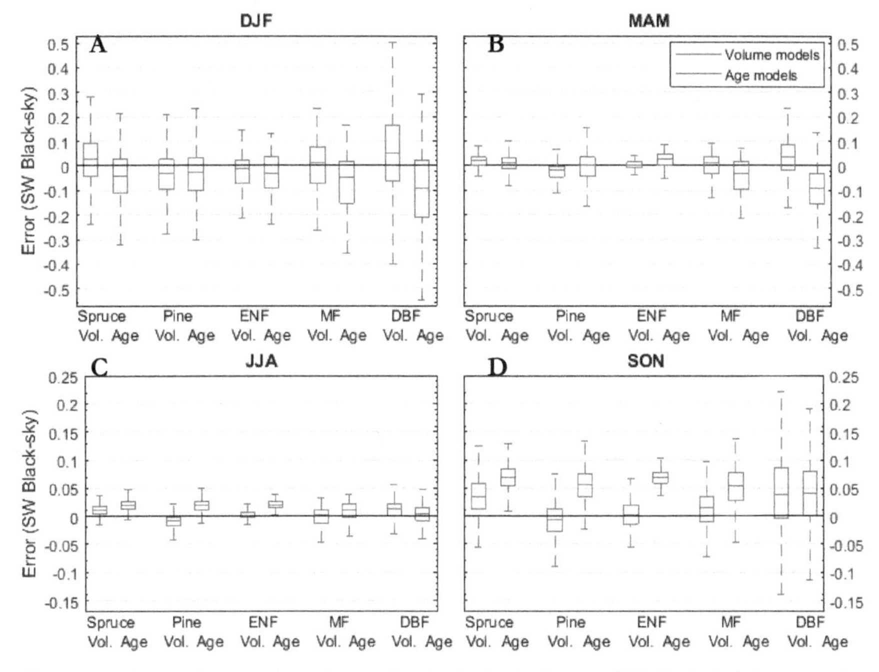

Figure 7. Seasonal mean error in predicted albedo in forests (SW black-sky) between the *Volume*-based models and the *Age*-based models of Bright et al. [27]: (**A**) December-January-February; (**B**) March-April-May; (**C**) June-July-August; and (**D**) September-October-November. Horizontal lines represent the medians, boxes represent the interquartile ranges and dashed whiskers represent the extent of the upper and lower quartiles.

Starting in winter ("DJF"; Figure 7A), compared to the *Age*-based models the *Volume*-based models presented in this work appeared to slightly improve the DJF accuracy in forests judging by the median errors, with the exception of pine. For pine, the median error using the *Volume* model was similar to that of the *Age* model. The spread in errors for the *Volume* models were similar to the *Age* models, although the error interquartile ranges appear to have been slightly reduced for pine, ENF and MF with the *Volume* models.

For spring ("MAM"; Figure 7B), the median errors for ENF, MF and DBF using the *Volume* models were notably reduced compared to the *Age* models. The error interquartile ranges were reduced for all forest types although for spruce and pine, the median errors were slightly higher (in absolute terms).

In summer ("JJA"; Figure 7C), the median prediction errors were noticeably reduced in most forest types compared to the *Age* models. The exception is DBF where the median error using the *Age* model was slightly lower than that of the *Volume* model.

The most notable improvements over the *Age*-based models were found in the autumn months ("SON"; Figure 7D), which demonstrates that while monthly mean T may be a good proxy for snow during winter and spring, it suffers as a proxy for snow during autumn. Relative to the *Age* models, the application of the *Volume* models led to large reductions in the median errors in all forest types with the exception of DBF, where the median errors were similar.

Shifting our attention to the non-forest endmembers and the limiting results presented henceforth of the full SW broad band, spreads in model errors were largest in DJF as expected (Figure 8, upper left panel). Median prediction errors were under 0.05 for all land cover types, except for "O-sv", "PB-f" and "O-nv". Median errors for "O-sv" and "O-nv" were negative for each season. Median error for "PB-f" was positive for all seasons although we note that the number of MCD43A3 retrievals containing "PB-f" homogeneity to ≥95% was limited to just three.

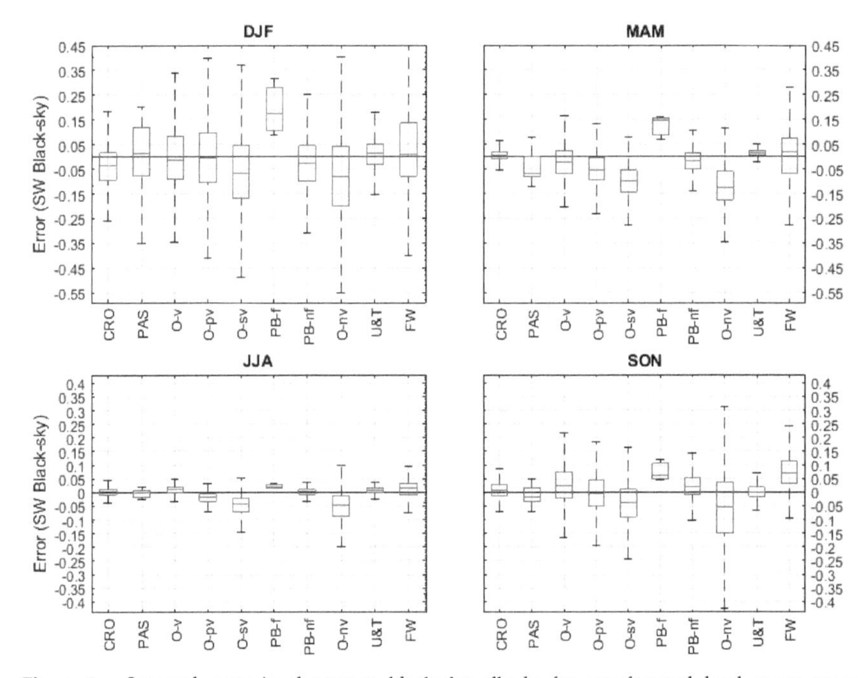

Figure 8. Seasonal error in shortwave black-sky albedo for non-forested land cover types (endmembers) for pixels with greater than 95% area (effective resolution) of one type (**Upper left**) December-January-February; (**Upper right**) March-April-May; (**Lower left**) June-July-August; and (**Lower right**) September-October-November. "CRO" = croplands (n = 3504); "PAS" = pasture (n = 11); "O-v" = Open, vegetated (n = 21,930); "O-pv" = Open, partly vegetated (n = 11,966); "O-sv" = Open, sparsely vegetated (n = 21,330); "PB-f" = Peat bog, forested (n = 3); "PB-nf" = Peat bog, non-forested (n = 1561); "O-nv" = Open, non-vegetated (n = 19,900); "U&T" = Urban & transport (n = 1498); and "FW" = freshwater (n = 17,573). Horizontal lines represent the medians, boxes represent the interquartile ranges and dashed whiskers represent the extent of the upper and lower quartiles.

In order to more rigorously assess whether these error patterns were systematic, we relaxed the homogeneity requirement described in Section 2.9 and recomputed the error statistics for all pixels using the sub-MODIS pixel modes from AR5/SR16 to define the majority land cover type. Figure 9 contains the results after normalizing the seasonal mean prediction error to the MCD43A3 retrievals, which reinforces the above-mentioned results that error patterns for "PB-f", "O-sv" and "O-nv" were indeed systematic.

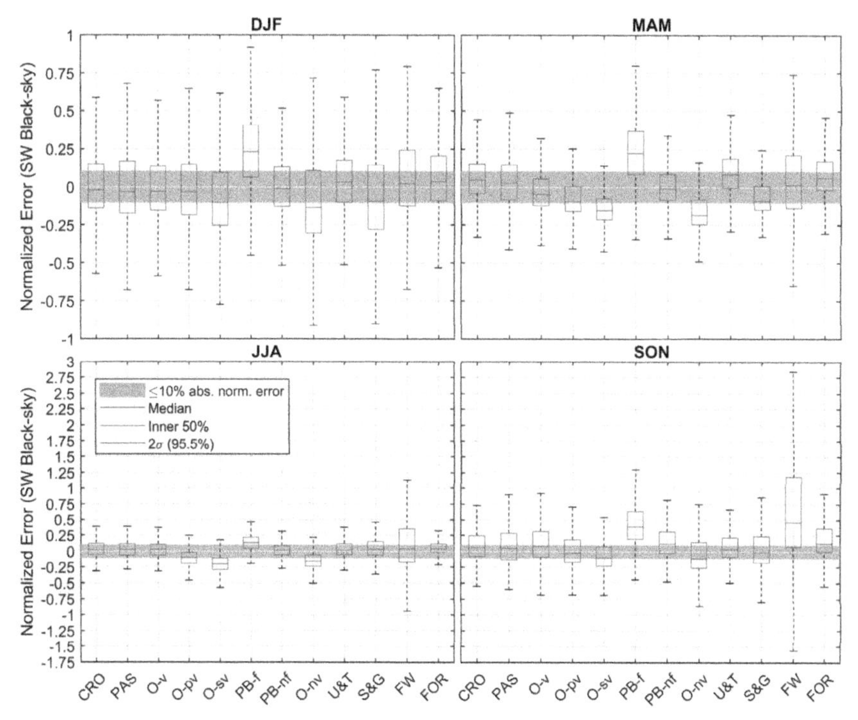

Figure 9. Normalized error (SW black-sky) by season and land cover (endmember) type; (**Upper left**) December-January-February; (**Upper right**) March-April-May; (**Lower left**) June-July-August; and (**Lower right**) September-October-November Pixels were allocated to individual land cover types according to the largest relative land cover fraction within the MCD43A3 v6 *effective* spatial resolution. Whiskers denote 2σ sigma confidence intervals, boxes denote interquartile ranges, and red horizontal lines denote medians. The gray shaded area represents the accepted error range of $\pm$0.1 ($\pm$10%).

Figure 9 also shows that the range in errors for freshwater ("FW") relative to the retrieved values is the largest among all land cover types, particularly in autumn (Figure 9 SON and JJA, whiskers). Given an acceptable median error threshold of $\leq$10%, Figure 9 shows that all models except "PB-f", "O-sv," and "O-nv" are accurate in spring (MAM) and summer (JJA). In winter, median normalized errors of all models except "O-nv" and "PB-f" meet this accuracy threshold. For autumn, median normalized errors of all models except "PB-f", "PB-nf," and "FW" meet this threshold. Additionally, the large spread in error for "Birch" (Figure 6) causes the median normalized error for forests as a whole ("FOR" in Figure 9) to slightly exceed the 10% accuracy threshold.

4. Discussion

We combined a five-year climatology of the surface albedo, near-surface air temperature, NDSI snow cover and detailed maps of Norwegian land cover and structure to yield a set of empirical models giving estimates of the monthly mean surface albedo as a function of Norwegian land cover classification. Although the endmember fraction and NDSI snow cover alone could explain 55–81% of the variance in the land surface albedo in our model training domain (depending on which albedo and broad band one looks at), the addition of both air temperature and forest structure (volume or biomass) further increased the explained variance to 85–88% (Table 1). We consider this a remarkable outcome given the multiple sources of error inherent in the various land cover products employed in the model training exercise.

For non-forested vegetated land covers, we found strong relationships between air temperature and albedo (all bands), which suggests that tgrowing season phenology was sufficiently synchronized with monthly mean air temperatures. With the exception of croplands ("CRO") and pastures ("PAS"), SW albedo (black-sky) decreased with increasing air temperature, suggesting that it was driven by increased vegetation masking of a higher albedo surface during the growing season. However, for cropland, pastures, and all forest endmembers, the SW albedo increased with increasing air temperature, suggesting either a larger role played by understory vegetation or by increased canopy masking of a lower albedo surface. During the snow season, we found that air temperature and surface albedo were negatively correlated. This is a relationship that is presumably driven by the influence of air temperature on snow metamorphosis and snow physical state [46,87]. When applied outside the training region, we found absolute normalized median errors to be $\leq$10% for most non-forest endmember models (SW black-sky). The non-forest endmember models on average performed best during spring (MAM) and summer (JJA) where the proportion of the total number of predictions agreeing within 10% of the MCD43A3 retrievals was around 35% and 45%, respectively (Figure S7, Supporting Information). These shares were weighed downward by the persistent positive error for forested peatbogs ("PB-f") and persistent negative error for non-vegetated open areas ("O-nv") whose proportions of the total number of predictions agreeing within 10% of the MCD43A3 retrievals were only 20% and 25% within the validation region, respectively (Figure S7). In general, these two land cover (endmember) types exhibit the largest variation in both geological attributes (exposed mineral composition) and vegetation attributes (vegetation cover fraction). These are two physical attributes important for the surface albedo which were not captured by the models. Furthermore, the model training region contained a disproportionately low share of "O-nv" relative to the validation region (Figure S1), and given the large variation in surface attributes within "O-nv", a larger "O-nv" sample during model fitting would likely have resulted in an improved performance by this endmember model.

Our models were least capable of explaining variance in the NIR albedos, which is unsurprising given the large share of forest cover in the region and the large variation in foliage reflectance in the shortwave infrared (1.5–2.5 μm) band among Fennoscandic tree species [88]. This variation is likely attributed to differences in leaf-level functional traits [19,20,89], and although seasonal trends in air temperature likely explained some of the albedo variance linked to trends in the leaf area [26], such functional controls could not be accounted.

For the forest endmembers, the maximum and minimum albedo predictions under both snow-covered and snow-free conditions (Figure 4) appeared well-constrained when benchmarked to observations reported in other Fennoscandic regions based on satellite remote sensing [32,33,40,41,81]. The dependency of the albedo on forest structure was more difficult to compare quantitatively to the results of other studies since the environmental state conditions (i.e., SC and T) were often not reported alongside the reported results. However, qualitatively, the general relationship between forest albedo (all endmembers, all bands) and stand volume under snow-free conditions (Figure 4B) appeared be consistent with the regression fits based on the Landsat observations reported in Kuusinen et al. [41]. However, we did not find as strong a dependency as Kuusinen et al. [41] of the VIS broad band albedo on stand total volume (cf. Figure 1 in Kuusinen et al. [41]).

Under snow-covered conditions, the use of an exponential function to model the albedo-structure relationship in forests appeared appropriate when compared to that reported for Finnish forests for age (c.f., Figure 1 in ref. [33]) or biomass (cf. Figure 9A in ref. [40]). However, the shape parameters of our volume (Figure 4A) and biomass models (Figures S2 and S3) seemed to differ from the age-based models of Kuusinen et al. [33] where the asymptotic SW albedo value for DBF (birch) and spruce forests was reached around age 50 in ref. [31]. This is an age that corresponds to a stand volume and biomass in forests of our training domain for which the asymptotic albedo has yet to be realized. The behavior of the SW albedo model for DBF under snow-covered conditions reported in this study appears to be more consistent with the age-based DBF model of Bright et al. [27], which is unsurprising given the partly-overlapping model training domains.

5. Conclusions

We presented a coherent empirical modeling framework based on linear spectral unmixing of optical satellite retrievals of surface albedo. The unmixing framework allowed us to sample a relatively large area within a finite region of Norway containing multiple land cover features spanning large climate gradients. The unmixing approach may be computationally less expensive than the approaches based on homogenous pixel sampling, which often requires a more extensive spatial domain to be sampled in order to allow for a sufficient number of satellite retrievals. The resulting land (or vegetation) cover-dependent models developed and presented here can be applied to create a spatially-explicit surface albedo dataset (or database) for use in regional land-climate research and in climate-oriented land management planning in Norway. The models can also facilitate comparisons to land-cover dependent albedo predictions made by land surface (climate) models, which are difficult to evaluate by direct comparisons to satellite retrievals.

The models required just two environmental state predictors: monthly mean air temperature and monthly mean NDSI snow cover (*SC*), where the latter was also provided by optical satellite remote sensing. Given a normalized error threshold of $\leq$10%, most of the models proved to be accurate when validated against high quality MODIS albedo retrievals in a region outside of the model training region (based on the medians reported in Figure 9). The exception was the performance for forested peat bogs ("Pb-f") and non-vegetated open areas ("O-nv") whose normalized absolute median prediction errors exceeded 10% in all seasons. Although we reported a large range in upper and lower error quartiles for all models during winter months (DJF; Figures 6–9), these larger errors may be deemed acceptable in most modeling contexts given the lower relevance of surface albedo during periods of low solar energy input (such is the case for Norway during DJF). Finally, although the MODIS NDSI-based *SC* can be higher than the 0.75 constraint shown in Figure 4, the higher values in our domain mostly coincided with the end of the snow season when the monthly mean *T* was also higher. We stress that great caution should be exercised when extrapolating the model behavior outside the extreme *SC* and *T* ranges presented in Figure 2.

Supplementary Materials: The following are available online at http://www.mdpi.com/2072-4292/11/7/871/s1: Table S1: Root mean squared errors of candidate models; Table S2: White-sky model parameters for forests with *volume* as structural predictor; Table S3: Black-sky model parameters for forests with *biomass* as structural predictor; Table S4: White-sky model parameters for forests with *biomass* as structural predictor; Table S5: White-sky model parameters for non-forest endmembers; Figure S1: Distribution of MODIS retrievals used in model fitting and validation by month and land cover type; Figure S2: White-sky model behavior for forests with *volume* as structural predictor; Figure S3: Black-sky model behavior for forests with *biomass* as structural predictor; Figure S4: White-sky model behavior for forests with *biomass* as structural predictor; Figure S5: Weighting scheme applied when fitting models to the *effective* spatial resolution; Figure S6: Distribution of Köppen-Geiger climate zones in Norway and within the study domain; Figure S7: Fraction of total predictions have $\leq$10% normalized absolute error (SW black-sky) by land cover type and season.

Author Contributions: R.M.B. conceptualized and designed the study, compiled and processed all data, carried out all modeling, analysis and validation, produced all figures and wrote the original manuscript draft. R.A. edited and commented on the original draft and R.M.B. and R.A. revised the final manuscript.

Funding: This research was funded by the Research Council of Norway (Norges Forskningsråd), grant number 255307/E10.

Conflicts of Interest: The authors declare no conflict of interest.

References

1. Mahmood, R.; Quintanar, A.I.; Conner, G.; Leeper, R.; Dobler, S.; Pielke, R.A.; Bonan, G.; Chase, T.; McNider, R.; McAlpine, C.; et al. Impacts of Land Use/Land Cover Change on Climate and Future Research Priorities. *Bull. Am. Meteorol. Soc.* **2010**, *91*, 37–46. [CrossRef]
2. Mahmood, R.; Pielke, R.A.; Loveland, T.R.; McAlpine, C.A. Climate Relevant Land Use and Land Cover Change Policies. *Bull. Am. Meteorol. Soc.* **2016**, *97*, 195–202. [CrossRef]
3. Stephens, G.L.; O'Brien, D.; Webster, P.J.; Pilewski, P.; Kato, S.; Li, J. The albedo of Earth. *Rev. Geophys.* **2015**, *53*, 141–163. [CrossRef]
4. Jackson, R.B.; Randerson, J.T.; Canadell, J.G.; Anderson, R.G.; Avissar, R.; Baldocchi, D.D.; Bonan, G.B.; Caldeira, K.; Diffenbaugh, N.S.; Field, C.B.; et al. Protecting climate with forests. *Environ. Res. Lett.* **2008**, *3*, 044006. [CrossRef]
5. Pielke, R.A., Sr.; Marland, G.; Betts, R.A.; Chase, T.N.; Eastman, J.L.; Niles, J.O.; Niyogi, D.d.S.; Running, S.W. The influence of land-use change and landscape dynamics on the climate system: Relevance to climate-change policy beyond the radiative effect of greenhouse gases. *Phil. Trans. R. Soc. Lond. A* **2002**, *360*, 1705–1719. [CrossRef] [PubMed]
6. Lutz, D.A.; Howarth, R. Valuing albedo as an ecosystem service: Implications for forest management. *Clim. Chang.* **2014**, *124*, 53–63. [CrossRef]
7. Lutz, D.A.; Burakowski, E.A.; Murphy, M.B.; Borsuk, M.E.; Niemiec, R.M.; Howarth, R.B.; Burakowski, E. Tradeoffs between three forest ecosystem services across the state of New Hampshire, USA: Timber, carbon, and albedo. *Ecol. Appl.* **2015**, *26*, 146–161. [CrossRef]
8. Favero, A.; Sohngen, B.; Huang, Y.; Jin, Y. Global cost estimates of forest climate mitigation with albedo: A new integrative policy approach. *Environ. Res. Lett.* **2018**, *13*, 125002. [CrossRef]
9. Thompson, M.P.; Adams, D.; Sessions, J. Radiative forcing and the optimal rotation age. *Ecol. Econ.* **2009**, *68*, 2713–2720. [CrossRef]
10. Matthies, B.D.; Valsta, L.T. Optimal forest species mixture with carbon storage and albedo effect for climate change mitigation. *Ecol. Econ.* **2016**, *123*, 95–105. [CrossRef]
11. He, T.; Liang, S.; Song, D.-X. Analysis of global land surface albedo climatology and spatial-temporal variation during 1981–2010 from multiple satellite products. *J. Geophys. Res. Atmos.* **2014**, *119*, 10281–10298. [CrossRef]
12. Gao, F.; He, T.; Wang, Z.; Ghimire, B.; Shuai, Y.; Masek, J.; Schaaf, C.; Williams, C. Multi-scale climatological albedo look-up maps derived from MODIS BRDF/albedo products. *J. Appl. Remote Sens.* **2014**, *8*, 083532-1. [CrossRef]
13. Gao, F.; Schaaf, C.B.; Strahler, A.H.; Roesch, A.; Lucht, W.; Dickinson, R. MODIS bidirectional reflectance distribution function and albedo Climate Modeling Grid products and the variablility of albedo for major global vegetation types. *J. Geophys. Res.* **2005**, *110*, 1–13. [CrossRef]
14. Zhao, K.; Jackson, R.B. Biophysical forcings of land-use changes from potential forestry activities in North America. *Ecol. Monogr.* **2014**, *84*, 329–353. [CrossRef]
15. Román, M.O.; Schaaf, C.B.; Woodcock, C.E.; Strahler, A.H.; Yang, X.; Braswell, R.H.; Curtis, P.S.; Davis, K.J.; Dragoni, D.; Goulden, M.L. The MODIS (Collection V005) BRDF/albedo product: Assessment of spatial representativeness over forested landscapes. *Remote. Sens. Environ.* **2009**, *113*, 2476–2498. [CrossRef]
16. Ni, W.; Woodcock, C.E. Effect of canopy structure and the presence of snow on the albedo of boreal conifer forests. *J. Geophys. Res. Phys.* **2000**, *105*, 11879–11888. [CrossRef]
17. Kung, E.C.; Bryson, R.A.; Lenschow, D.H. Study of a continental surface albedo on the basis of flight measurements and structure of the earth's surface cover over north america. *Mon. Weather. Rev.* **1964**, *92*, 543–564. [CrossRef]
18. Betts, A.K.; Ball, J.H. Albedo over the boreal forest. *J. Geophys. Res.* **1997**, *102*, 28901–28909. [CrossRef]
19. Richardson, A.D.; Black, T.A.; Ciais, P.; Delbart, N.; Friedl, M.A.; Gobron, N.; Hollinger, D.Y.; Kutsch, W.L.; Longdoz, B.; Luyssaert, S.; et al. Influence of spring and autumn phenological transitions on forest ecosystem productivity. *Philos. Trans. R. Soc. B Boil. Sci.* **2010**, *365*, 3227–3246. [CrossRef]

20. Ollinger, S.V.; Richardson, A.D.; Martín, M.E.; Hollinger, D.Y.; Frolking, S.E.; Reich, P.B.; Plourde, L.C.; Katul, G.G.; Munger, J.W.; Oren, R.; et al. Canopy nitrogen, carbon assimilation, and albedo in temperate and boreal forests: Functional relations and potential climate feedbacks. *Proc. Natl. Acad. Sci. USA* **2008**, *105*, 19336–19341. [CrossRef]
21. Duveiller, G.; Hooker, J.; Cescatti, A. A dataset mapping the potential biophysical effects of vegetation cover change. *Sci. Data* **2018**, *5*, 180014. [CrossRef]
22. Duveiller, G.; Hooker, J.; Cescatti, A. The mark of vegetation change on Earth's surface energy balance. *Nat. Commun.* **2018**, *9*, 679. [CrossRef] [PubMed]
23. Seneviratne, S.I.; Phipps, S.J.; Pitman, A.J.; Hirsch, A.L.; Davin, E.L.; Donat, M.G.; Hirschi, M.; Lenton, A.; Wilhelm, M.; Kravitz, B. Land radiative management as contributor to regional-scale climate adaptation and mitigation. *Nat. Geosci.* **2018**, *11*, 88–96. [CrossRef]
24. Miller, J.N.; Vanloocke, A.; Gomez-Casanovas, N.; Bernacchi, C.J.; Gomez-Casanovas, N. Candidate perennial bioenergy grasses have a higher albedo than annual row crops. *GCB Bioenergy* **2015**, *8*, 818–825. [CrossRef]
25. Wickham, J.; Barnes, C.; Nash, M.; Wade, T. Combining NLCD and MODIS to create a land cover-albedo database for the continental United States. *Remote. Sens. Environ.* **2015**, *170*, 143–152. [CrossRef]
26. Leonardi, S.; Magnani, F.; Nolè, A.; Van Noije, T.; Borghetti, M. A global assessment of forest surface albedo and its relationships with climate and atmospheric nitrogen deposition. *Glob. Chang. Boil.* **2014**, *21*, 287–298. [CrossRef]
27. Bright, R.M.; Astrup, R.; Strømman, A.H. Empirical models of monthly and annual albedo in managed boreal forests of interior Norway. *Clim. Chang.* **2013**, *120*, 183–196. [CrossRef]
28. Rechid, D.; Raddatz, T.J.; Jacob, D. Parameterization of snow-free land surface albedo as a function of vegetation phenology based on MODIS data and applied in climate modelling. *Theor. Appl. Climatol.* **2009**, *95*, 245–255. [CrossRef]
29. Song, J. Phenological influences on the albedo of prairie grassland and crop fields. *Int. J. Biometeorol.* **1999**, *42*, 153–157. [CrossRef]
30. Granhus, A.; Hylen, G.; Nilsen, J.-E.Ø. *Statistics of Forest Conditions and Resources in Norway, in Ressursoversikt fra Skog og Landskap 03/12*; Norwegian Forest and Landscape Institute: Ås, Norway, 2012.
31. Larsson, J.Y.; Hylen, G. Skogen i Norge: Statistikk over Skogforhold og Skogressurser i Norge Registrert i Perioden 2000–2004 [Forest in Norway: Forest Resource Statistics for the Period 2000–2004]. Norwegian Forest and Landscape Institute: Ås, Norway, 2007; p. 95. Available online: https://brage.bibsys.no/xmlui/bitstream/handle/11250/2508185/SoL-Viten-2007-01.pdf?sequence=1&isAllowed=y (accessed on 21 October 2018).
32. Kuusinen, N.; Tomppo, E.; Berninger, F. Linear unmixing of MODIS albedo composites to infer subpixel land cover type albedos. *Int. J. Appl. Earth Obs. Geoinf.* **2013**, *23*, 324–333. [CrossRef]
33. Kuusinen, N.; Tomppo, E.; Shuai, Y.; Berninger, F. Effects of forest age on albedo in boreal forests estimated from MODIS and Landsat albedo retrievals. *Remote. Sens. Environ.* **2014**, *145*, 145–153. [CrossRef]
34. Bright, R.M.; Eisner, S.; Lund, M.T.; Majasalmi, T.; Myhre, G.; Astrup, R. Inferring Surface Albedo Prediction Error Linked to Forest Structure at High Latitudes. *J. Geophys. Res. Atmos.* **2018**, *123*, 4910–4925. [CrossRef]
35. Bioucas-Dias, J.M.; Plaza, A.; Dobigeon, N.; Parente, M.; Du, Q.; Gader, P.; Chanussot, J. Hyperspectral Unmixing Overview: Geometrical, Statistical, and Sparse Regression-Based Approaches. *IEEE J. Sel. Top. Appl. Earth Obs. Remote Sens.* **2012**, *5*, 354–379. [CrossRef]
36. Qu, X.; Hall, A. On the persistent spread in snow-albedo feedback. *Clim. Dyn.* **2014**, *42*, 69–81. [CrossRef]
37. Essery, R. Large-scale simulations of snow albedo masking by forests. *Geophys. Res. Lett.* **2013**, *40*, 5521–5525. [CrossRef]
38. Lukeš, P.; Stenberg, P.; Rautiainen, M. Relationship between forest density and albedo in the boreal zone. *Ecol. Model.* **2013**, *261*, 74–79. [CrossRef]
39. Bright, R.M.; Myhre, G.; Astrup, R.; Antón-Fernández, C.; Strømman, A.H. Radiative forcing bias of simulated surface albedo modifications linked to forest cover changes at northern latitudes. *Biogeosciences* **2015**, *12*, 2195–2205. [CrossRef]
40. Lukeš, P.; Rautiainen, M.; Manninen, T.; Stenberg, P.; Mõttus, M. Geographical gradients in boreal forest albedo and structure in Finland. *Remote. Sens. Environ.* **2014**, *152*, 526–535. [CrossRef]
41. Kuusinen, N.; Stenberg, P.; Korhonen, L.; Rautiainen, M.; Tomppo, E. Structural factors driving boreal forest albedo in Finland. *Remote. Sens. Environ.* **2016**, *175*, 43–51. [CrossRef]

42. Loranty, M.M.; Berner, L.T.; Goetz, S.J.; Jin, Y.; Randerson, J.T. Vegetation controls on northern high latitude snow-albedo feedback: Observations and CMIP5 model simulations. *Glob. Chang. Biol.* **2014**, *20*, 594–606. [CrossRef] [PubMed]
43. Kuusinen, N.; Kolari, P.; Levula, J.; Porcar-Castell, A.; Stenberg, P.; Berninger, F. Seasonal variation in boreal pine forest albedo and effects of canopy snow on forest reflectance. *Agric. For. Meteorol.* **2012**, *164*, 53–60. [CrossRef]
44. Pomeroy, J.W.; Parviainen, J.; Hedstrom, N.; Gray, D.M. Coupled modelling of forest snow interception and sublimation. *Hydrol. Process.* **1998**, *12*, 2317–2337. [CrossRef]
45. Hedstrom, N.R.; Pomeroy, J.W. Measurements and modelling of snow interception in the boreal forest. *Hydrol. Process.* **1998**, *12*, 1611–1625. [CrossRef]
46. Wiscombe, W.J.; Warren, S.G. A model for the spectral albedo of snow. I. Pure Snow. *J. Atmos. Sci.* **1980**, *37*, 2712–2733. [CrossRef]
47. Aoki, T.; Hachikubo, A.; Hori, M. Effects of snow physical parameters on shortwave broadband albedos. *J. Geophys. Res. Phys.* **2003**, *108*, 4616. [CrossRef]
48. Brun, E. Investigation on Wet-Snow Metamorphism in Respect of Liquid-Water Content. *Ann. Glaciol.* **1989**, *13*, 22–26. [CrossRef]
49. Painter, T.H.; Rittger, K.; McKenzie, C.; Slaughter, P.; Davis, R.E.; Dozier, J. Retrieval of subpixel snow covered area, grain size, and albedo from MODIS. *Remote. Sens. Environ.* **2009**, *113*, 868–879. [CrossRef]
50. Essery, R.; Morin, S.; Lejeune, Y.; Ménard, C.B. A comparison of 1701 snow models using observations from an alpine site. *Adv. Resour.* **2013**, *55*, 131–148. [CrossRef]
51. Pedersen, C.A.; Winther, J.-G. Intercomparison and validation of snow albedo parameterization schemes in climate models. *Clim. Dyn.* **2005**, *25*, 351–362. [CrossRef]
52. Olsson, C.; Jönsson, A.M. Process-based models not always better than empirical models for simulating budburst of Norway spruce and birch in Europe. *Glob. Chang. Boil.* **2014**, *20*, 3492–3507. [CrossRef] [PubMed]
53. Zohner, C.M.; Benito, B.M.; Svenning, J.-C.; Renner, S.S. Day length unlikely to constrain climate-driven shifts in leaf-out times of northern woody plants. *Nat. Clim. Chang.* **2016**, *6*, 1120–1123. [CrossRef]
54. Rutter, N.; Essery, R.; Pomeroy, J.; Altimir, N.; Andreadis, K.; Baker, I.; Barr, A.; Bartlett, P.; Boone, A.; Deng, H.; et al. Evaluation of forest snow processes models (SnowMIP2). *J. Geophys. Res. Phys.* **2009**, *114*, 06111. [CrossRef]
55. Bjørdal, I.; Bjørkelo, K. *AR5 Klassifikasjonssystem—Klassifikasjon av arealressurser, in Håndbok fra Skog og Landskap—01/2006*; Norwegian Forest and Landscape Institute: Ås, Norway, 2006; p. 26.
56. Gjertsen, A.K.; Angeloff, M.; Strand, G.-H. Arealressurskart over fjellområdene. *Kart og Plan* **2012**, *71*, 45–51.
57. Mathiesen, H.F. *Arealstatistikk—Fulldyrka Jord og Dyrkbar Jord*; Norwegian Forest and Landscape Institute: Ås, Norway, 2014; p. 43.
58. Astrup, R.; Rahlf, J.; Bjørkelo, K.; Debella-Gilo, M.; Gjertsen, Ar.; Breidenbach, J. Forest information at multiple scales: Development, evaluation and application of the Norwegian Forest Resources Map. *Scand. J. For. Res.* **2018**. under review. [CrossRef]
59. Blaschke, T.; Lang, S.; Hay, G.J. *Object-Based Image Analysis: Spatial Concepts for Knowledge-Driven Remote Sensing Applications*; Springer: Berlin/Heidelberg, Germany, 2008.
60. Gjertsen, A. Accuracy of forest mapping based on Landsat TM data and a kNN-based method. *Remote. Sens. Environ.* **2007**, *110*, 420–430. [CrossRef]
61. Gjertsen, A.K.; Nilsen, J.-E.Ø. *SAT-SKOG: Et skogkart basert på tolking av Satellittbilder [SAT-SKOG: A Forest Map Based on Interpretation of Satellite Imagery]*; Norwegian Forest and Landscape Institute: Ås, Norway, 2012; p. 54. Available online: http://www.skogoglandskap.no/filearchive/rapport_23_12_sat_skog_skogkart_basert_pa_tolking_av_satellittbilder.pdf (accessed on 11 February 2014). (In Norwegian)
62. Kotchenova, S.Y.; Vermote, E.F.; Matarrese, R.; Klemm, J.F.J. Validation of a vector version of the 6S radiative transfer code for atmospheric correction of satellite data Part I: Path radiance. *Appl. Opt.* **2006**, *45*, 6762. [CrossRef] [PubMed]
63. Kotchenova, S.Y.; Vermote, E.F. Validation of a vector version of the 6S radiative transfer code for atmospheric correction of satellite data. Part II. Homogeneous Lambertian and anisotropic surfaces. *Appl. Opt.* **2007**, *46*, 4455–4464. [CrossRef]

64. Schaaf, C.B.; Gao, F.; Strahler, A.H.; Lucht, W.; Li, X.; Tsang, T.; Strugnell, N.C.; Zhang, X.; Jin, Y.; Muller, J.-P.; et al. First operational BRDF, albedo nadir reflectance products from MODIS. *Remote. Sens. Environ.* **2002**, *83*, 135–148. [CrossRef]
65. Lucht, W.; Lewis, P. Theoretical noise sensitivity of BRDF and albedo retrieval from the EOS-MODIS and MISR sensors with respect to angular sampling. *Int. J. Remote. Sens.* **2000**, *21*, 81–98. [CrossRef]
66. Schaaf, C.; Strahler, A.; Lucht, W. An algorithm for the retrieval of albedo from space using semiempirical BRDF models. *IEEE Trans. Geosci. Remote. Sens.* **2000**, *38*, 977–998.
67. Schaaf, C.; Wang, Z. *MCD43A3 MODIS/Terra + Aqua BRDF/Albedo Daily L3 Global—500m V006 [Data set]*; NASA EOSDIS LP DAAC: Sioux Falls, SD, USA, 2015. [CrossRef]
68. Wang, Z.; Schaaf, C.B.; Strahler, A.H.; Chopping, M.J.; Román, M.O.; Shuai, Y.; Woodcock, C.E.; Hollinger, D.Y.; Fitzjarrald, D.R. Evaluation of MODIS albedo product (MCD43A) over grassland, agriculture and forest surface types during dormant and snow-covered periods. *Remote. Sens. Environ.* **2014**, *140*, 60–77. [CrossRef]
69. Wang, Z.; Schaaf, C.B.; Sun, Q.; Shuai, Y.; Román, M.O. Capturing rapid land surface dynamics with Collection V006 MODIS BRDF/NBAR/Albedo (MCD43) products. *Remote. Sens. Environ.* **2018**, *207*, 50–64. [CrossRef]
70. Schaaf, C.; Wang, Z. *MCD43A2 MODIS/Terra+Aqua BRDF/Albedo Quality Daily L3 Global—500m V006 [Data set]*; NASA LP DAAC: Sioux Falls, SD, USA, 2015. [CrossRef]
71. Hall, D.K.; Riggs, G.A. MODIS/Terra Snow Cover Daily L3 Global 500m Grid, Version 6 [Data set]. Available online: https://doi.org/10.5067/MODIS/MOD10A1.006 (accessed on 5 November 2018).
72. Salomonson, V.; Appel, I. Estimating fractional snow cover from MODIS using the normalized difference snow index. *Remote. Sens. Environ.* **2004**, *89*, 351–360. [CrossRef]
73. Riggs, G.A.; Hall, D.K.; Román, M.O. MODIS Snow Products Collection 6 User Guide (Version 1.0). Available online: https://modis-snow-ice.gsfc.nasa.gov/?c=userguides2016 (accessed on 7 October 2018).
74. Riggs, G.A.; Hall, D.K.; Román, M.O. Overview of NASA's MODIS and Visible Infrared Imaging Radiometer Suite (VIIRS) snow-cover Earth System Data Records. *Earth Syst. Sci. Data* **2017**, *9*, 765–777. [CrossRef]
75. Masson, T.; Dumont, M.; Mura, M.; Sirguey, P.; Gascoin, S.; Dedieu, J.-P.; Chanussot, J. An Assessment of Existing Methodologies to Retrieve Snow Cover Fraction from MODIS Data. *Remote Sens.* **2018**, *10*, 619. [CrossRef]
76. Dong, J.; Ek, M.; Hall, D.; Peters-Lidard, C.; Cosgrove, B.; Miller, J.; Riggs, G.; Xia, Y. Using Air Temperature to Quantitatively Predict the MODIS Fractional Snow Cover Retrieval Errors over the Continental United States. *J. Hydrometeorol.* **2013**, *15*, 551–562. [CrossRef]
77. Lussana, C.; Tveito, O.E.; Uboldi, F. Three-dimensional spatial interpolation of 2 m temperature over Norway. *Q. J. R. Meteorol. Soc.* **2018**, *144*, 344–364. [CrossRef]
78. Daeseong, K.; Hyung-sup, J.; Jeong-cheol, K. Comparison of Snow Cover Fraction Functions to Estimate Snow Depth of South Korea from MODIS Imagery. *Korean J. Remote Sens.* **2017**, *33*, 401–410.
79. Cescatti, A.; Marcolla, B.; Vannan, S.K.S.; Pan, J.Y.; Román, M.O.; Yang, X.; Ciais, P.; Cook, R.B.; Law, B.E.; Matteucci, G.; et al. Intercomparison of MODIS albedo retrievals and in situ measurements across the global FLUXNET network. *Remote Sens. Environ.* **2012**, *121*, 323–334. [CrossRef]
80. Campagnolo, M.L.; Sun, Q.; Liu, Y.; Schaaf, C.; Wang, Z.; Román, M.O. Estimating the effective spatial resolution of the operational BRDF, albedo, and nadir reflectance products from MODIS and VIIRS. *Remote Sens. Environ.* **2016**, *175*, 52–64. [CrossRef]
81. Hovi, A.; Lindberg, E.; Lang, M.; Arumäe, T.; Peuhkurinen, J.; Sirparanta, S.; Pyankov, S.; Rautiainen, M. Seasonal dynamics of albedo across European boreal forests: Analysis of MODIS albedo and structural metrics from airborne LiDAR. *Remote Sens. Environ.* **2019**, *224*, 365–381. [CrossRef]
82. Yamazaki, T. A One-dimensional Land Surface Model Adaptable to Intensely Cold Regions and its Applications in Eastern Siberia. *J. Meteorol. Soc. Jpn. Ser. II* **2001**, *79*, 1107–1118. [CrossRef]
83. Jordan, R. *A One-Dimensional Temperature Model for a Snow Cover: Technical Documentation for SNTHERM89*; Cold Regions Research and Engineering Laboratory: Hanover, NH, USA; Engineer Research and Development Center: Vicksburg, MS, USA, 1991; Available online: https://erdc-library.erdc.dren.mil/xmlui/handle/11681/11677 (accessed on 11 December 2018).
84. Rautiainen, M.; Nilson, T.; Lükk, T. Seasonal reflectance trends of hemiboreal birch forests. *Remote. Sens. Environ.* **2009**, *113*, 805–815. [CrossRef]

85. Kuusk, A.; Lang, M.; Kuusk, J.; Lükk, T.; Nilson, T.; Mõttus, M.; Rautiainen, M.; Eenmäe, A. *Database of Optical and Structural Data for the Validation of Radiative Transfer Models*; Technical Report Version 04.2015; Tartu Observatory: Tõravere, Estonia, 2015; p. 60. Available online: http://www.aai.ee/bgf/jarvselja_db/jarvselja_db.pdf (accessed on 2 December 2018).
86. Borden, J.H.; Campbell, S.A. Bark reflectance spectra of conifers and angiosperms: Implications for host discrimination by coniferophagous bark and timber beetles. *Can. Èntomol.* **2005**, *137*, 719–722.
87. Warren, S.G. Optical properties of snow. *Rev. Geophys.* **1982**, *20*, 67–89. [CrossRef]
88. Hovi, A.; Raitio, P.; Rautiainen, M. A spectral analysis of 25 boreal tree species. *Silva Fenn.* **2017**, *51*. [CrossRef]
89. Hollinger, D.Y.; Ollinger, S.V.; Richardson, A.D.; Meyers, T.P.; Dail, D.B.; Martín, M.E.; Scott, N.A.; Arkebauer, T.J.; Baldocchi, D.D.; Clark, K.L.; et al. Albedo estimates for land surface models and support for a new paradigm based on foliage nitrogen concentration. *Glob. Chang. Boil.* **2010**, *16*, 696–710. [CrossRef]

Article

Time Series High-Resolution Land Surface Albedo Estimation Based on the Ensemble Kalman Filter Algorithm

Guodong Zhang [1,2], Hongmin Zhou [1,*], Changjing Wang [1,2], Huazhu Xue [2,*], Jindi Wang [1] and Huawei Wan [3]

1 State Key Laboratory of Remote Sensing Science, Beijing Engineering Research Center for Global Land Remote Sensing Products, Faculty of Geographical Science, BNU, Beijing 100875, China; 201704020045@home.hpu.edu.cn (G.Z.); 201704020046@home.hpu.edu.cn (C.W.); wangjd@bnu.edu.cn (J.W.)
2 School of Surveying & Land Information Engineering, Henan Polytechnic University, Henan 454000, China
3 Satellite Environment Center, Ministry of Environmental Protection, Beijing 100094, China; wanhw@secmep.cn
* Correspondence: zhouhm@bnu.edu.cn (H.Z.); xhz@hpu.edu.cn (H.X.); Tel.: +86-10-58806011 (H.Z.); +86-391-3987661 (H.X.)

Received: 6 March 2019; Accepted: 26 March 2019; Published: 28 March 2019

Abstract: Continuous, long-term sequence, land surface albedo data have crucial significance for climate simulations and land surface process research. Sensors such as the Moderate-Resolution Imaging Spectroradiometer (MODIS) and Visible Infrared Imaging Radiometer (VIIRS) provide global albedo product data sets with a spatial resolution of 500 m over long time periods. There is demand for new high-resolution albedo data for regional applications. High-resolution observations are often unavailable due to cloud contamination, which makes it difficult to obtain time series albedo estimations. This paper proposes an "amalgamation albedo" approach to generate daily land surface shortwave albedo with 30 m spatial resolution using Landsat data and the MODIS Bidirectional Reflectance Distribution Functions (BRDF)/Albedo product MCD43A3 (V006). Historical MODIS land surface albedo products were averaged to obtain an albedo estimation background, which was used to construct the albedo dynamic model. The Thematic Mapper (TM) albedo derived via direct estimation approach was then introduced to generate high spatial-temporal resolution albedo data based on the Ensemble Kalman Filter algorithm (EnKF). Estimation results were compared to field observations for cropland, deciduous broadleaf forest, evergreen needleleaf forest, grassland, and evergreen broadleaf forest domains. The results indicated that for all land cover types, the estimated albedos coincided with ground measurements at a root mean squared error (RMSE) of 0.0085–0.0152. The proposed algorithm was then applied to regional time series albedo estimation; the results indicated that it captured spatial and temporal variation patterns for each site. Taken together, our results suggest that the amalgamation albedo approach is a feasible solution to generate albedo data sets with high spatio-temporal resolution.

Keywords: land surface albedo; time series; high spatio-temporal resolution; EnKF

---

1. Introduction

Land surface albedo, which is defined as the fraction of incident solar radiation (0.3–5.0 μm) reflected by land surfaces [1], is widely used in ground energy balance analysis, weather climate prediction, and climate change research [2,3]. Existing global albedo products include the Advanced Very High-resolution Radiometer (AVHRR) [4–6], Earth Radiation Budget Experiment (ERBE) [7], Moderate-Resolution Imaging Spectroradiometer (MODIS) [8], Multi-angle Imaging

SpectroRadiometer (MISR) [9], POLarization and Directionality of the Earth's Reflectances (POLDER) [10–12], Meteorological Satellites (Meteosat) [13], and Global Land Surface Satellite (GLASS) [14], are all conventional production. Their spatial resolution ranges from several hundred meters to several kilometers, which is much larger than the characteristic patch size for forest management, the typical field size of global agriculture, or the dominant extent of impervious surfaces in urban areas [15]. There is yet demand for new land surface albedo estimation methodologies which span longer time periods without sacrificing high spatio-temporal resolution.

Previous researchers have proposed various techniques for estimating high spatial resolution albedo with satellite imagery. Liang [16], for example, used an extensive radiative transfer model as a conversion formula for calculating total shortwave albedo, total-, direct-, and diffuse-visible, and near-infrared broadband albedos for several narrowband sensors. He [17,18] later applied the algorithm to the Chinese environment and disaster monitoring and forecasting small satellite constellation (HJ) data and Landsat series data to derive 30 m resolution albedo data. The results were validated at Surface Radiation (SURFRAD), AmeriFlux, Baseline Surface Radiation Network (BSRN), and Greenland Climate Network (GC-Net) sites; the direct estimation algorithm did provide accurate albedo estimations for different land cover types with root mean squared errors (RMSEs) ranging from 0.022 to 0.034 for snow-free surfaces. Shuai et al. [19] proposed an algorithm for generating land surface albedo at 30 m resolution using Landsat and the anisotropy information from Moderate-Resolution Imaging Spectroradiometer (MODIS) observations. Their estimated albedos showed an absolute accuracy of $\pm$ 0.02–0.05, RMSE less than 0.03, and a bias less than 0.02 by comparison against field measurements. Zhang [20] used the Spatial and Temporal Adaptive Reflectance Fusion Model (STARFM) to blend spatial information from fine-resolution shortwave albedo images and temporal information from coarse-resolution shortwave albedo images to successfully estimate high spatio-temporal resolution albedos. These methods all center on the use of Landsat imagery to obtain high spatial resolution albedos. However, Landsat data sets are affected by the long satellite return cycle and cloud contamination, and do not readily provide surface albedo data with high temporal resolution or time series albedo data.

Data assimilation is an effective approach to time series land surface parameter estimation. It is a mechanism to integrate various direct or indirect observation information types of varying resolution and from different sources to automatically adjust the model trajectory which provides accurate dynamic model state and model predictions. The Kalman filter algorithm was first developed in 1960 and later used to estimate time series soil moisture, leaf area index (LAI), and other surface parameters [21,22]. Li [23] proposed the Dual ensemble Kalman Filter (Dual EnKF) to estimate a time series LAI; Dual EnKF represents an updated LAI estimation with more sensitive parameters (LAI, Markov parameter, weight of the first price function, and weight of the second price function) in the dynamic model. Zhou [24] built a data-based mechanistic assimilation technique by coupling a revised universal data-based mechanistic model (LAI_UDBM) with a vegetation canopy radiative transfer model (PROSAIL). The Ensemble Kalman filtering algorithm was applied to enhance the LAI estimation accuracy.

Many other researchers have achieved notable results in this field. Shi [25] used the China Land Soil Moisture Date Assimilation System (CLSMDAS) to perform an assimilation experiment on soil moisture based on the EnKF and Land Surface Process Model; their results reasonably reflected the spatial and temporal distribution of soil moisture. Jin [26] used the EnKF algorithm coupled with the canopy radiative transfer model (ACRM) to accurately predict time series LAI data from a phenological model. Xu [27] proposed an algorithm that integrated Direct Insertion (DI) and Deterministic Ensemble Kalman Filter (DEnKF) methods to assimilate snow depth with surface albedo; the solution was shown to improve the precision of snow depth simulations. Although the EnKF assimilation method has been widely used in the inversion of many surface parameters, it has not yet been applied to the estimation of land surface albedos.

In this paper, we propose a time series high-resolution land surface albedo estimation algorithm based on the ensemble Kalman filter. The albedo dynamic model was established based on historical MODIS albedo products and Thematic Mapper (TM) observations, and albedo data was recursively updated via the EnKF method.

2. Study Areas and Data

2.1. Study Areas

We selected 18 FluxNet sites as our research areas. Among them, 15 sites were used for estimating and verifying the single-point time series albedo for five different land cover types (cropland; deciduous broadleaf forest; evergreen needleleaf forest; grassland; evergreen broadleaf forest). Five sites were selected as the central pixel for estimating and verifying regional time series albedos for each land cover type.

Due to the long satellite reentry cycle and cloud contamination, satellite data and ground data did not always match temporally. We also needed to match ground data with TM data in the time series to conduct this study, so the year of data used for each site in the experiment was inconsistent . Table 1 provides further information.

Table 1. Ground stations used for validation.

Site Name	Network	Latitude (°)	Longitude (°)	Land Cover Type	Data Year
CA-Oas *	FluxNet	53.6289N	106.1978W	DBF	2009
IT-Col *	FluxNet	41.8494N	13.5881E	DBF	2009
IT-Ro2 *	FluxNet	42.3903N	11.9209E	DBF	2010
US-Bar +	AmeriFlux	44.0646N	71.2881W	DBF	2009
FR-Pue *	FluxNet	43.7414N	3.5958E	EBF	2009
MY-Pso *	FluxNet	2.9730N	102.3062E	EBF	2009
AU-Wac *+	FluxNet	37.4259S	145.1878E	EBF	2007
CH-Dav *	FluxNet	46.8153N	9.8559E	ENF	2009
FI-Hyy *	FluxNet	61.8474N	24.2948E	ENF	2009
NL-Loo *	FluxNet	52.1666N	5.7436E	ENF	2010
IT-Sro +	FluxNet	43.7279N	10.2844E	ENF	2009
DE-Kli *	FluxNet	50.8931N	13.5224E	CRO	2009
IT-Bci *	FluxNet	40.5238N	14.9574E	CRO	2010
FR-Gri *	FluxNet	48.8442N	1.9519E	CRO	2010
US-Arm +	AmeriFlux	36.6058N	97.4888W	CRO	2010
AU-DaP *	FluxNet	14.0633S	131.3181E	GRA	2008
US-Ib2 *	AmeriFlux	41.8406N	88.2410W	GRA	2010
AU-Stp *+	FluxNet	17.1507S	133.3502E	GRA	2009

Cropland (CRO); deciduous broadleaf forest (DBF); evergreen needleleaf forest (ENF); grassland (GRA); evergreen broadleaf forest (EBF). Sites with * are used for estimating and verifying single-point time series albedo, sites with + are selected as the central pixel for estimating and verifying regional time series albedo.

2.2. Ground Verification Data

FluxNet is a global network of micrometeorological flux measurement sites that measure the exchanges of carbon dioxide, water vapor, and energy between the biosphere and atmosphere. It was established based on other observation networks including AmeriFlux, CarboEurope, AsiaFlux, OzFlux, and a few independent sites. At present, over 140 Flux tower stations are operating on a long-term and continuous basis. Data and site information are available online at the FluxNet website, http://fluxnet.fluxdata.org/. Land surface types include temperate conifer and broadleaf (deciduous and evergreen) forests, tropical and boreal forests, crops, grasslands, chaparral, wetlands, and tundra. Sites exist on five continents and their latitudinal distribution ranges from 70 °N to 30 °S. The FluxNet site records upwelling and downwelling shortwave radiative flux data with a half-hour observation period. For our purposes, we selected observation data from 0.5 hours before and after noon. The shortwave albedo of the site was calculated as the ratio of upwelling radiation and downwelling radiation. After eliminating invalid observations (filled data) and data with albedos less than 0 or greater than 1, each selected site was associated with sufficient time series ground albedo data for validation.

2.3. Landsat Satellite Data

Landsat sensors have continuously imaged the land surface since the 1970s [28]. The TM onboard Landsat 4 and 5 satellites with seven spectral bands sampled the shortwave range at a spatial resolution of 30 m from 1984 to 2011. TM images are an important remote sensing data source for earth resources and environments as per their high spatial resolution, spectral resolution, and positioning accuracy. We downloaded all available Landsat data during the ground measurement period. The distribution of available high-quality TM data corresponding to each FluxNet site within one year is shown in Figure 1. Abundant cloud-free TM data can be obtained for individual pixels; TM data are prone to contamination over larger areas. As shown in Figure 1, all of the sites have more than five available cloud-free TM images within one year. This provides reliable observation data for the effective operation of the EnKF algorithm.

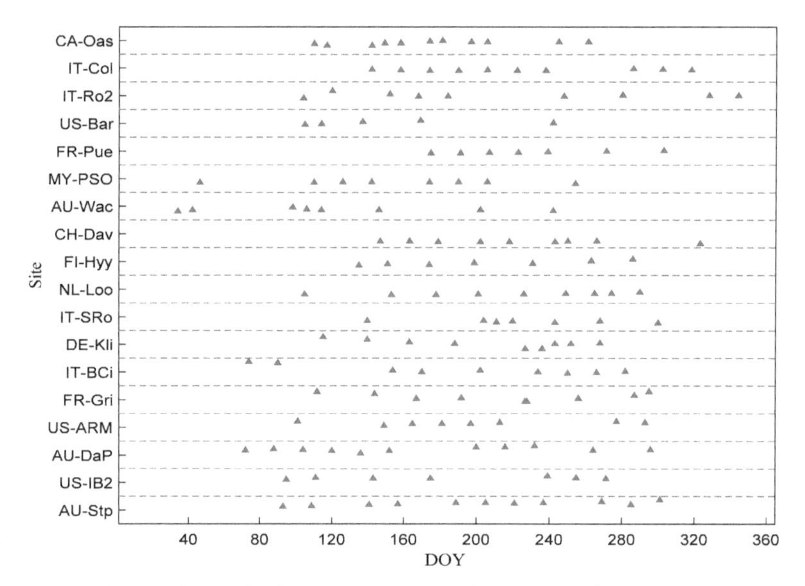

Figure 1. Distribution of available Thematic Mapper (TM) data corresponding to Flux sites (one year).

2.4. MCD43A3 BRDF/Albedo Product

MCD43A3 (Version 6) (V006) is the latest version of the MODIS Bidirectional Reflectance Distribution Functions (BRDF)/Albedo product. It includes bi-hemispherical reflectance (white-sky albedo) and directional-hemispherical reflectance (black-sky albedo) with a 500 m spatial resolution. The daily albedo is composed of 16-day multi-angle observations, where the Julian date of each specific file represents the 9th day of the 16-day retrieval period. MCD43A3 was produced based on the Algorithm for Model Bidirectional Reflectance Anisotropies of the Land Surface (AMBRALS), which uses all atmospherically corrected, high-quality, cloudless surface reflectance over the course of 16 days to achieve the best-fit surface bidirectional reflectivity via Ross–Li kernel models [29,30]. The Ross–Li kernel model is expressed as follows:

$$R(\theta, \vartheta, \phi, \Lambda) = f_{iso}(\Lambda) + f_{vol}(\Lambda)K_{vol}(\theta, \vartheta, \phi) + f_{geo}(\Lambda)K_{geo}(\theta, \vartheta, \phi) \quad (1)$$

where $R(\theta, \vartheta, \phi, \Lambda)$ is the frontal reflectivity with a solar zenith angle θ, observed zenith angle ϑ, relative azimuth ϕ, and wavelength band Λ. $f_{iso}(\Lambda)$ is the proportion of uniform scattering in all directions, $f_{vol}(\Lambda)$ is the proportion of body scattering, $f_{geo}(\Lambda)$ is the proportion of geometric optical scattering, $K_{vol}(\theta, \vartheta, \phi)$ is the RossThick kernel, and $K_{geo}(\theta, \vartheta, \phi)$ is the LiSparse kernel.

MODIS V006 products provide shortwave black-sky albedo and white-sky albedo [31], which must be converted into blue-sky shortwave albedo according to the proportion of sky scattered light [32].

$$\alpha(\theta_{\rm i}, \lambda) = (1 - {\rm s}(\theta_{\rm i}\tau(\lambda)))\alpha_{\rm bs}(\theta_{\rm i}, \lambda) + {\rm s}(\theta_{\rm i}\tau(\lambda))\alpha_{\rm ws}(\theta_{\rm i}, \lambda) \quad (2)$$

where $\alpha(\theta_{\rm i}, \lambda)$ is the blue-sky albedo of the band λ at a solar zenith angle of θ, $\alpha_{\rm bs}(\theta_{\rm i}, \lambda)$ is the black-sky albedo, $\alpha_{\rm ws}(\theta_{\rm i}, \lambda)$ is the white-sky albedo, and ${\rm s}(\theta_{\rm i}, \tau(\lambda))$ is the fraction of diffuse skylight when the solar zenith angle is θ, which is a function of aerosol optical depth and can be calculated using a predetermined look-up table (LUT) based on the 6S atmospheric radiative transfer code [33].

There was some data missing in MCD43A3 due to retrieval failure, so we used a gap filling algorithm to construct a continuous albedo data set. Missing data was filled with the average of all pixel values of the same land type in a given window (10 $\times$ 10 pixels) centered on the target pixel.

3. Methods

Our data assimilation methodology included an EnKF which recursively updated the albedo by coupling the direct estimation approach with a dynamic model. A flow chart of this process is shown in Figure 2. The existing high quality multi-year MODIS albedo product data is averaged to compute albedo climatology, then a simple dynamic model is established based on the climatology to evolve albedo over time and to forecast short-range albedos. A direct estimation approach is used to generate high-resolution TM albedo data. The EnKF technique is used to estimate real-time albedos by combining the predictions from the dynamic model and the high-resolution TM data.

The proposed method was essentially a three-step process:

(1) Obtain MODIS albedo data for the historical period of interest, correct them geometrically, convert black-sky albedo (BSA) and white-sky albedo (WSA) to blue-sky albedo using the scatter ratio, resample to a 30 m resolution, and average the albedo of the historical period to obtain the time-series shortwave albedo climatology;

(2) Transform the cloudless TM image into a 30 m spatial resolution TM shortwave albedo via direct estimation algorithm; and

(3) Use the EnKF method with the MODIS time series albedo as the background field and input the TM high spatial resolution albedo data to estimate the high spatial resolution albedo.

This method effectively integrated the time change information from MODIS data and spatial Landsat data to resolve issues with the low spatial resolution, sparsity, and long return cycle of MODIS data, as well as the sparsity of Landsat data to ultimately obtain high spatial and temporal resolution albedo data.

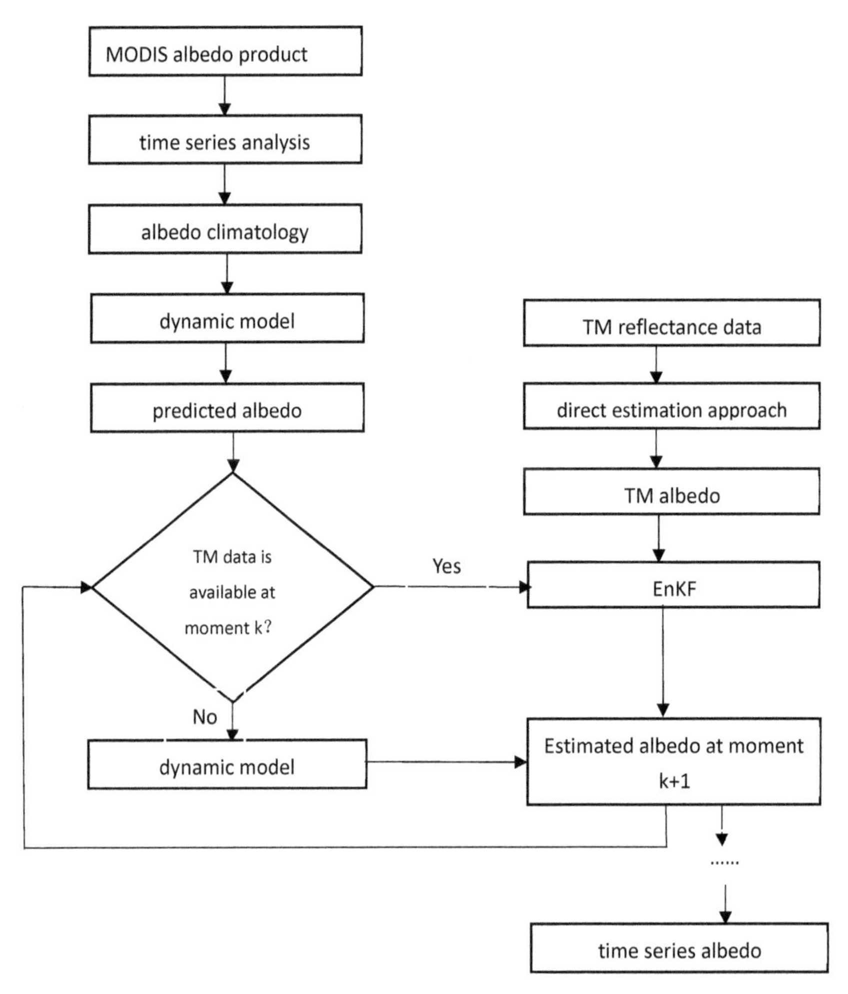

Figure 2. Real-time albedo inversion based on ensemble Kalman Filter (EnKF).

3.1. Validation Site Land Surface Heterogeneity

Previous researchers have verified the consistency of global land surface albedo products against ground measurement data at MODIS 500 m spatial scales at SURFRAD, Atmospheric Radiation Measurement Southern Great Plains (ARM/SGP) [34], and Cloud and Radiation Testbed-Southern Great Plains (CART/SGP) sites [35]. The MODIS albedo product has high accuracy with RMSE of 0.013–0.018. However, these verifications are based on homogeneous sites using point observations. In heterogeneous surface areas, site observations do not represent the MODIS pixels area, and thus the remote sensing products cannot be fully verified. As such, we must analyze site representativeness before validation with remote sensing data [36]. The heterogeneity of surface space describes the extent of surface heterogeneity within a certain range and reflects the heterogeneity in the area around FluxNet sites [37,38]. In this study, we used variogram model parameters and the relative coefficient of variation (R_{CV}) to analyze the effect of heterogeneity in the area around FluxNet sites according to Roman [39].

The isotropic spherical variogram model was used to evaluate landscape heterogeneity with 1.0 km^2, 1.5 km^2, and 2.0 km^2 TM albedo subsets. The TM albedo was obtained using the direct estimation approach proposed by He [18]. The spherical model formula is:

$$\gamma_{sph}(h) = \begin{cases} c_0 + c\cdot(1.5 \cdot \frac{h}{a} - 0.5(\frac{h}{a})^3) & \text{for } 0 \le h \le a \\ c_0 + c & \text{for } h > a \end{cases} \quad (3)$$

The range a defines the distance from a point beyond which there is no further correlation of a biophysical property associated with that point. It has also been described as the average patch size of the landscape when the curve reaches the high-level distance [40]. The data are correlated within the range of a. Otherwise, the data are not correlated with each other; that is, the observations outside the range do not affect the estimation results. For a 1.0 km^2 subset a_{max} = 690 m, for a 1.5 km^2 subset a_{max} = 1050 m, and for a 2.0 km^2 subset a_{max} = 1410 m. The sill c is the variation of ordinates when the abscissa is greater than the variable range. The nugget c_0 is the variation of the abscissa at 0, which describes the variability of the data at the microcosmic level.

The coefficient of variation (CV) is the ratio of the standard deviation to the mean. It is independent of spatial scale and can provide an estimate of the overall variability of data. The relative CV is:

$$R_{cv} = \frac{CV_{1.5x} - CV_x}{CV_x}; x = 1.0\ km^2 \quad (4)$$

where R_{cv} is the relative CV, $CV_{1.5x}$ is the CV calculated from a 1.5 km^2 TM albedo subset at the center of a given observation station, and CV_x is the CV calculated from a 1.0 km^2 TM albedo subset at the center of the given observation station. If a site is more representative, it has a weaker spatial heterogeneity, a similar landscape around the site, and R_{cv} closer to 0.

3.2. Albedo Dynamic Model Construction

Albedo climatology is determined by the average of albedo over historical period. In this study, we calculated albedo climatology from high-quality multi-year MODIS albedo data as follows:

$$ALB_t' = \frac{1}{n}\sum_{k=1}^{n} ALB_t(k) \quad (5)$$

where ALB_t' is the albedo value corresponding to time t on the background field curve, $ALB_t(k)$ is the MODIS albedo value at time t, and n is the year.

The albedo climatology describes the general change tendency of albedo in a one-year period. There may be some deviation from the ground measurements due to precipitation, anthropogenic activities, and other reasons. Albedo climatology simulates the general trend of land surface albedo and supplies the albedo estimation background. When new observations are available, the EnKF updates the estimation to produce a final value. We constructed our dynamic model based on the climatology used to forecast the short-range albedo:

$$ALB_t = F_t \times ALB_{t-1} \quad (6)$$

where ALB_t represents the current estimated albedo, ALB_{t-1} represents the estimated albedo at the preceding time step, and F_t is:

$$F_t = 1 + \frac{1}{ALB_t + \varepsilon} \times \frac{dALB_t}{dt} \quad (7)$$

where $\varepsilon = 10^{-3}$ prevents negative denominators and $\frac{dALB_t}{dt}$ is the growth rate of albedo at time t. The dynamic model was adopted from Samain et al. [41] and Xiao et al. [42].

3.3. High-Resolution Albedo Estimation

Surface albedo can be obtained directly from Top of Atmosphere (TOA) reflectance without requiring atmospheric correction [43]. The apparent surface albedo was separated from the inherent surface albedo, and the empirical relationship between the simulated surface albedo and TOA reflectance was built based on extensive radiative transfer simulations under a variety of atmospheric conditions. To mitigate errors from nonlinearities, a new statistical relationship was extended to generate albedo measurements from the spectral bands and three broadbands [18], including the visible (300–700 nm), near-infrared (NIR; 700–3000 nm), and total shortwave ranges (300–3000 nm), based on the following linear equation:

$$\alpha_\lambda = \sum \rho_i^{TOA} \cdot c_i + c_0 \tag{8}$$

where α_λ is the surface albedo for the spectral range of λ, ρ_i^{TOA} is the TOA reflectance for spectral band i, and c_i and c_0 are the regression coefficients.

The direct estimation approach simulates TOA reflectance from radiative transfer information. Here, the sensor spectral response functions obtained from the USGS and the MODIS BRDF database were used as inputs for the 6S radiative transfer code, aerosol types, water vapor, ozone, and CO_2. The linear regression coefficients for each of the geometrical combinations were precalculated and stored in LUTs for operational use. TOA reflectance was simulated using the regression models in the LUT. Cloud-free observations were selected from the quality control document, then the TM surface albedo was obtained from TOA reflectance data. The direct estimation approach has been successfully applied to Landsat Multispectral Scanner (MSS), Thematic Mapper (TM), Enhanced Thematic Mapper Plus (ETM+), and Operational Land Imager (OLI) sensors [18]; extensive validations have also been carried out at SURFRAD, AmeriFlux, BSRN, and GC-Net sites. The direct estimation approach can generate reliable surface albedo estimates with accuracy of 0.022 to 0.034 in terms of RMSEs over snow-free surfaces [44,45]. The direct estimation approach has also been used in OLI and Gaofen-1 satellite (GF-1) surface albedo inversion [46,47].

3.4. Ensemble Kalman Filter

In the 1960s, Kalman [48,49] and Bucy [48] developed the Kalman filter algorithm for optimal estimation of system states according to a linear system state equation and combined input and output system observation data. The traditional Kalman filter algorithm is often applied to linear problems [50,51]. It can estimate the state of a dynamic system from a series of data with measurement noise when the measurement variance is known, and is a common component in communication, navigation, guidance, and control applications. The dynamic model and observer in a Kalman filter are typically expressed as follows:

$$x_{k+1} = M_k x_k + \omega_{k+1} \tag{9}$$

$$z_k = H_k x_k + v_k \tag{10}$$

where x_k is the dynamic model state vector; $M_k \in R^{n \times n}$ is the system matrix, which changes with time; z_k represents observed values, and H_k is a time-varying observation system which transforms the state variable into the same value as the observed value. In this study, the observed values and state variable were land surface albedos. ω_k is the model error and v_k is observed noise. The albedo has different levels of inversion accuracy over time.

The predicted value of the state variable of the dynamic model at time k is defined as x_k^f. The data assimilation step serves to calculate the optimal estimate x_k^a of the state variables at moment k by combining the predicted value x_k^f and the observed value z_k of the model state variables at time k, x_k^a, the optimal estimate of x_k, is a linear function of x_k^f and z_k.

$$x_k^a = x_k^f + K_k[z_k - H_k x_k^f] \tag{11}$$

where $z_k - H_k x_k^f$ is observed incrementally (innovation) and K_k is the Kalman gain matrix. Data assimilation is conducted to minimize the variance of x_k^a by determining K_k. The update equations for state variables and covariance are:

$$x_k^a = x_k^f + K_k[z_k - H_k x_k^f] = x_k^f + P_k^f H_k^T (H_k P_k^f H_k^T + R_k)^{-1} (z_k - H_k x_k^f) \tag{12}$$

$$P_k^a = P_k^f - K_k H_k P_k^f = P_k^f - P_k^f H_k^T (H_k P_k^f H_k^T + R_k)^{-1} H_k P_k^f \tag{13}$$

where x_k^a is the updated state variable value, x_k^f is the predicted state variable, P_k^f is the predicted covariance, and P_k^a is the posterior covariance of state variables.

The ensemble Kalman filter algorithm is a complex sequential data assimilation method which produces non-linear models within a Kalman gain scheme. In the ensemble Kalman filter algorithm, x is defined as an n-dimensional model state vector and $A = (x_1, x_2, \ldots x_N) \in \Re^{n \times N}$ is composed of a collection of N model state vectors. The ensemble mean is stored in each column of $\overline{A} \in \Re^{n \times N}$. The ensemble perturbation matrix is defined as:

$$A' = A - \overline{A} \tag{14}$$

The ensemble covariance matrix is:

$$P_e = \frac{A'(A')^T}{N-1} \tag{15}$$

Given a vector of measurements $d \in \Re^m$, where m is the number of measurements, the observation matrix is expressed as follows:

$$D = (d_1, d_2, \ldots d_N) \in \Re^{m \times N} \tag{16}$$

Then, the standard analysis equation of the Kalman filter is:

$$A^a = A + A'A'^T H^T (HA'A'^T H + R)^{-1} (D - HA) \tag{17}$$

where $H \in \Re^{n \times N}$ (m is the number of measurements) is the measurement operator relating the model state to the observations; $R \in \Re^{m \times m}$ is the observational error covariance matrix, and $D \in \Re^{m \times N}$ is the disturbance observation matrix set. H is a linear operator which is not suitable in cases when the operator is nonlinear. By augmenting the model state vector, $\overline{x}^T = [x^T, h^T(x)]$. $\overline{Y} \in \Re^{n \times N}$ is defined as the set matrix of augmented state vectors and $\overline{Y'} \in \Re^{n \times N}$ is the set disturbance matrix of state variables. As such, the analytical equation is:

$$A^a = A + A'\overline{Y'}^T \overline{H'}^T (\overline{HY'}\overline{Y'}^T \overline{H'}^T + R)^{-1} (D - \overline{HY}) \tag{18}$$

where $\overline{H}$ is a new observation operator. The analytical equation can solve the data assimilation problem for which the observation operator is nonlinear.

In this study, we calculated the background error covariance from albedo background and field observations. We calculated observation error covariance values from TM albedo and field observations over the time series described above. At each time point, we generated N random noises with 0 as the mean and 0.01 as the variance. N is the ensemble number, which affects estimation efficiency and accuracy: A larger N has a lower calculation efficiency and higher accuracy. We set the ensemble number to 100 and added random noise values to the background field values of the current date to obtain N new background field values [52]. Finally, we calculated the standard deviation between the N background field values and the ground field observations as the background field errors of the current date. Observation errors were simulated in the same way. Figure 3 shows the error setting for the background field, TM albedo, and field observation for the AU-DaP site in 2008.

The "background error" and "TM error" labels represent simulated background errors and TM albedo errors, respectively. For regional use, the error of the area was set based on the error of the center pixel; a normal distribution noise with a mean of 0 and a variance of 0.01 was added for each pixel.

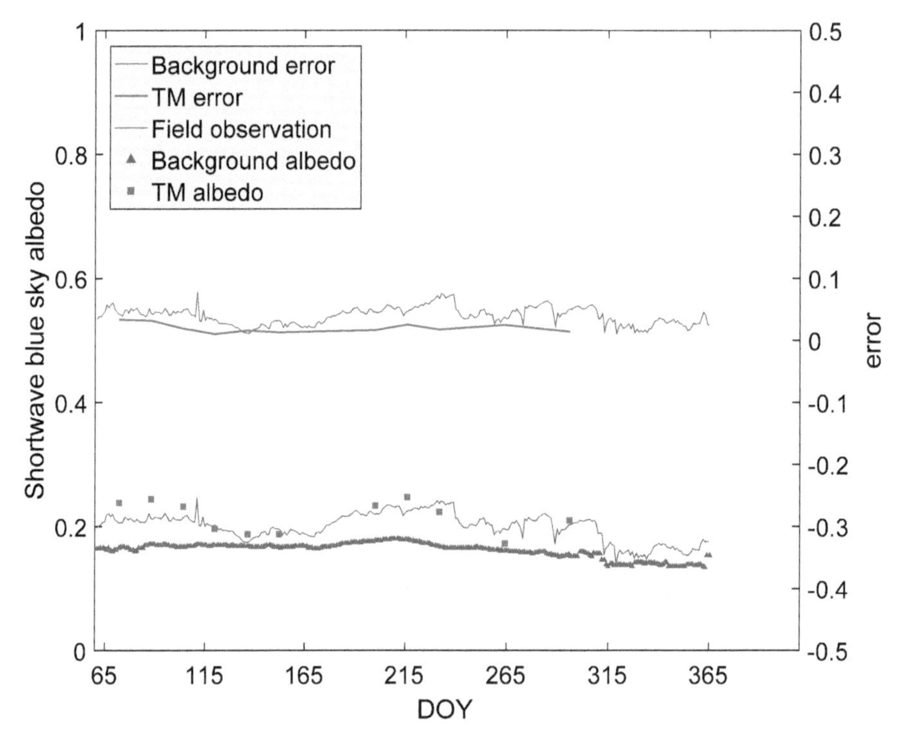

Figure 3. Error setting for albedo background and TM data.

Surface albedo is the only model state here, so the size of the model state vector n is equal to 1. We used the albedo value derived by the direct estimation approach as the albedo observation data, so the sum of the size of the model state vector n and the number of measurement equivalents added to the original model state vector was equal to 2. The simulated albedo was included in the state variables of the dynamic model at the given observation time. We emphasize the observation error covariance R here as per its significant influence on the A^a value. In our experiments, all the errors were simulated at site AU-DaP.

3.5. Validation

We used the time series field-observed local noon albedos at FluxNet sites to validate the albedos estimated with the EnKF algorithm. Spatially continuous albedos derived by the direct estimation approach were selected to validate the accuracy of estimated albedo at the regional scale. Single-point verification was performed at 15 FluxNet sites as shown in Table 1; the spatial representativeness of each FluxNet site was investigated according to the method described in Section 3.1. Three statistical parameters, coefficient of determination (R^2), RMSE and Bias, were used to evaluate the results. If the maximum error between the estimated result and the validated value did not exceed 0.05, the estimated result was considered accurate enough for scientific use [1,53].

4. Results

4.1. Land Surface Heterogeneity at FluxNet Sites

Figure 4a shows an 18 km subset TM albedo for IT-Col and a magnified image of the measurement point and spatial boundaries of 1.0 km^2 and 1.5 km^2, which provides a detailed visual representation of the landscape heterogeneity at the site. Figure 4b shows the variogram estimator (point values) for the TM albedo subsets. The spatial variability is more obvious over the larger squared regions (1.5 km^2 and 2.0 km^2), and different land cover types are more likely to be found at larger separation distances. This variogram estimator reaches an asymptote or constant variance between spatial uncorrelated samples near the sample variance. We calculated the range and R_{cv} for each Flux site to determine the heterogeneity of MODIS pixels as summarized in Table 2.

When the TM footprint increased from 1.0 km to 1.5 km, R_{cv} increased 0%–10% over CA-Oas, IT-Col, IT-Ro2, FR-Pue, AU-Wac, CH-Dav, NL-Loo, DE-Kli, IT-BCi, FR-Gri, US-ARM, AU-DaP, and US-IB2. These sites differ less from the surrounding landscape; when the TM footprint changes, R_{cv} only changes slightly relative to the smaller landscapes in the surrounding regions. The size of R_{cv} depends on the degree of landscape similarity within the given range. When surface heterogeneity around the site is weak, the overall degree of stationarity between regions is similar, the overall variability is small, and the value of R_{cv} is relatively low. R_{cv} increased 32.46% over HI-Fyy and 86.38% over IT-Sro as the TM footprint increased from 1.0 km to 1.5 km. Based on the range of FI-Hyy beyond the 1 km^2 limit, the landscapes around the two sites are obviously different; the small lakes near FI-Hyy are not within the 1 km^2 boundary, IT-Sro is near the sea, and internal ($CV_{1\ km}$) regions do not include the sea. At the US-Bar, MY-Pso, and AU-Stp stations, the results for R_{CV} were moderate (10%–20%). There are two types of vegetation around the three sites which include mixtures of grass and trees. There are some grasslands located approximately 1.0 km north of the US-Bar tower and southeast of the MY-Pso tower. A broadleaf forest is located southwest of the AU-Stp station, but the internal ($CV_{1\ km}$) does not include the forest. The overall fluctuation between regions was not high—the three sites were moderately heterogeneous.

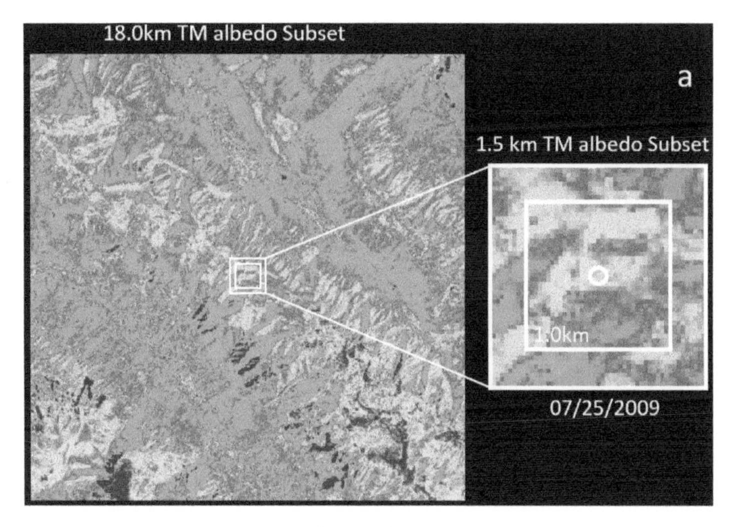

Figure 4. *Cont.*

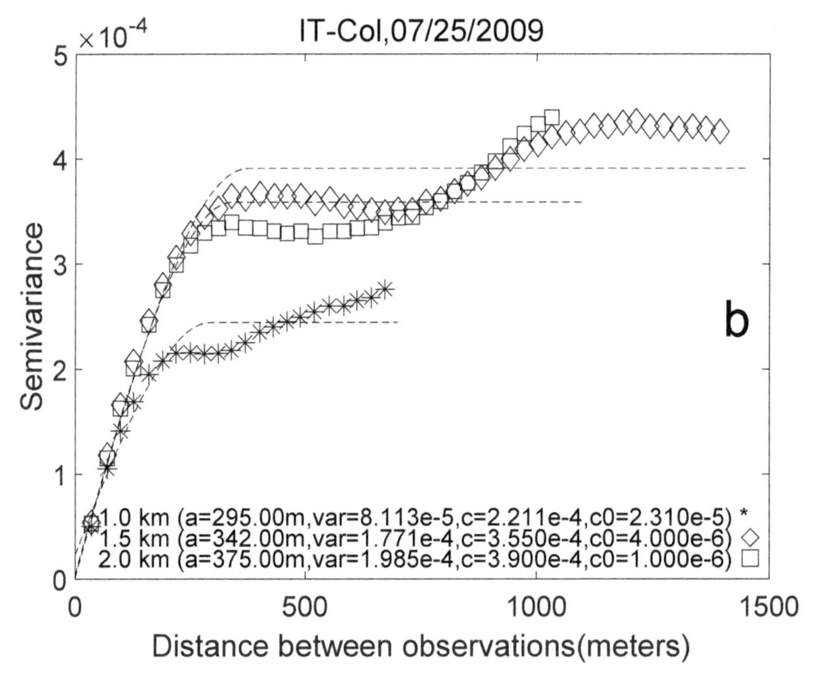

Figure 4. (**a**) Subset of 18.0 km TM albedo at IT-Col (measurement point) and spatial boundaries of 1.0 km^2 and 1.5 km^2. (**b**) Variogram plot, spherical model (dotted curves), and sample variance obtained via TM albedo on 25 July, 2009 for 1.0 km^2 (asterisks), 1.5 km^2 (diamonds), and 2.0 km^2 (squares) regions.

Table 2. R_{cv} for selected ground sites.

Site	TM Overpass Time	1 km Range	1.5 km Range	R_{cv}
CA-Oas	10-Aug-09	71.00 m	55.00 m	5.26%
IT-Col	25-Jul-09	295.00 m	342.00 m	7.44%
IT-Ro2	03-Jul-09	260.00 m	129.00 m	5.38%
US-Bar	17-May-10	168.00 m	353.00 m	16.87%
FR-Pue	26-Jul-09	624.00 m	332.00 m	7.8%
MY-PSO	11-Sep-09	45.00 m	644.00 m	14.38%
AU-Wac	16-Apr-07	109.00 m	112.00 m	3.40%
CH-Dav	06-Aug-09	1453.00 m	396.00 m	9.11%
FI-Hyy	31-May-09	877.00 m	155.00 m	32.46%
NL-Loo	06-Sep-10	70.00 m	34.00 m	5.33%
IT-SRo	23-Jul-09	304.00 m	641.00 m	86.38%
DE-Kli	24-Aug-09	304.00 m	193.00 m	7.25%
IT-BCi	09-Oct-10	234.00 m	182.00 m	8.94%
FR-Gri	24-May-10	114.00 m	87.00 m	7.44%
US-ARM	16-Jul-11	584.00 m	393.00 m	5.53%
AU-DaP	23-Aug-09	257.00 m	422.00 m	6.74%
US-IB2	12-Sep-10	216.00 m	243.00 m	4.30%
AU-Stp	25-Aug-09	470.00 m	341.00 m	16.77%

4.2. Estimating and Verifying Single-Point Time Series Albedo

We included five vegetated land surface types in this study: Deciduous broadleaf forest, evergreen broad-leaved forest, evergreen coniferous forest, grassland, and farmland. Three sites were subjected to validation for each land surface type. The results are shown in Figure 5.

The algorithm starts with cloudless TM albedo data spanning the whole year. The starting days for each site are different, so the MODIS albedo data for different land surface types have different degrees of overestimation and underestimation; however, they are relatively continuous in time and all reflect the variation characteristics of surface albedo in snow-free periods. The EnKF algorithm combines the advantages of MODIS and TM sensors. The estimated results were very close to the directly calculated TM albedo data when adequate TM data were included. When TM data is absent, the albedo background is used as the final estimation to maintain a complete time series. Our results also showed that with the continuous introduction of TM observations, the errors induced by MODIS background data could be corrected by EnKF. The albedo background is generated by resampling MODIS albedo data.

We also evaluated the heterogeneity of the test sites (Section 4.1) within the entire MODIS 500 m spatial scale. The results indicated that the proposed algorithm can adapt to heterogeneous surfaces (Figure 5e,h,o).

Figure 6 shows scatter plots of the estimated albedos and ground measurements. Different colors represent the three different sites for each land surface type.

As shown in Figure 6, RMSEs of the estimated albedos in deciduous broadleaf forest (DBF), evergreen broadleaf forest (EBF), evergreen needleleaf forest (ENF), cropland (CRO), and grassland (GRA) were 0.0152, 0.0085, 0.0087, 0.0152, and 0.0109; the R^2 values were 0.5980, 0.7061, 0.9190, 0.8191, and 0.6911, respectively. The RMSEs for all land surface types were less than 0.02, which meets the requirements of global and regional climate models [53]. The correlation coefficients between the estimated surface albedos and the surface measurements were high ($R^2 > 0.6$). The estimation results were close to the line $y = x$. The accuracies of ENF and CRO were the highest among all land surface types with R^2 values of 0.9190 and 0.8191, respectively. The lowest R^2 value was 0.5980 for DBF, which can be attributed to the dispersion of TM data (Figure 1). At ENF and CRO sites, the distribution of TM data was more uniform across the whole time series. At the DBF site, there was a slightly longer period of missing data which prevented the albedo from being updated in a timely manner. The estimation accuracy depended on the temporal distribution of the available TM data at a certain degree, as temporally uniform distributed TM data led to higher estimation accuracy. Nevertheless, the proposed algorithm greatly improves the estimation accuracy compared to field observation alone.

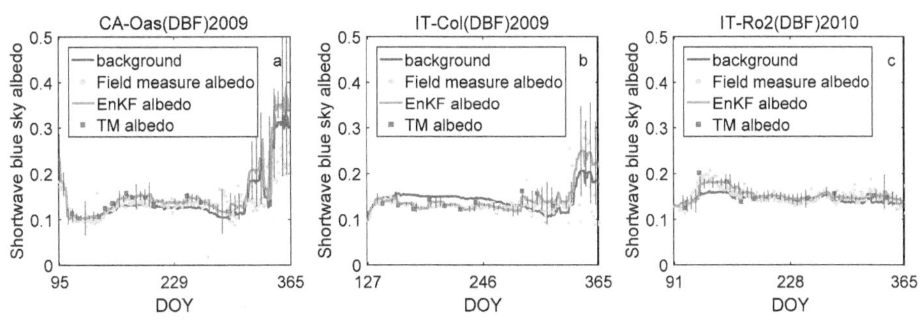

Figure 5. *Cont.*

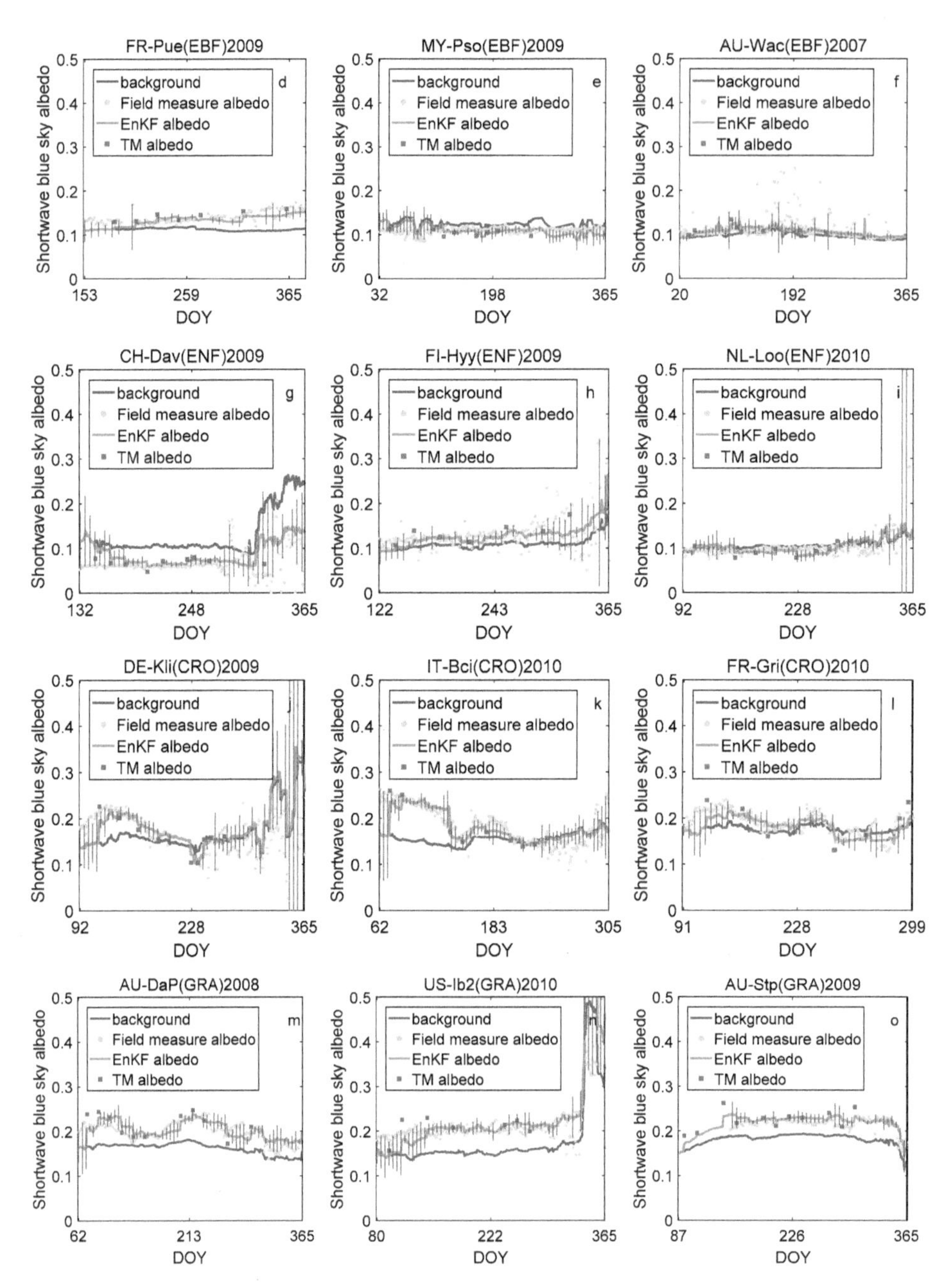

Figure 5. Time series albedo estimations based on EnKF for DBF (**a**–**c**), EBF (**d**–**f**), ENF (**g**–**i**), CRO (**j**–**l**), GRA (**m**–**o**); three sites for each surface type. Grey vertical line represents error between estimated and measured values.

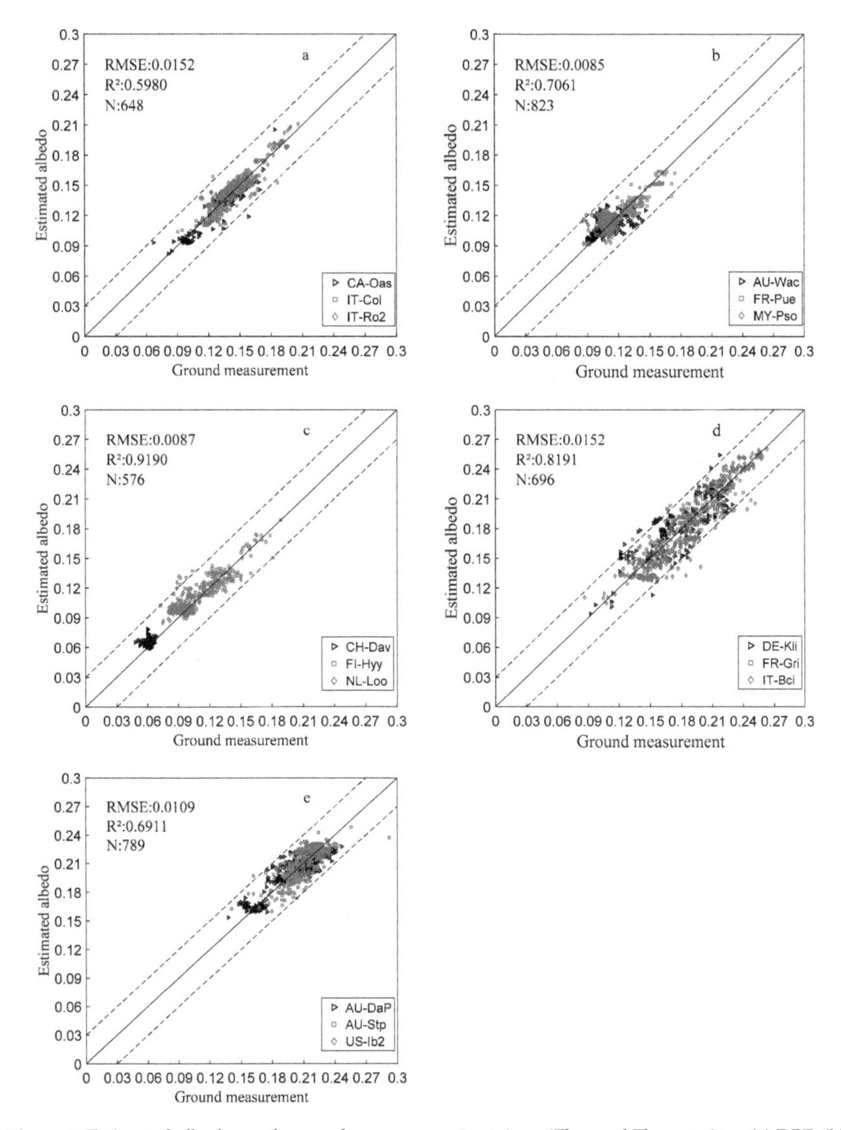

Figure 6. Estimated albedos and ground measurements at AmeriFlux and Flxunet sites: (**a**) DBF; (**b**) EBF; (**c**) ENF; (**d**) CRO; (**e**) GRA. Different colors represent different sites.

4.3. Regional Timing Albedo Estimation and Verification

We validated the proposed assimilation algorithm on five types of land surface areas, each area containing a ground-based station. The five stations, as mentioned above, were US-Bar, AU-Wac, IT-Sro, US-Arm, and AU-Stp (Table 1). The results are shown in Figure 7. The daily estimation result was difficult to display and the beginning date for each site was different, so we used nine days-worth of estimation results from each site.

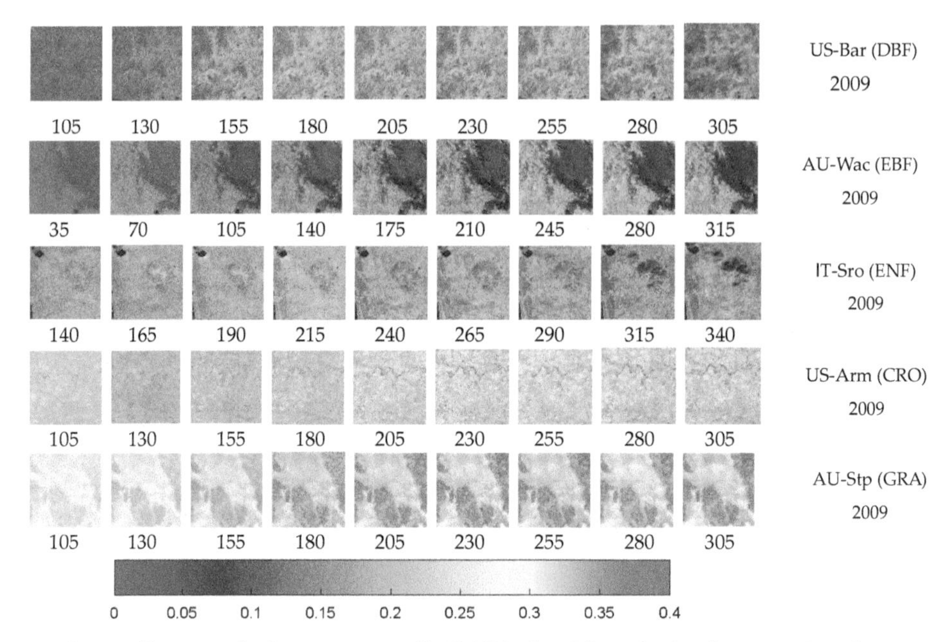

Figure 7. Shortwave albedo maps generated by EnKF for five different land surface types. Day of year (DOY) indicated for each albedo map. Estimations have one-day temporal resolution starting with the first cloud-free TM image (only a portion of the results are shown).

The EnKF assimilation method uses TM albedo as observation data. The estimation is strongly dependent on the quantity of TM albedo values. To obtain time-continuous, high spatial resolution albedo data, the EnKF algorithm updates the background field from the first cloudless TM image; therefore, assimilation results will be improved as more TM images become available. Figure 7 shows the estimated time series albedos for five different land cover types. Again, the starting point of each time series was not exactly the same as was determined by the availability of high-quality TM data. Starting with the first cloudless TM image, the algorithm can obtain an albedo dataset with a temporal resolution of one day and a spatial resolution of 30 m (shown here as 25-day or 35-day intervals), thereby reflecting albedo fluctuations within a one-year period.

As verification, we compared the albedo generated by the EnKF algorithm and the TM albedo directly estimated on the same date. Figure 8 shows that the EnKF and TM albedo results are consistent. The scatter plot has RMSE values between 0.0031 and 0.0112, and R^2 values between 0.8584 and 0.9494. One exception was the farmland area (US-Arm site), where regional estimation and TM albedo verification results had R^2 lower than the other four surface types, and RMSE higher than other landforms. This may be due to the fact that farmland surface vegetation is more affected by anthropogenic factors than other land surface types; the jump in surface albedo makes the deviation of farmland albedo estimations larger than others. In this case, inducing more high-resolution images into the assimilation process could effectively improve the albedo estimation accuracy.

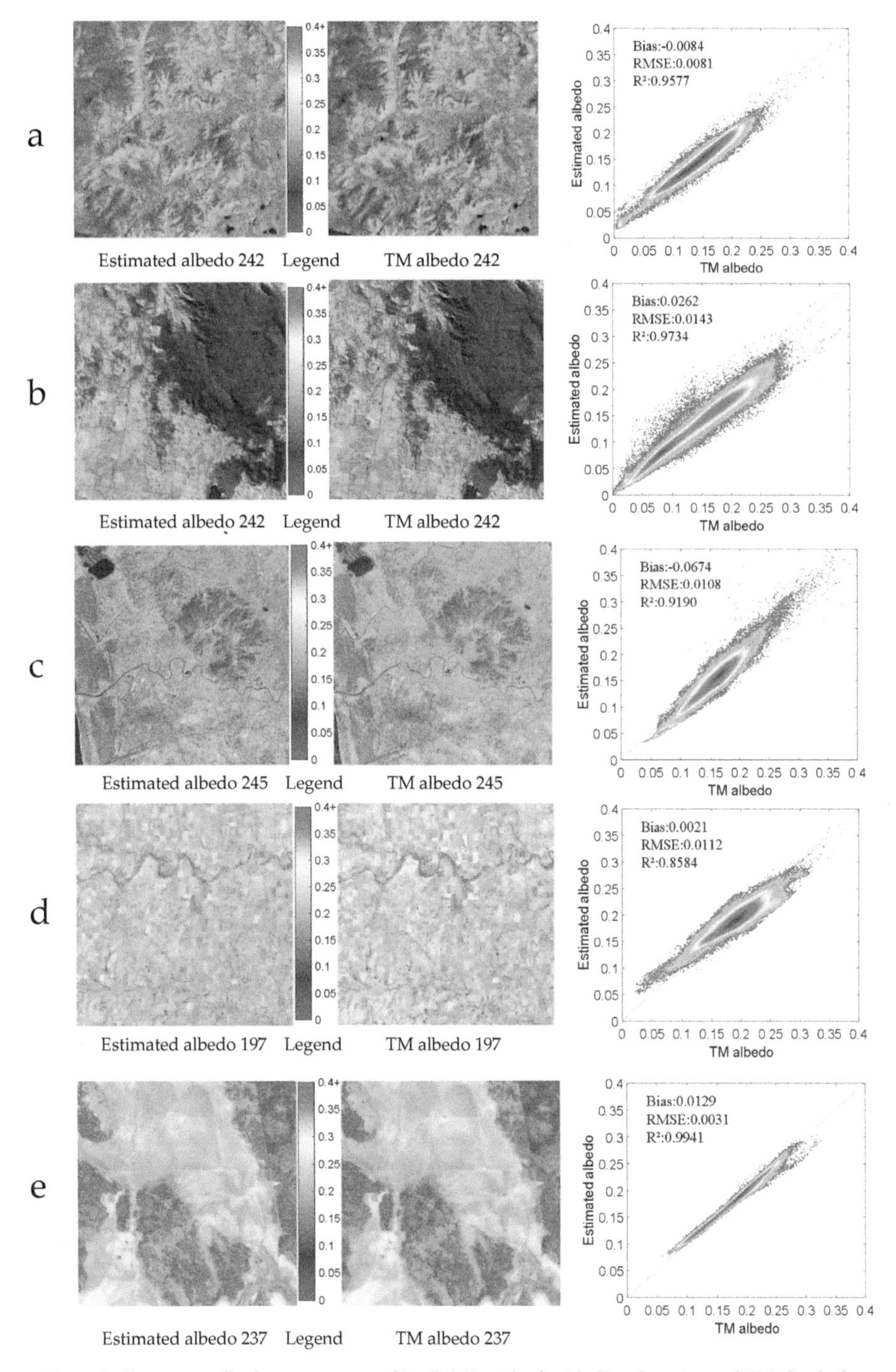

Figure 8. Shortwave albedo map generated by EnKF method with directly estimated TM albedo for five different land surface types: (**a**) DBF at US-Bar site (2009); (**b**) EBF at AU-Wac site (2007); (**c**) ENF at IT-Sro site (2009); (**d**) CRO at US-Arm site (2010); (**e**) GRA at AU-Stp site (2009). Right-hand scatter plots show consistency between the two sets of results, as number of dots gradually decreases from red to blue.

The DBF, EBF, and CRO sites selected in this study are heterogeneous research areas and have RMSE values less than 0.01 and R^2 values greater than 0.9, which indicate that the algorithm is well-suited to albedo estimation in heterogeneous areas.

5. Discussion

5.1. Accuracy of the Albedo Background

In the data assimilation algorithm, the background field is a preliminary estimate of target parameters, which is obtained from historical and empirical data, and reflects the general variation trend of the estimator [54]. Our EnKF algorithm results were obtained based on the optimal estimation of the current time. The background field trend introduces an observation increment in the optimal weight, so its accuracy is very important. We conducted spatial registration before constructing the background; after processing, the projection mode and resolution of MODIS data were consistent with Landsat. MODIS data actually provides background fields and dynamic models (preliminary estimates), so errors do exist, but they are allowed under the assimilation algorithm. Surface albedo presents a certain regularity in the growing season. We averaged the historical MODIS data, instead of the contemporary MODIS data, as the albedo background field because MODIS data were missing across our study area [55]. Consider the AU-DaP site as an example: The background field albedo has the same change trend as the original MODIS albedo, but with no missing data and an overall smoother structure. To this effect, the background field is more suitable for data assimilation.

The background field encompasses periodic albedo characteristics, but the albedo is overestimated or underestimated due to the accuracy of MODIS products (especially in heterogeneous pixels) [56]. Under the assimilation scheme, however, the background field only provides an initial prediction and the weight of background data is determined according to its accuracy. When the background accuracy is low, the weight is decreased and the weight of the TM observation is increased. Therefore, the estimation results gradually approach the field observations as the quantity of TM observations increases; using the mean albedo of the time series as a background field is feasible and effective.

5.2. Errors Induced by TM Albedo Estimation

In this study, TM albedos derived by direct estimation were utilized as albedo observation data. The TM albedos were used to adjust model-based predictions to bring them as close as possible to actual measured values. The quality of the TM data directly affects the accuracy of the final estimations. Previous researchers have thoroughly validated TM albedos derived by direct estimation [18]. In this study, we also compared the TM albedo with the ground measured albedo for several FluxNet sites. Figure 9 shows that the surface albedo estimated from TM images is highly accurate (maximum error < 0.03) and can be used as the input observation value for the data assimilation algorithm.

The accuracy of inversion is affected by the accuracy, quantity, and time distribution of TM data, where quantity and time distribution complement each other. In this study, we used L1T data with less than 10% cloud contamination for our estimations. There were approximately 5–11 TM data points available at the selected sites, which was sufficient for albedo estimation by data assimilation. If the quantity of available TM data in one year falls below 5 points, the albedo during the time series may not be updated for long periods of time, and the accuracy of the estimations degrades. Several high-quality observations are needed to ensure accurate estimations. Other high-resolution satellite images such as HJ1A/B or Sentinel-2 can be supplemented to resolve this problem for those areas where high-resolution observations are limited.

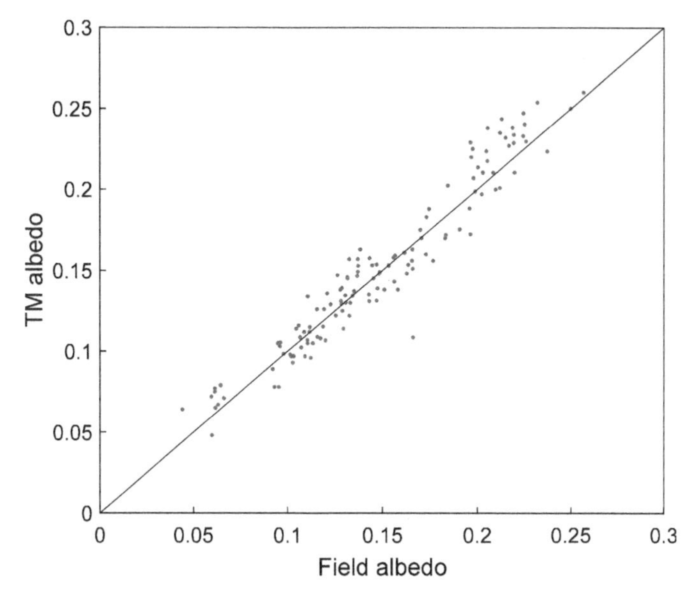

Figure 9. Directly estimated TM albedos and field albedos including all available TM images used in this study.

5.3. Error Setting in the Data Assimilation Algorithm

In the data assimilation system, the observation-model error is critical for land surface albedo estimation. We calculated observation error from field observations and TM data and calculated model error from field observations and the albedo background. Field observation data are key in terms of error determination. For each site, we used the field observation data to generate errors resulting in high estimation accuracy. When extending the model to regional use where no field observation data was available, we attempted to generate a "common" error by averaging the error of the 18 ground stations and inputting it to the assimilation scheme. The test results on the AU-DaP site are shown in Figure 10. Errors caused by TM data are relatively small and uniform, and using the mean errors instead of the original observation error did not affect the estimation results significantly. In effect, the common error from different sites represents the averaged field condition and can be used to account for large areas.

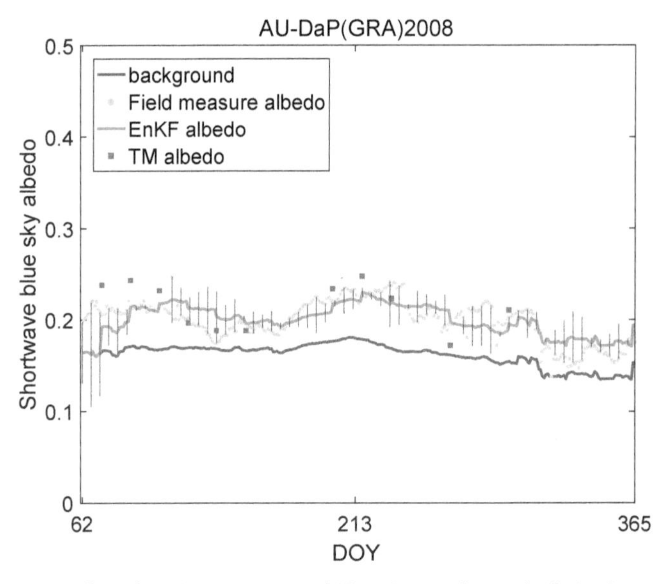

Figure 10. Estimated results using mean error of 18 stations as data assimilation input at AU-DaP. Replacing original observation error with mean error does not significantly affect the estimation results.

5.4. *Capability of Capturing Abrupt Variations in Land Surface Albedo*

Abrupt changes in surface albedo are often due to drastic changes in surface characteristics over a short period, which are difficult to capture using remote sensing data [57]. The background field as obtained from MODIS historical period data can reflect abrupt changes to a certain extent, but the effect is not obvious. If TM data are available during the period of abrupt change, the model prediction is updated and the estimated results are closer to the observed values; the abrupt change in surface albedo can be captured effectively.

In our time-series albedo estimation for farmland stations, surface albedo jump occurred on the 230th day at the DE-Kli site, the 150th day at the IT-Bci site, and the 230th and 250th days at the FR-Gri site. There are TM data available before and after these days, and the trajectory of the estimated results was corrected in time after adding the observed data. On the 170th day at the DE-Kli site, the 140th day at the IT-Bci site, and the 245th day at the FR-Gri site, there were no TM data available to update the model prediction and only the background field trajectory could be used for forward predictions. The abrupt variation in surface albedo is difficult to show and the estimated results contain deviations. The jump is more severe during snowy periods (especially snowfall and melting), where the surface albedo deviation may be higher than 0.5 and the EnKF algorithm cannot be applied effectively without support from TM data. To operate the surface albedo data assimilation algorithm, it is necessary to secure more high-quality satellite data as input observations. With support from more observations, the data assimilation algorithm is better able to capture abrupt variations and produce accurate estimations.

6. Conclusions

High spatio-temporal resolution albedo products are essential for climate simulations, as well as for various agricultural and environmental monitoring applications. In this study, we developed a novel method based on the ensemble Kalman filter algorithm to integrate MODIS high temporal resolution albedo and Landsat high spatial resolution albedo data to estimate high spatial-temporal resolution albedos.

We constructed an albedo background field using MODIS historical surface albedo data, from which the initial predicted albedo was obtained from a dynamic model. TM albedos derived by direct estimation were then used as observation data to update the initial predictions. The error of the dynamic model and observations was generated from field observations. When extending the model to regions where no field observation data were available, we created a "common" error by averaging the error of all test stations. The results were compared with ground measurement data for cropland, deciduous broadleaf forest, evergreen needleleaf forest, grassland, and evergreen broadleaf forest sites. We found that estimation accuracy is high (RMSE < 0.0152) for all land cover types. When applied to large areas, the proposed algorithm also shows high estimation accuracy both for homogeneous and heterogeneous regions. The estimated and TM albedos derived by direct estimation were in accordance with RMSE values of 0.003 to 0.0112, and R^2 values of 0.8584 to 0.9964.

The proposed algorithm has four distinct advantages over other similar methodologies. (1) It can generate reliable estimates of land surface albedo with high spatio-temporal resolution. (2) It can be used for all manner of snow-free vegetation surfaces, even heterogeneous surfaces. (3) It compensates for return-cycle and cloud contamination problems with high spatial resolution satellite image data. (4) The algorithm can be easily extended to other fine-resolution data similar to Landsat data (e.g., Sentinel-2).

The proposed algorithm does still have some drawbacks. For instance, the EnKF starts with the first available observation data to update the dynamic model, so the start time of high accuracy estimation depends on the availability of cloudless Landsat satellite images. The quality and quantity of observation data are crucial to the accuracy of the assimilation results. To further improve estimation accuracy, it is necessary to obtain more, higher-quality albedo data.

Author Contributions: Conceptualization, H.Z. and H.X.; Methodology, H.Z. and G.Z.; Software, H.Z. and G.Z.; Validation, G.Z.; Formal Analysis, H.Z. and G.Z.; Investigation, G.Z. and C.W.; Resources, H.Z., H.W. and J.W.; Data Curation, G.D.; Writing-Original Draft Preparation, G.Z.; Writing-Review & Editing, H.Z., J.W. and H.X.; Visualization, G.D. and C.W.; Supervision, H.Z. and H.X.; Project Administration, H.Z.; Funding Acquisition, H.Z.

Funding: This research was funded by the National Natural Science Foundation of China grant number 41801242, 41801366, the Key research and development program of China under grant 2016YFB0501502, the Chinese 973 Program under grant 2013CB733403, the Key Scientific and Technological Project of Henan Province under grant 172102110268.

Conflicts of Interest: The authors declare no conflict of interest.

References

1. Dickinson, R.E. Land surface processes and climate surface albedos and energy-balance. *Adv. Geophys.* **1983**, *25*, 305–353.
2. Ollinger, S.V.; Richardson, A.D.; Martin, M.E.; Hollinger, D.Y.; Frolking, S.E.; Reich, P.B.; Plourde, L.C.; Katul, G.G.; Munger, J.W.; Oren, R. Canopy nitrogen, carbon assimilation, and albedo in temperate and boreal forests: Functional relations and potential climate feedbacks. *Proc. Natl. Acad. Sci. USA* **2008**, *105*, 19336–19341. [CrossRef] [PubMed]
3. Amut, A.; Lu, G.; Yuan, Z. Spatial distributions of surface albedo from satellite data in arid oasis. *Proc. SPIE Int. Soc. Opt. Eng.* **2007**, *6679*, 66791V–66796V.
4. Csiszar, I.; Gutman, G. Mapping global land surface albedo from NOAA AVHRR. *J. Geophys. Res. Atmos.* **1999**, *104*, 6215–6228. [CrossRef]
5. Strugnell, N.C.; Lucht, W.; Schaaf, C. A global albedo data set derived from AVHRR data for use in climate simulations. *Geophys. Res. Lett.* **2001**, *28*, 191–194. [CrossRef]
6. Strugnell, N.C.; Lucht, W. An algorithm to infer continental-scale albedo from AVHRR data, land cover class, and field observations of typical BRDFs. *J. Clim.* **1999**, *14*, 1360–1376. [CrossRef]
7. Li, Z.; Garand, L. Estimation of surface albedo from space: A parameterization for global application. *J. Geophys. Res. Atmos.* **1994**, *99*, 8335–8350. [CrossRef]
8. Lucht, W.; Schaaf, C.B.; Strahler, A.H. An algorithm for the retrieval of albedo from space using semiempirical BRDF models. *IEEE Trans. Geosci. Remote Sens.* **2002**, *38*, 977–998. [CrossRef]

9. Diner, D.J.; Beckert, J.C.; Reilly, T.H.; Bruegge, C.J.; Conel, J.E.; Kahn, R.A.; Martonchik, J.V.; Ackerman, T.P.; Davies, R.; Gerstl, S.A.W. Multi-angle Imaging Spectro Radiometer (MISR) instrument description and experiment overview. *IEEE Trans. Geosci. Remote Sens.* **2007**, *98*, 1072–1087.
10. Leroy, M.; Deuzé, J.L.; Bréon, F.M.; Hautecoeur, O.; Herman, M.; Buriez, J.C.; Tanré, D.; Bouffiès, S.; Chazette, P.; Roujean, J.L. Retrieval of atmospheric properties and surface bidirectional reflectances over land from POLDER/ADEOS. *J. Geophys. Res. Atmos.* **1997**, *102*, 17023–17037. [CrossRef]
11. Hautecœur, O.; Leroy, M.M. Surface bidirectional reflectance distribution function observed at global scale by POLDER/ADEOS. *Geophys. Res. Lett.* **1998**, *25*, 4197–4200. [CrossRef]
12. Hautecoeur, O.; Roujean, J.L. Validation of POLDER surface albedo products based on a review of other satellites and climate databases. In Proceedings of the IEEE International Geoscience & Remote Sensing Symposium, Barcelona, Spain, 23–28 July 2007.
13. Pinty, B.; Roveda, F.; Verstraete, M.M.; Gobron, N.; Govaerts, Y.; Martonchik, J.V.; Diner, D.J.; Kahn, R.A. Surface albedo retrieval from Meteosat: 1. Theory. *J. Geophys. Res. Atmos.* **2000**, *105*, 18113–18134. [CrossRef]
14. Liang, S.; Zhao, X.; Liu, S.; Yuan, W.; Cheng, X.; Xiao, Z.; Zhang, X.; Liu, Q.; Cheng, J.; Tang, H. A long-term Global LAnd Surface Satellite (GLASS) data-set for environmental studies. *Int. J. Digit. Earth* **2013**, *6*, 5–33. [CrossRef]
15. Barnes, C.A.; Roy, D.P. Radiative forcing over the conterminous United States due to contemporary land cover land use albedo change. *Geophys. Res. Lett.* **2009**, *35*, 148–161. [CrossRef]
16. Liang, S. Narrowband to broadband conversions of land surface albedo I: Algorithms. *Remote Sens. Environ.* **2001**, *76*, 213–238. [CrossRef]
17. Tao, H.; Liang, S.; Wang, D.; Chen, X.; Song, D.X.; Bo, J. Land surface albedo estimation from Chinese HJ satellite data based on the direct estimation approach. *Remote Sens.* **2015**, *7*, 5495–5510.
18. He, T.; Liang, S.; Wang, D.; Cao, Y.; Gao, F.; Yu, Y.; Feng, M. Evaluating land surface albedo estimation from Landsat MSS, TM, ETM +, and OLI data based on the unified direct estimation approach. *Remote Sens. Environ.* **2018**, *204*, 181–196. [CrossRef]
19. Shuai, Y.; Masek, J.G. An algorithm for the retrieval of 30-m snow-free albedo from Landsat surface reflectance and MODIS BRDF. *Remote Sens. Environ.* **2011**, *115*, 2204–2216. [CrossRef]
20. Zhang, K.; Zhou, H.; Wang, J.; Xue, H. Estimation and validation of high temporal and spatial resolution albedo. *J. Remote Sens.* **2014**, *18*, 497–517.
21. Lü, H.; Yu, Z.; Zhu, Y.; Drake, S.; Hao, Z.; Sudicky, E.A. Dual state-parameter estimation of root zone soil moisture by optimal parameter estimation and extended Kalman filter data assimilation. *Adv. Water Res.* **2011**, *34*, 395–406. [CrossRef]
22. Zhao, Y.; Chen, S.; Shen, S. Assimilating remote sensing information with crop model using Ensemble Kalman Filter for improving LAI monitoring and yield estimation. *Ecol. Modell.* **2013**, *270*, 30–42. [CrossRef]
23. Li, X.; Xiao, Z.; Wang, J.; Song, J. Simultaneous estimation of LAI and dynamic model parameters using dual EnKF from time series MODIS data. In Proceedings of the International Conference on Multimedia Technology, Ningbo, China, 29–31 October 2010.
24. Zhou, H.; Chen, P.; Wang, J.; Liang, S.; Guo, L.; Zhang, K. A data-based mechanistic assimilation method to estimate time series LAI. In Proceedings of the Geoscience and Remote Sensing Symposium, Melbourne, Australia, 21–26 July 2013.
25. Shi, C.X.; Xie, Z.H.; Hui, Q.; Liang, M.L.; Yang, X.C. China land soil moisture EnKF data assimilation based on satellite remote sensing data. *Sci. China Earth Sci.* **2011**, *54*, 1430–1440. [CrossRef]
26. Jin, H.; Wang, J.; Xiao, Z.; Fu, Z. Leaf area index estimation from MODIS data using the ensemble Kalman smoother method. In Proceedings of the IEEE International Geoscience & Remote Sensing Symposium, Honolulu, HI, USA, 25–30 July 2010.
27. Xu, J.; Shu, H. Assimilating MODIS-based albedo and snow cover fraction into the Common Land Model to improve snow depth simulation with direct insertion and deterministic ensemble Kalman filter methods. *J. Geophys. Res. Atmos.* **2015**, *119*, 10, 684–610, 701. [CrossRef]
28. Chander, G.; Helder, D.L.; Markham, B.L.; Dewald, J.D.; Kaita, E.; Thome, K.J.; Micijevic, E.; Ruggles, T.A. Landsat-5 TM reflective-band absolute radiometric calibration. *IEEE Trans. Geosci. Remote Sens.* **2004**, *42*, 2747–2760. [CrossRef]
29. Roujean, J.L.; Leroy, M.; Deschamps, P.Y. A bidirectional reflectance model of the Earth's surface for the correction of remote sensing data. *J. Geophys. Res. Atmos.* **1992**, *97*, 20455–20468. [CrossRef]

30. Schaaf, C.B.; Gao, F.; Strahler, A.H.; Lucht, W.; Li, X.; Tsang, T.; Strugnell, N.C.; Zhang, X.; Jin, Y.; Muller, J.P. First operational BRDF, albedo nadir reflectance products from MODIS. *Remote Sens. Environ.* **2002**, *83*, 135–148. [CrossRef]
31. Svacina, N.A.; Duguay, C.R.; King, J.M.L. Modelled and satellite-derived surface albedo of lake ice—Part II: Evaluation of MODIS albedo products. *Hydrol. Processes* **2014**, *28*, 4562–4572. [CrossRef]
32. Privette, J.L.; Mukelabai, M.; Zhang, H.; Schaaf, C.B. Characterization of MODIS land albedo (MOD43) accuracy with atmospheric conditions in Africa. In Proceedings of the IGARSS 2004 IEEE International Geoscience and Remote Sensing Symposium, Anchorage, AK, USA, 20–24 September 2004.
33. Zhou, H.; Wang, J.; Liang, S. Design of a novel spectral albedometer for validating the MODerate resolution Imaging Spectroradiometer Spectral Albedo Product. *Remote Sens.* **2018**, *10*, 101. [CrossRef]
34. Salomon, J.G.; Schaaf, C.B.; Strahler, A.H.; Gao, F.; Jin, Y. Validation of the MODIS bidirectional reflectance distribution function and albedo retrievals using combined observations from the aqua and terra platforms. *IEEE Trans. Geosci. Remote Sens.* **2006**, *44*, 1555–1565. [CrossRef]
35. Jin, Y.; Schaaf, C.B.; Woodcock, C.E.; Gao, F.; Li, X.; Strahler, A.H.; Lucht, W.; Liang, S. Consistency of MODIS surface bidirectional reflectance distribution function and albedo retrievals: 2. Validation. *J. Geophys. Res. Atmos.* **2003**, *108*, 4159. [CrossRef]
36. Cescatti, A.; Marcolla, B.; Vannan, S.K.S.; Pan, J.Y.; Román, M.O.; Yang, X.; Ciais, P.; Cook, R.B.; Law, B.E.; Matteucci, G. Intercomparison of MODIS albedo retrievals and in situ measurements across the global FLUXNET network. *Remote Sens. Environ.* **2012**, *121*, 323–334. [CrossRef]
37. Moon, S.K.; Ryu, Y.R.; Lee, D.H.; Kim, J.; Lim, J.H. Quantifying the spatial heterogeneity of the land surface parameters at the two contrasting KoFlux Sites by Semivariogram. *J. Agric. For. Meteorol.* **2007**, *9*, 140–148. [CrossRef]
38. Tittebrand, A. Analysis of the spatial heterogeneity of land surface parameters and energy flux densities. *Atmos. Chem. Phys.* **2009**, *9*, 2075–2087. [CrossRef]
39. Román, M.O.; Schaaf, C.B.; Woodcock, C.E.; Strahler, A.H.; Yang, X.Y.; Braswell, R.H.; Curtis, P.S.; Davis, K.J.; Dragoni, D.; Goulden, M.L. The MODIS (Collection V005) BRDF/albedo product: Assessment of spatial representativeness over forested landscapes. *Remote Sens. Environ.* **2009**, *113*, 2476–2498. [CrossRef]
40. Cooper, S.D.; Barmuta, L.; Sarnelle, O.; Kratz, K.; Diehl, S. Quantifying spatial heterogeneity in streams. *J. N. Am. Benthol. Soc.* **1997**, *16*, 174–188. [CrossRef]
41. Samain, O.; Roujean, J.-L.; Geiger, B. Use of a Kalman filter for the retrieval of surface BRDF coefficients with a time-evolving model based on the ECOCLIMAP land cover classification. *Remote Sens. Environ.* **2008**, *112*, 1337–1346. [CrossRef]
42. Xiao, Z.; Liang, S.; Wang, J.; Jiang, B.; Li, X. Real-time retrieval of Leaf Area Index from MODIS time series data. *Remote Sens. Environ.* **2011**, *115*, 97–106. [CrossRef]
43. Liang, S.; Strahler, A.; Walthall, C. Retrieval of land surface albedo from satellite observations: A simulation study. In Proceedings of the 1998 IEEE International Geoscience and Remote Sensing Symposium IGARSS '98, Seattle, WA, USA, 6–10 July 1998; Volume 1283, pp. 1286–1288.
44. Shuai, Y.; Masek, J.G.; Gao, F.; Schaaf, C.B.; He, T. An approach for the long-term 30-m land surface snow-free albedo retrieval from historic Landsat surface reflectance and MODIS-based a priori anisotropy knowledge. *Remote Sens. Environ.* **2014**, *152*, 467–479. [CrossRef]
45. Franch, B.; Vermote, E.; Claverie, M. Intercomparison of Landsat albedo retrieval techniques and evaluation against in situ measurements across the US SURFRAD network. *Remote Sens. Environ.* **2014**, *152*, 627–637. [CrossRef]
46. He, T.; Liang, S. Mapping surface Albedo from the complete landsat archive since the 1980s and its cryospheric application. In Proceedings of the IGARSS 2018 IEEE International Geoscience and Remote Sensing Symposium, Valencia, Spain, 22–27 July 2018.
47. Zhou, H.; Hu, N.; He, T.; Liang, S.; Wang, J. High resolution Albedo estimation with Chinese GF-1 WFV data. In Proceedings of the IGARSS 2018 IEEE International Geoscience and Remote Sensing Symposium, Valencia, Spain, 22–27 July 2018.
48. Kalman, R.E.; Bucy, R.S. New results in linear filtering and prediction theory. *ASME J. Basic Eng.* **1961**, *83*, 95–108. [CrossRef]
49. Kalman, R.E. A new approach to linear filtering and prediction problems. *ASME J. Basic Eng.* **1960**, *82*, 35–45. [CrossRef]

50. Angus, J. Forecasting, structural time series and the Kalman filter. *J. Oper. Res. Soc.* **1991**, *34*, 496–497. [CrossRef]
51. Lefferts, E.J.; Markley, F.L.; Shuster, M.D. Kalman Filtering for spacecraft attitude estimation. *J. Guid. Control Dynam.* **1982**, *5*, 536–542. [CrossRef]
52. Yin, X.; Xiao, Z. Optimal integration of MODIS and MISR albedo products. In Proceedings of the International Symposium on Image and Data Fusion, Tengchong, China, 9–11 August 2011.
53. Sellers, P.J.; Meeson, B.W.; Hall, F.G.; Asrar, G.; Murphy, R.E.; Schiffer, R.A.; Bretherton, F.P.; Dickinson, R.E.; Ellingson, R.G.; Field, C.B. Remote sensing of the land surface for studies of global change: Models—Algorithms—Experiments. *Remote Sens. Environ.* **1995**, *51*, 3–26. [CrossRef]
54. Xiao, Z.; Liang, S.; Wang, J.; Wu, X. Use of an ensemble Kalman filter for real-time inversion of leaf area index from MODIS time series data. In Proceedings of the Geoscience and Remote Sensing Symposium, Honolulu, HI, USA, 25–30 July 2010.
55. Friedl, M.A.; Mciver, D.K.; Hodges, J.C.F.; Zhang, X.Y.; Muchoney, D.; Strahler, A.H.; Woodcock, C.E.; Gopal, S.; Schneider, A.; Cooper, A. Global land cover mapping from MODIS: Algorithms and early results. *Remote Sens. Environ.* **2002**, *83*, 287–302. [CrossRef]
56. Serbin, S.P.; Ahl, D.E.; Gower, S.T. Spatial and temporal validation of the MODIS LAI and FPAR products across a boreal forest wildfire chronosequence. *Remote Sens. Environ.* **2013**, *133*, 71–84. [CrossRef]
57. Bolten, J.D.; Gupta, M.; Gatebe, C.K.; Ichoku, C.M. Regional land surface hydrology impacts from fire-induced surface Albedo darkening in Northern Sub-Saharan Africa. In Proceedings of the AGU Fall Meeting, San Francisco, CA, USA, 14–18 December 2015.

Article

Mapping Climatological Bare Soil Albedos over the Contiguous United States Using MODIS Data

Tao He [1,2,*], Feng Gao [3], Shunlin Liang [1,4] and Yi Peng [1]

1 School of Remote Sensing and Information Engineering, Wuhan University, Wuhan 430079, China; sliang@umd.edu (S.L.); ypeng@whu.edu.cn (Y.P.)
2 State Key Laboratory of Information Engineering in Surveying, Mapping and Remote Sensing, Wuhan University, Wuhan 430079, China
3 USDA-ARS Hydrology and Remote Sensing Laboratory, Beltsville, MD 20705, USA; feng.gao@ars.usda.gov
4 Department of Geographical Sciences, University of Maryland, College Park, MD 20742, USA
* Correspondence: taohers@whu.edu.cn; Tel.: +86-27-68778913

Received: 30 January 2019; Accepted: 16 March 2019; Published: 19 March 2019

Abstract: Surface bare soil albedo is an important variable in climate modeling studies and satellite-based retrievals of land-surface properties. In this study, we used multiyear 500 m albedo products from the Moderate Resolution Imaging Spectroradiometer (MODIS) to derive the bare soil albedo for seven spectral bands and three broadbands over the contiguous United States (CONUS). The soil line based on red and green spectral signatures derived from MODIS data was used as the basis to detect and extract bare soil albedo. A comparison against bare soil albedo derived from 30 m Landsat data has been made, showing that the MODIS bare soil albedo had a bias of 0.003 and a root-mean-square-error (RMSE) of 0.036. We found that the bare soil albedo was negatively correlated with soil moisture from the Advanced Microwave Scanning Radiometer-Earth Observing System (AMSR-E), with a relatively stable exponential relationship reflecting the darkening effect that moisture has on most soils. However, quantification of the relationship between bare soil albedo and soil moisture still needs to be improved through simultaneous and instantaneous measurements at a finer spatial resolution. Statistics of the multiyear climatological bare soil albedos calculated using soil types and the International Geosphere-Biosphere Programme (IGBP) land cover types suggest that: Land cover type is a better indicator for determining the magnitude of bare soil albedos for the vegetated areas, as the vegetation density is correlated with soil moisture; and soil type is a better indicator for determining the slope of soil lines over sparsely vegetated areas, as it contains information of the soil texture, roughness, and composition. The generated bare soil albedo can be applied to improve the parameterization of surface energy budget in climate and remote sensing models as well as the retrieval accuracy of some satellite products.

Keywords: bare soil albedo; MODIS albedo; contiguous United States; soil line; Landsat albedo; soil moisture

---

1. Introduction

Bare soil albedo has been widely used in climate models and remote sensing estimates of surface energy balance as one key component of the surface albedo by determining the amount of solar radiation reflected and absorbed at the Earth's surface [1,2]. Bare soil albedo has also been widely used in ecological research as a controlling factor in algorithms for deriving leaf area index (LAI) and fraction of Photosynthesis Active Radiation (fPAR) from satellite observations, which requires fine resolution data for satellite applications, e.g., [3,4]. A limited number of prescribed bare soil albedo values are assigned based on global soil color maps [5–7]. However, soil reflectivity varies both temporally and spatially. Besides soil color, bare soil albedo is also a function of moisture content [8,9], organic matter [10], texture/roughness [11], and other surface characteristics. Oversimplified soil albedo parameterization has been reported to introduce uncertainties both in climate modeling [12,13] and satellite LAI/fPAR products [14–17]. Significant differences in soil albedo were found in comparisons of land surface models and satellite products [12], which lead to substantially different estimations of surface energy balance and hydrologic budget partitioning [18].

Efforts have been made towards generating global or regional soil albedo datasets that are independent of coarse resolution soil color maps. Zhou et al. [19] analyzed the Moderate Resolution Imaging Spectroradiometer (MODIS) 1 km albedo products over desert areas using principle component analysis and found the extracted spatial pattern could improve the soil albedo parameterization in climate models. In a recent study, a method has been proposed to estimate soil albedo by removing impacts of solar zenith angle and soil moisture empirically from MODIS albedo products [20]. In their method, coarse resolution (~15–100 km) soil moisture data from the North American and Global Land Data Assimilation System (LDAS) were used.

To develop a bare soil albedo dataset over densely vegetated area and without the support of reanalysis data, a method has been developed based on an empirical relationship between broadband albedos and the Normalized Difference Vegetation Index (NDVI) [21]. In their method, broadband albedo of bare soil over densely vegetated area was predicted through extrapolation of linear logarithmic relationship between albedo and NDVI. It was proposed that bare soil albedos were estimated at NDVI = 0.09 globally, based on the empirical relationship. However, we found that this empirical relationship could generate larger values for visible albedo than near infrared albedo due to lack of training data for extrapolation purposes. In addition, previous studies have proved that NDVI may not be a good quantitative indicator of surfaces with sparse vegetation coverage because of mutual shadow effects on surface anisotropy [22,23].

Pisek and Chen [16] proposed a method using multi-angular satellite observations to map background spectral reflectance in forested pixels and found significant variations between coniferous and deciduous forests, particularly in the near infrared wavelengths.

In some recent studies, MODIS LAI and fPAR products were introduced to calculate vegetation fraction used in a linear regression to estimate soil albedo and the difference of vegetation and soil albedos based on the assumption that these two variables were invariant within a certain period [24,25]. This was later improved by using a Kalman filter to generate dynamic albedo on a daily basis [26]. However, MODIS LAI/fPAR products were derived assuming a prescribed bare soil albedo based on a soil type map, and these products were believed to have lower reliability during the vegetation growing season [14,15].

Existing bare soil albedo datasets are subject to several limitations. First, most suffer from uncertainties in ancillary input products used in the albedo retrieval algorithm due to, for example residual cloud contamination, inaccurate soil maps, or scale differences between model inputs and the retrieval scale. Second, most of the existing global bare soil albedo datasets are only available at a spatial resolution coarser than 5 km, which cannot satisfy the increasing demand for high resolution soil parameterizations, especially in agricultural and ecological applications, e.g., [27]. Third, validation of these datasets has been generally limited to inter-comparison against model-simulated results.

In addition to mapping the broadband albedo for bare soil, there is also a need to generate spatially dynamic background spectral reflectance/albedo to improve the LAI/fPAR estimates from remote sensing data [16]. Thus, it is important to develop an approach that can generate both spectral and broadband albedos at a fine spatial resolution (e.g., 30–500 m) to satisfy the needs of climate modeling and ecosystem monitoring purposes. Significant impacts of soil moisture content on soil reflectivity have been reported based on laboratory measurements [28,29], ground measurements [8,9,30], and satellite products [31–33]. However, few studies have demonstrated the relationship over a large spatial domain, which is critical for land surface energy balance and hydrological modeling purposes.

The concept of a "soil line", a linear relationship between red and near infrared (NIR) reflectances typically observed over bare soils, has been widely used to discriminate vegetation from soil using spectral information [34–39]. The soil line has been commonly used to develop the vegetation indices, such as the simple ratio (SR), NDVI, Enhanced Vegetation Index (EVI), and Soil-Adjusted Vegetation Index (SAVI), to identify dense vegetation. However, this concept had mostly been applied to extracting information about the vegetation canopy rather than the understory and soil.

The purpose of this study is to build a 500 m bare soil albedo dataset over the contiguous United States (CONUS) on a pixel-basis from multi-year MODIS data based on the soil line feature and to explore the relationship between bare soil albedo and other surface properties such as soil moisture, soil type, and vegetation type. A description of the data and methodology is given in Section 2. Comparison results against Landsat data are presented in Section 3 and are followed by the evaluation of soil albedo with regards to soil moisture content, soil type, and land cover type.

2. Materials and Methods

2.1. MODIS Albedo Anisotropy Products

The MODIS albedo and bidirectional reflectance distribution function (BRDF) products (MCD43A) [40] are available at a 500 m resolution globally using a 16-day temporal acquisition window. Of the available albedo anisotropy products, this study primarily used three datasets: Spectral/broadband albedos, nadir view corrected surface spectral reflectance, and quality control (QC) flags (overall accuracy and snow flag). Two types of albedos were included in the MODIS products: Black-sky albedo (BSA), also called directional-hemispherical albedo, and white-sky albedo (WSA), also called bi-hemispherical albedo. Datasets covering CONUS were chosen for the 13-year period from 2000 to 2012. Information of the datasets used in this study is summarized in Table 1.

To assist in the removal of possible vegetation, water, snow, and residual cloud shadow contaminated albedo values, the NDVI and normalized difference water index (NDWI) [41] were used, calculated from nadir BRDF-adjusted reflectances (NBAR) for the spectral bands (Equations (1) and (2)).

$$NDVI = \frac{\rho_{b2} - \rho_{b1}}{\rho_{b2} + \rho_{b1}}, \quad (1)$$

$$NDWI = \frac{\rho_{b2} - \rho_{b5}}{\rho_{b2} + \rho_{b5}}, \quad (2)$$

where ρ_{b1}, ρ_{b2}, and ρ_{b5} are the spectral reflectances for MODIS band 1 (620–670 nm), band 2 (841–876 nm), and band 5 (1230–1250 nm), respectively.

Table 1. Information of the datasets used in this study.

Variable	Dataset	Temporal Coverage	Spatial Resolution
Surface albedo	MCD43A	2001–2012	500 m
Soil type	Natural Resources Conservation Service (NRCS) soil suborder map	N/A	4000 m
Land cover	MCD12Q	2006	500 m
Land cover	National Land Cover Dataset (NLCD)	2006	30 m
Soil moisture	AMSR-E L3	2002–2011	25 km

2.2. Soil Type Map

Current climate and ecological modeling applications generally use maps of soil type to assign albedo values to certain locations [42] because soil type is believed to be the dominant factor controlling soil reflectivity. The Natural Resources Conservation Service (NRCS) in the U.S. Department of Agriculture (USDA) provides a general soil taxonomy map for the CONUS (Figure 1). In the soil taxonomy map, a dominant soil type is given for each pixel at 2 arcmin resolution (~4 km) from the 12 soil types, including: Alfisols, Andisols, Aridisols, Entisols, Gelisols, Histosols, Inceptisols, Mollisols, Oxisols, Spodosols, Ultisols, and Vertisols [43]. This soil type map was used in the soil albedo derivation methodology described in the following sections.

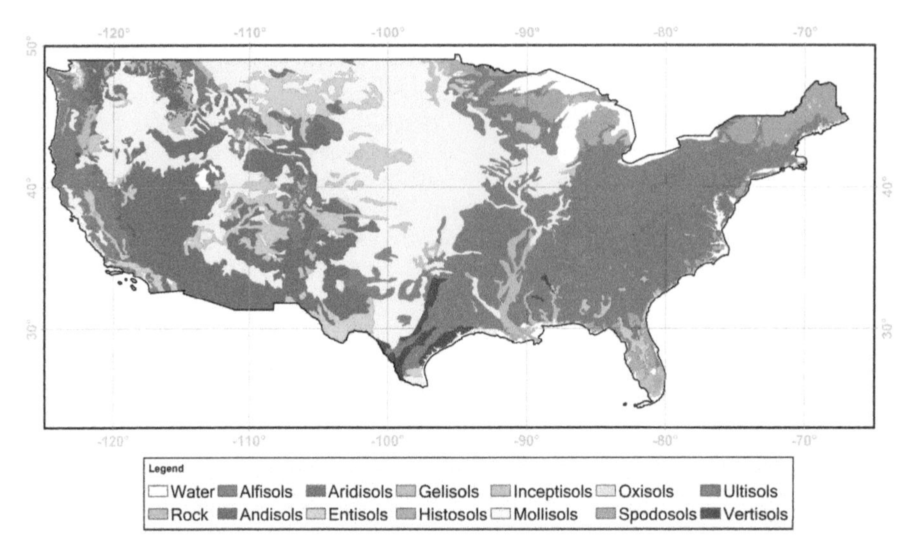

Figure 1. Dominant soil type map over the contiguous United States (CONUS) extracted from U.S. Department of Agriculture (USDA) soil survey data.

2.3. Using the Soil Line to Detect Bare Soil

In the solar shortwave range, bare soil surfaces normally have small reflectivity in shorter wavelength, with reflectance gradually increasing with wavelength [44,45]. In contrast, a vegetation canopy has very strong absorption in the visible spectrum and back scattering in the near infrared spectrum. The soil line is one of the key and stable spectral signatures of soil and has been examined in many experiments [34–36]. The soil line describes the linear feature (the "envelope") of surface reflectivity in multi-band space over a landscape of a given soil type. For example, the soil line is the lower envelope in the red-NIR space and has been commonly used to identify areas with dense vegetation. However, in practice, the red-NIR soil line can exhibit significant variation because both residual vegetation (leaf litter) and soil moisture may affect the NIR reflectivity greatly. This can cause problems for deriving a universal threshold to separate soil from vegetation, especially when sparse vegetation cover is not available for certain surfaces. In contrast, the green-red based soil line is very stable [35,46]. To verify the green-red soil line feature, surface spectra from Advanced Spaceborne Thermal Emission and Reflection Radiometer (ASTER) and US Geological Survey (USGS) spectral libraries [44,45] were collected and used to identify soil line features with different combinations of spectral albedo and vegetation/soil related indices including NDVI, EVI, SR, NDWI, and Visible Atmospherically Resistant Index (VARI). Of all combinations tested, the green-red band combination could detect soil line with the highest R^2 of 0.92, while the others ranged from 0.6 to 0.8.

In this study, thirteen years of MODIS albedo/BRDF data (2000–2012) were used to identify bare soil albedo characteristics over the CONUS using the following procedure. First, the soil line was generated for each of the 12 major soil types in the NRCS classification using MODIS data in 2005. To reduce uncertainties and minimize effects of vegetation and snow, only high quality snow-free albedo values (based on the QC flag in MODIS albedo product) collected between November and March were used in the soil line generation. MODIS NDVI and NDWI products were used to further exclude possible vegetation, snow/ice, water, and residual cloud/shadow pixels. Once the soil line was detected, the second step was to calculate statistics describing the distance of the bare soil albedo from the soil line for each pixel over the whole period 2000–2012. Considering the accuracy of the estimated soil line and the uncertainty in MODIS data products, the soil albedos were considered as candidates if their distance to the soil line was smaller than 20% of the magnitude of their albedo values. Our sensitivity test showed that the 20% threshold did not introduce any significant difference in the soil albedo climatology results, and, at the same time, it significantly increased the number of samples that were used in the calculation. The bare soil albedos (both spectral and broadband) were generated at a 16-day interval. The third step was to calculate the mean and standard deviation (SDEV) of the 16-day bare soil albedos over the entire thirteen years for each location. Figure 2 illustrates the procedure used to generate the multiyear mean bare soil albedos from the MODIS datasets. In the final bare soil albedo map, the soil albedo represents the average state of the soil condition, regardless of the differences in moisture content and organic matter. The SDEV maps contain information regarding impacts of variation in soil moisture, organic matter, and residual vegetation on the apparent soil reflectivity.

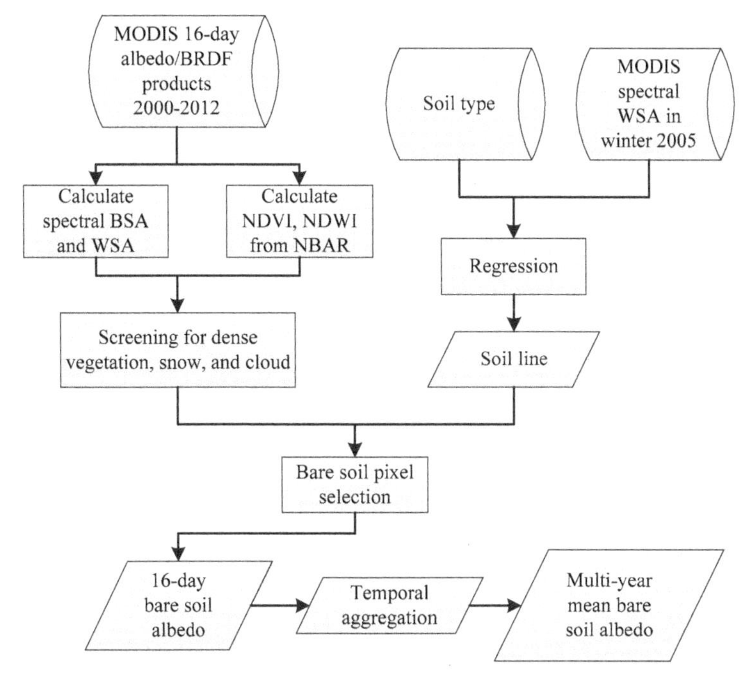

Figure 2. Flowchart for mapping bare soil albedo from multiyear Moderate Resolution Imaging Spectroradiometer (MODIS) data based on soil line.

3. Results and Discussion

3.1. Charateristics of Soil Lines Derived from MODIS Albedo

It has been demonstrated that the visible spectral space is useful in estimating vegetation fraction and indicating crop development stages tested at multiple crop sites with various soil backgrounds [35,46]. Based on field data, the green-red soil line feature has been proven to be a robust method for separating bare soil and vegetated surface conditions [35]. However, this feature has not yet been well documented using long-term satellite data. Figure 3 shows the soil line feature in green-red space from MODIS albedo samples over different land cover types based on MODIS land cover product during 2000–2012 using red and green bands.

In both crop and forest MODIS-derived samples, there was an obvious linear-line feature with relatively high green and red reflectance values in the spectral space (black line shown in Figure 3), which was in accordance with in-situ observations [46]. Based on the observations, this line was occupied mostly by samples in winter when vegetation had not sprouted. Such samples tend to be bare soil samples with low NDVI and positive NDWI. The variations in surface reflectivity at this stage were likely due to changes of soil moisture [28].

There are several reasons why data obtained in winter are preferable for deriving the soil line, as demonstrated in Figure 3. First, the effect of vegetation is minimized during the fall and winter dormant months. As vegetation grows such as in spring, surface albedos for green and red bands rapidly decrease due to the strong absorption by chlorophyll pigment in leaves, while the red band has a larger decrease rate because chlorophyll absorbance coefficient is much higher in red than in green [47]. Samples were moving downwards in spring when vegetation began to grow and ended up by a short "vegetation line" with red reflectance almost invariant occupied mostly by samples in summer when vegetation density was quite high. During the fall at the onset of vegetation senescence, green and red reflectance increased due to less absorption by the degraded chlorophyll pigments. As such, the samples at this stage were moving upwards from "vegetation line" back to "soil line" (red points in Figure 3). Thus, we can see from Figure 3 that albedos in winter are on the soil line and those in summer are on the vegetation line. For spring and fall, the albedo values lie between these extremes due to partial (sub-pixel) vegetation cover.

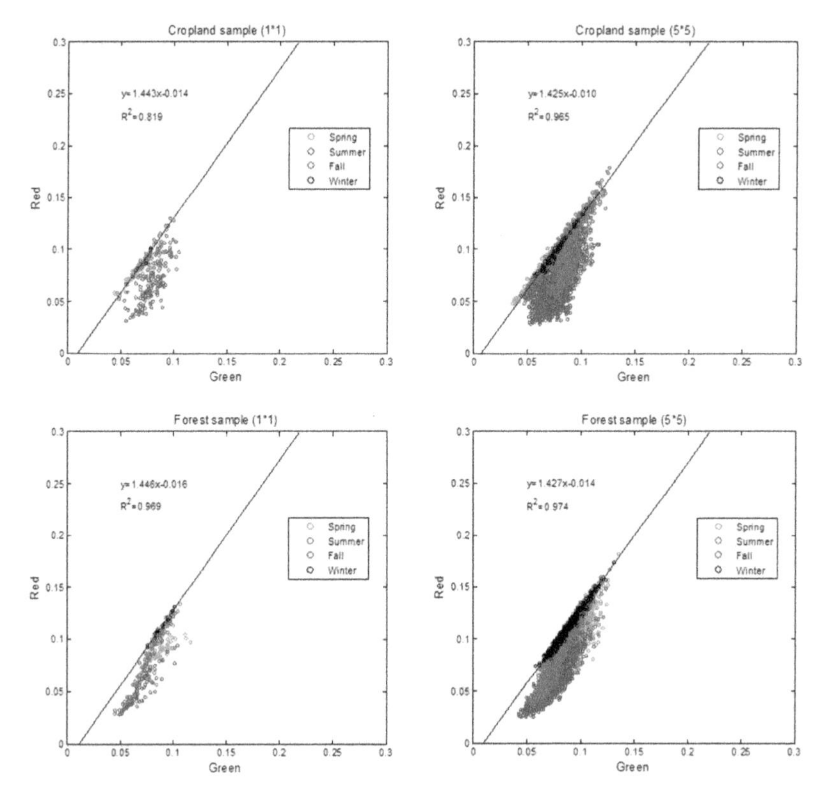

Figure 3. MODIS white-sky albedo (or diffuse reflectance) for the red and green bands over crop and forest samples in MODIS tile h11v04 during 2000–2012. The left panel shows the albedo values for one pixel for the time period; the right panel shows albedo values for the pixels in the 5-by-5 surrounding spatial window with the same land cover type to that in the left panel.

Second, soil albedo varies with soil moisture. During drier seasons (fall and winter) soil albedo tends to have a higher value, while during wetter seasons (spring and summer), it decreases as soil moisture increases. As soil moisture is usually small in winter (excluding the snow albedo), a small increase/decrease in moisture can cause large decrease/increase in albedo. Thus, a large range of soil albedo would produce robust soil line estimation.

Therefore, the soil line features were estimated by fitting the albedo values obtained in winter after excluding the data with large NDVI (>0.3) and/or negative NDWI to reduce the impacts of dense vegetation, clouds, and snow/ice.

The scatter plots in the left panel in Figure 3 are based on data at a single 500 m pixel; the plots in the right panel include data from the 5-by-5 surrounding pixels with the same land cover type. Because soil albedo varies both seasonally and inter-annually as the averaging area increases, including more surrounding pixels of the same land cover type as shown in the right panel, the slope of the soil line does not change much, but the R^2 increases. This procedure was tested over other locations with different land cover types, yielding results similar to those presented in Figure 3.

Soil line characteristics estimated from MODIS data from 2005 are listed in Table 2. The variability in these soil lines indicates that a global unified soil line may not be good enough to extract bare soil albedo. Most of the soil line estimations have an R^2 greater than 0.9 except for Alfisols, which demonstrates that the soil type information can support soil line estimations and other soil properties (e.g., soil moisture, organic matter, and surface texture/roughness) may have a secondary impact to the soil line [34].

Table 2. Parameters of soil lines estimated for the major soil types over the CONUS

Soil Type	Soil Line: $\alpha_{Red}=\alpha_{Green}\cdot a+b$				
	a	b	R^2	RMSE	N of Pixels
Alfisols	1.4830 ± 0.0057	−0.0142 ± 0.0006	0.8407	0.0114	190525
Andisols	1.3557 ± 0.0075	−0.0053 ± 0.0008	0.9393	0.0063	16682
Aridisols	1.3394 ± 0.0014	−0.0028 ± 0.0002	0.9047	0.0132	134894
Entisols	1.3195 ± 0.0016	−0.0027 ± 0.0002	0.9223	0.0112	99206
Gelisols	1.3058 ± 0.0599	−0.0053 ± 0.0053	0.9888	0.0024	1472
Histosols	1.2052 ± 0.0204	−0.0062 ± 0.0014	0.9400	0.0070	8340
Inceptisols	1.3661 ± 0.0089	−0.0116 ± 0.0016	0.9145	0.0084	43997
Mollisols	1.3242 ± 0.0014	−0.0017 ± 0.0002	0.9209	0.0077	262389
Oxisols	1.0827 ± 0.1283	0.0139 ± 0.0107	0.9394	0.0056	457
Spodosols	1.2062 ± 0.0265	−0.0101 ± 0.0019	0.9559	0.0086	120682
Ultisols	1.4115 ± 0.0088	−0.0177 ± 0.0007	0.9449	0.0079	105382
Vertisols	1.2849 ± 0.0082	−0.0030 ± 0.0008	0.9017	0.0067	15343

* Statistics are based on 95% confidence level.

Following the procedure described in Figure 2, bare soil albedo over the CONUS was first generated on a 16-day interval including black-sky (direct) and white-sky (diffuse) albedos for the seven spectral bands and three broadbands, and then aggregated temporally to generate the climatological mean and SDEV for each of the bands. Statistics for broadband white-sky albedos are presented in Figure 4. Based on visual comparison, a general agreement can be reached in terms of magnitude and spatial pattern between the bare soil shortwave albedo map generated in this study and the data aggregated from MODIS data during 2001–2010 in a recent study [26]. The shortwave albedo for soil is lower in the East and Northwest CONUS (around 0.15~0.18) where dense vegetation is located with a wetter climate and higher in the central and Southwest CONUS from 0.20 to more than 0.40 where much of the land is covered with crops, grass, and bare soil with a drier climate. A more detailed spatial pattern in the soil albedo over the CONUS is provided in this study.

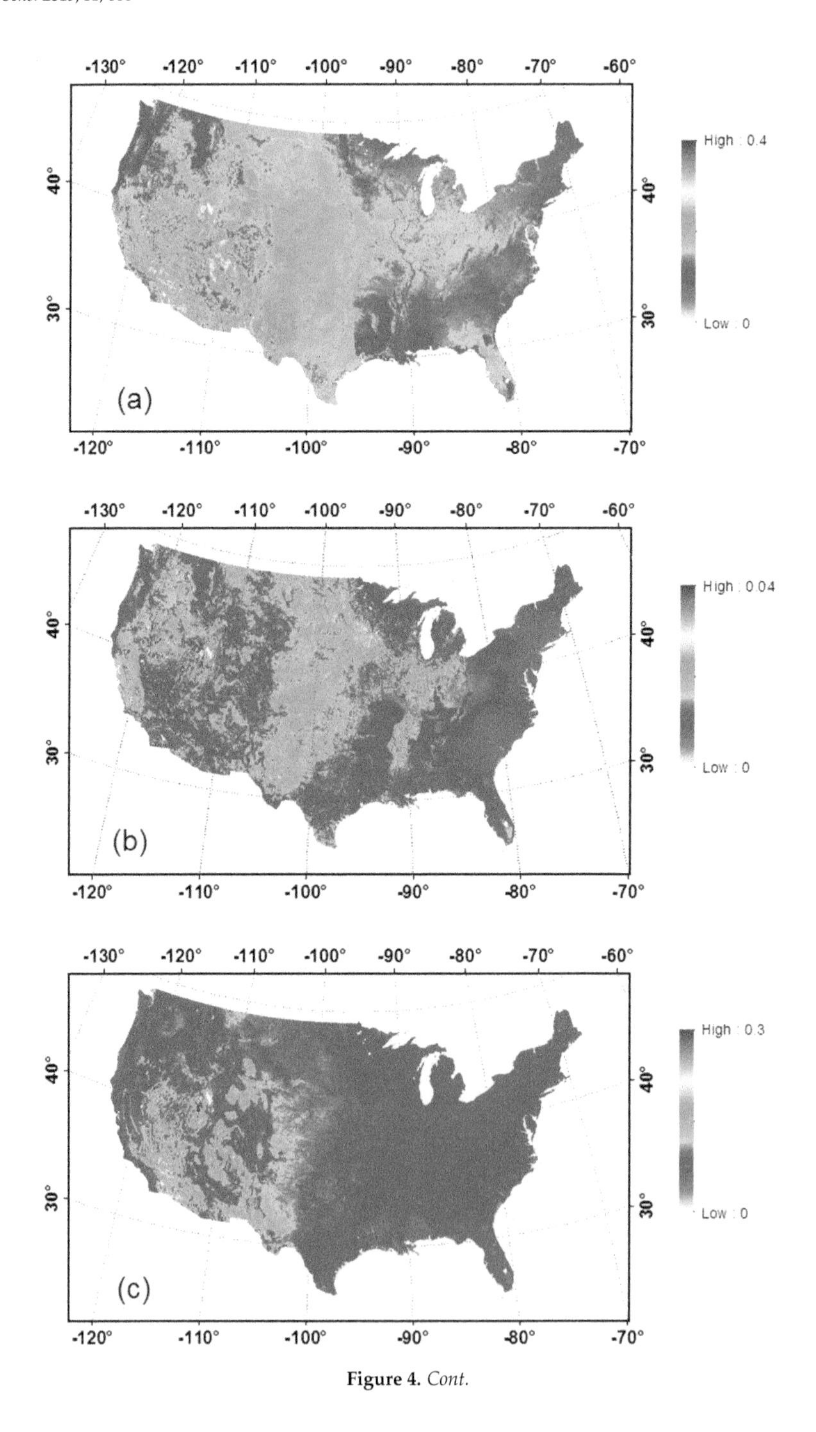

Figure 4. *Cont.*

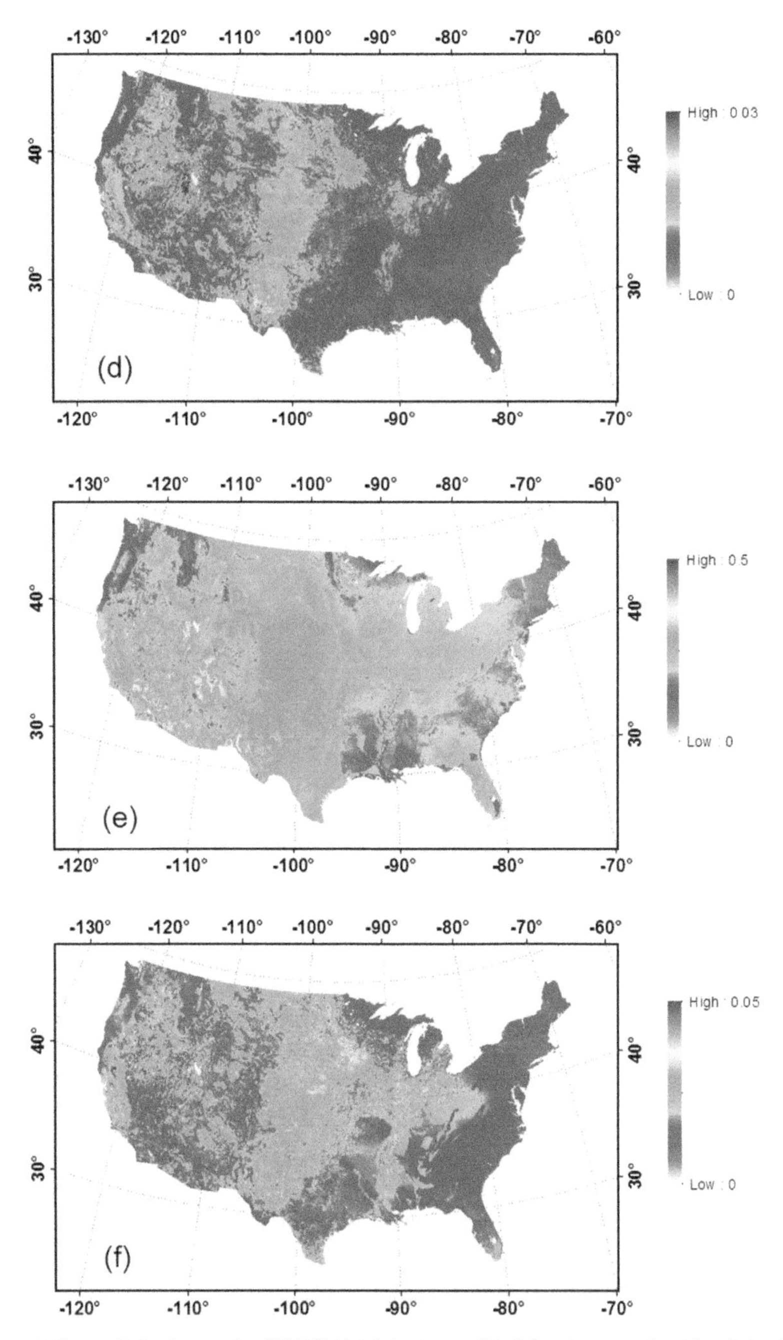

Figure 4. Bare soil albedo over the CONUS: Total shortwave (0.3–5.0 μm) mean value (**a**) and standard deviation (SDEV) (**b**); visible (0.3–0.7 μm) mean value (**c**) and SDEV (**d**); and near infrared (NIR) (0.7–5.0 μm) mean value (**e**) and SDEV (**f**).

3.2. *Validation of MODIS Bare Soil Albedo*

As bare soil albedo ground measurements are not widely available, validation of the bare soil albedo extracted from MODIS data at 500 m resolution is difficult. In previous studies, bare soil albedo was either directly compared with model inputs or indirectly validated against MODIS albedo products using pure vegetation albedo and vegetation fraction as inputs to calculate soil/vegetation mosaic albedos. In this study, we proposed an approach using finer resolution satellite data to help verify the 500 m MODIS-based bare soil albedo estimations.

Landsat data are available from the USGS at 16-day intervals and at a spatial resolution of 30 m. Though it is usually difficult to find pure bare soil pixels at the MODIS resolution, they are considerably easier to identify at the Landsat resolution. The finer resolution offers a unique capability to observe the bare soil directly, which is otherwise mixed with vegetation at MODIS resolution. In this study, it was assumed that if the Landsat pixels that belong to a MODIS pixel with a 500m-by-500m nominal spatial coverage can be identified as bare soil, the averaged soil albedo value of the Landsat pixels will represent the bare soil albedo for the MODIS pixel. The Landsat shortwave broadband albedo was estimated following the procedure described by He et al. [48] for the atmospheric correction and narrow-to-broadband conversion. Clouds in Landsat data were screened using the Landsat Ecosystem Disturbance Adaptive Processing System (LEDAPS) tool [49].

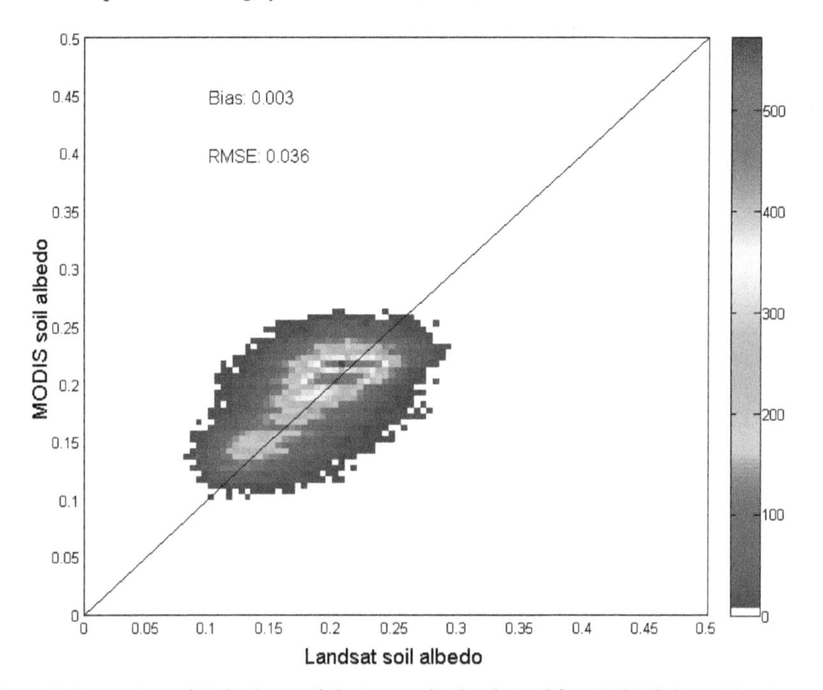

Figure 5. Comparison of 16-day bare soil shortwave albedos derived from MODIS data with estimates from Landsat data.

Potential soil pixels were extracted from the Landsat datasets based on land cover class designated in the National Land Cover Database 2006 (NLCD2006). Only pixels classified as sparse vegetation–including developed, barren, and shrubland–were used [50]. Both NDVI and NDWI were calculated for each Landsat scene to exclude cloud/shadow, water, snow, and vegetation pixels, similar to the MODIS data processing discussed in previous sections. In this study, we used the Landsat scenes (path 042, row 034) centered at 119.0°W, 37.5°N for validation, which have a variety of land cover types, including forest, grassland, and cropland. There are four major soil types in this region, including Alfisols, Aridisols, Entisols, and Ultisols. All the available Landsat 5 TM scenes for the period 2005–2007 with nominal cloud coverage less than 30% were used. As the NLCD class and vegetation indices thresholds (NDVI<0.3 same as in Section 3.1) were applied to the Landsat data, it was assumed that the generated Landsat soil albedo can represent the bare soil surfaces with/without vegetation cover in a larger area, e.g., a MODIS pixel. As demonstrated in Figure 5, a general agreement can be found for bare soil shortwave black-sky albedos derived from MODIS and Landsat data with a bias of 0.003, a root-mean-square-error (RMSE) of 0.036, and an R^2 of 0.316. There are several possible reasons for the differences between these two datasets found in the comparison: Landsat albedo estimations are instantaneous and thus more sensitive to soil moisture changes; Landsat and MODIS surface reflectance may not fit to 1-to-1 line exactly due to the differences in sensor characteristics and data processing; and Landsat pixels may not match MODIS pixels exactly due to the different registration accuracy and adjacency effects in the coarser resolution data.

3.3. *Impacts of Soil Moisture Content on Soil Albedo*

The water content in the top layer of soil is believed to have great impacts on the soil albedo. Previous studies have tried to establish the relationship between soil moisture content and soil reflectivity based on various datasets. Some studies reported that the relationship was linear, while others found an exponential equation provided a better fit (Equation (3)):

$$\alpha = A\exp(-B\theta) + C, \tag{3}$$

where α is soil albedo and θ is soil moisture. A, B, and C are regression coefficients.

Gascoin et al. [9] found that the best fit coefficients (A= 0.31, B =12.7 and C =0.15) with an RMSE of 0.030 using various soil samples from field measurements, and Wang et al. [30] reported a lower value for the B coefficient (B = 3.52, clay soil type). Gascoin et al. [9] also pointed it out that the parameters must be carefully examined before application to other regions.

In this study, a relationship between bare soil albedo and soil moisture was derived using remote sensing data over a large spatial domain. White-sky albedo was used exclusively to minimize the solar zenith impacts on albedo data. In addition, the Advanced Microwave Scanning Radiometer-Earth Observing System (AMSR-E) Level 3 daily soil moisture products were used as estimates of soil moisture in the top ~1 cm of soil averaged over the retrieval footprint [51,52]. The 16-day averaged soil moisture values were generated from the daily AMSR-E products to match the temporal resolution of MODIS albedo during 2002–2011 (due to demise of AMSR-E in 2011) over the CONUS. The MODIS soil albedo data were rescaled to 25 km to match AMSR-E soil moisture products. Comparisons between the soil moisture and broadband white-sky albedos for bare soil in the Southwest U.S. (MODIS tile h08v05) indicate that soil albedo generally decreases with an increase in soil moisture (Figure 6). The rate of decrease for all three broadband albedos are very similar, which suggests that the change of soil moisture is likely the major factor of albedo change and that the soil line approach is resistant to soil moisture changes. Variation in NIR broadband albedo is larger than that in visible and total shortwave albedos, which is likely due to the fact that: (1) NIR albedo is larger than visible, and thus the absolute values are more variant; and (2) soil moisture may cause a larger variation in the longer wavelength part of the NIR spectral domain [34].

The relationship between bare soil albedo and soil moisture on a 16-day interval was found to be relatively stable across different seasons and locations over the CONUS ($0.18 < A < 0.24$; $9.86 < B < 21.33$; $0.12 < C < 0.14$; $0.031 < RMSE < 0.037$). However, as the range of the 16-day averaged soil moisture variation is relatively small (most soil moisture values are between 0.08 and 0.12 g/cm^3), the exponential relationship between these two variables cannot be verified without simultaneous instantaneous small soil moisture values (<0.05 g/cm^3) and large bare soil albedos (>0.3). Nevertheless, the relationship found in this study is very close to the results presented in Gascoin et al. [9], considering that their soil moisture samples were taken at a ~5 cm depth from the surface.

Soil moisture is among the most important factors that can affect soil reflectivity, which should be considered in bare soil albedo estimations. However, soil moisture varies both spatially and temporally. Thus, it is difficult to remove the impacts of soil moisture in the final 500 m bare soil albedo map without the support of soil moisture products available at a similar resolution.

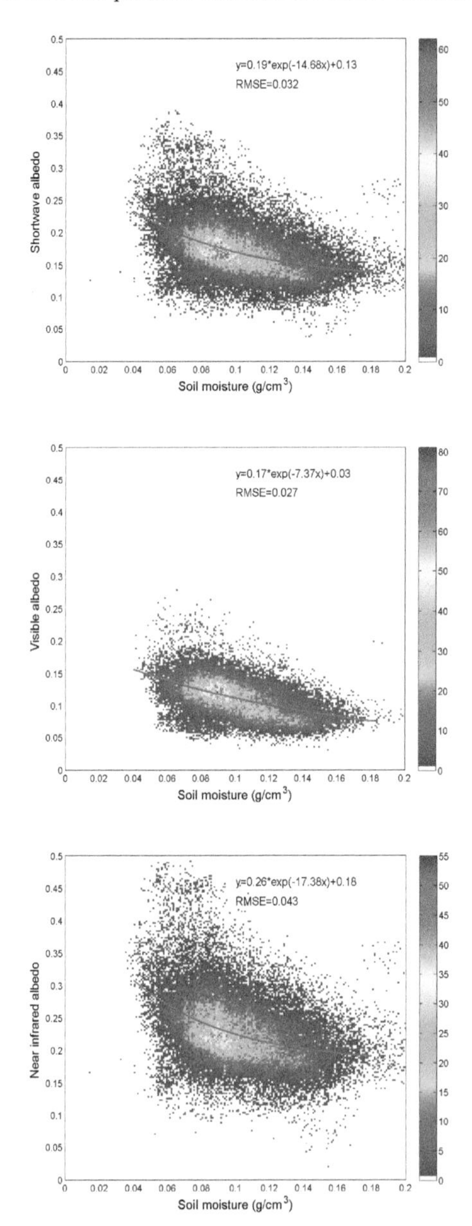

Figure 6. Relationship between bare soil broadband white-sky albedos and soil skin moisture over MODIS tile h08v05 for January 2002–2011. Red line is the best fit curve; its regression coefficients and RMSEs are shown above.

3.4. Relationship of Bare Soil Albedo with Major Soil Types and Land Cover Types

Traditional methods directly relate soil albedo with soil types, which are currently used by many land surface models to simplify the parameterization of soil albedo besides correcting the illumination angle effects [20].

The climatological bare soil white-sky albedos for different soil types derived for the period 2000–2012 (Table 3) were compared between each other. The Aridisols were found to have the largest soil albedo values because they tend to be the driest among the 12 major soil types—too dry for mesophytic plants to grow. Entisols, many of which are found in steep and rocky settings, have the second largest albedo values. Since the soil reflectivity is very sensitive to moisture changes under dry conditions, albedos of both Aridisols and Entisols have the largest variations. Among three broadband albedos, the near infrared albedos have the highest mean and SDEV values, and the visible albedos have the lowest mean and SDEV values for each soil type.

Table 3. Statistics of broadband soil albedo distribution for each of the major soil types.

Soil Type	Visible		Near Infrared		Shortwave	
	Mean	SDEV	Mean	SDEV	Mean	SDEV
Alfisols	0.0796	0.0152	0.2060	0.0375	0.1479	0.0251
Andisols	0.0738	0.0190	0.1914	0.0457	0.1382	0.0314
Aridisols	0.1196	0.0376	0.2493	0.0553	0.1858	0.0433
Entisols	0.1013	0.0295	0.2330	0.0449	0.1695	0.0345
Gelisols	0.0801	0.0159	0.1804	0.0322	0.1357	0.0227
Histosols	0.0619	0.0158	0.1604	0.0465	0.1174	0.0299
Inceptisols	0.0704	0.0170	0.1820	0.0449	0.1310	0.0296
Mollisols	0.0915	0.0160	0.2258	0.0327	0.1623	0.0230
Oxisols	0.0674	0.0157	0.2138	0.0427	0.1460	0.0258
Spodosols	0.0691	0.0183	0.1702	0.0462	0.1257	0.0304
Ultisols	0.0739	0.0179	0.2004	0.0486	0.1429	0.0320
Vertisols	0.0800	0.0112	0.2248	0.0261	0.1550	0.0167

Table 4. p-value of two tailed t-test on the soil albedos among different soil types.

Soil Type	Andisols	Aridisols	Entisols	Gelisols	Histosols	Inceptisols	Mollisols	Oxisols	Spodosols	Ultisols	Vertisols
Alfisols	***	***	***	0.78	***	***	***	0.39	***	***	***
Andisols		***	***	***	***	***	***	0.55	**	***	***
Aridisols			***	***	***	***	***	0.36	***	***	0.40
Entisols				0.66	***	***	***	0.40	***	***	***
Gelisols					***	***	0.31	0.34	***	***	***
Histosols						0.28	***	0.49	***	***	***
Inceptisols							***	0.49	***	***	***
Mollisols								0.39	***	***	***
Oxisols									0.52	0.38	0.35
Spodosols										***	***
Ultisols											*

*: $p < 0.05$; **: $p < 0.01$; ***: $p < 0.001$.

Histosols have the lowest reflectivity, probably because of their high organic matter content. Oxisols and Spodosols also have low albedo values. Possible reasons are that the Oxisols are mainly distributed in tropical and subtropical areas where precipitation increases soil moisture, and the Spodosols are acid soils mainly under forest.

The two tailed t-test results shown in Table 4 suggest that differences between soil albedo values are statistically significant for most soil types. However, there are several exceptions. Oxisols are quite inseparable from all the other types. It is also difficult to separate the albedos of Gelisols from those of Alfisols and Entisols. Their small sample size (Oxisols have less than 0.004% of the total samples; Gelisols have 0.147% of the total samples) is likely to have resulted in the difficulty in the separation of their soil albedos because a 4 km soil map grid may contain multiple soil types.

Bare soil albedo can also be linked with the type of vegetation above the ground because that the soil moisture–precipitation feedback [53] and vegetation–soil moisture feedback [54] will determine the relationship between vegetation type and magnitude of soil albedo (because of soil moisture). Table 5 lists the climatological values of bare soil albedo for each of the International Geosphere-Biosphere Programme (IGBP) land cover types from MODIS land cover data. In this study, it was found that bare soil albedo decreases as the vegetation density increases from open shrubland to evergreen needle leaf forest.

Table 5. Statistics of broadband soil albedo distribution for each of the land cover classes.

IGBP Land Cover	Visible		Near Infrared		Shortwave	
	Mean	SDEV	Mean	SDEV	Mean	SDEV
ENF	0.0458	0.0147	0.1122	0.0296	0.0847	0.0210
EBF	0.0526	0.0193	0.1237	0.0460	0.0938	0.0304
DNF	0.0593	0.0139	0.1308	0.0272	0.1006	0.0191
DBF	0.0633	0.0124	0.1627	0.0265	0.1192	0.0184
MIX	0.0541	0.0134	0.1354	0.0330	0.1009	0.0222
CSH	0.0664	0.0161	0.1692	0.0334	0.1217	0.0234
OSH	0.1096	0.0315	0.2386	0.0512	0.1764	0.0393
WSV	0.0669	0.0142	0.1780	0.0338	0.1283	0.0234
SAV	0.0723	0.0144	0.1911	0.0368	0.1366	0.0252
GRA	0.0987	0.0192	0.2318	0.0341	0.1682	0.0245
WET	0.0489	0.0135	0.1081	0.0301	0.0847	0.0204
CRO	0.0893	0.0136	0.2255	0.0311	0.1610	0.0218
URB	0.0868	0.0191	0.1937	0.0324	0.1428	0.0238
CRC	0.0814	0.0127	0.2090	0.0291	0.1504	0.0200
GLA	0.0925	0.0519	0.1590	0.0663	0.1290	0.0548
BRN	0.1939	0.0551	0.3267	0.0813	0.2561	0.0627

ENF: Evergreen needleleaf forest; EBF: Evergreen broadleaf forest; DNF: Deciduous needleleaf forest; DBF: Deciduous broadleaf forest; MIX: Mixed forest; CSH: Closed shrubland; OSH: Open shrublands; WSV: Woody savannas; SAV: Savannas; GRA: Grasslands; WET: Permanent wetlands; CRO: Croplands; URB: Urban and built-up; CRC: Cropland/natural vegetation mosaic; GLA: Snow and ice; BRN: Barren or sparsely vegetated.

Compared with the statistics of albedo–soil types in Tables 2 and 3, bare soil albedos have smaller within-class SDEVs for the albedo–land cover types, except for open shrublands and barren surface. This suggests that land cover type might be a better indicator to quantify the magnitude of bare soil albedo than soil type for the vegetated areas. For sparsely vegetated areas, the soil type map provides spatial variation that is necessary to characterize the relationship between bare soil albedo and soil type.

4. Conclusions

Bare soil albedo is an important ancillary dataset used in both climate and energy balance models and satellite LAI/fPAR retrieval procedures, but is usually prescribed based on soil type maps at very coarse spatial resolution. This fixed classification is unable to account for variability observed in soil albedo due to soil moisture and other soil properties particularly in the spatial domain. Long-term satellite observations provide a great opportunity to extract the bare soil albedo information at a much finer resolution. Methods that rely mostly on a single vegetation index (NDVI, LAI, fPAR, etc.) have been widely practiced in producing the existing soil albedo datasets from satellite data, the results of which, however, may suffer from the uncertainties in the upstream vegetation index product. To overcome this problem, a novel method was proposed in this study for the detection and extraction of bare soil albedo from thirteen years of MODIS albedo product and USDA soil type over the CONUS based on the soil line concept, NDVI, and NDWI derived from the MODIS spectral bands. The soil line concept turned out to be very effective in extracting the bare soil information with minimized impact from vegetation.

The validation of the bare soil albedo is quite challenging over a large spatial domain as ground measurements are not widely available. In this study, bare soil pixels were extracted from Landsat data using the classification map from NLCD. Good agreement has been found between bare soil albedo from Landsat and our estimations from MODIS data. Further work is expected to make comparisons using more Landsat and MODIS data over different regions and soil types.

In most cases, the proposed soil line method in mapping the bare soil albedo is very efficient and effective. However, it is not always effective, especially when the soil is covered by dense evergreen vegetation canopy. Though the NDVI and NDWI thresholds can help exclude the observations with dense vegetation canopy coverage, it is very possible that the regressed soil line detects the observation with least green canopy coverage rather than pure bare soil background. In other words, the derived climatological bare soil albedo in the evergreen vegetation covered area is not as reliable as those over land surface with sparser vegetation. Such an issue also exists in other current bare soil albedo products, which draws attention in applying them over areas covered by evergreen vegetation. Compared with the existing datasets, the maps generated in this study have reduced uncertainty because more background soil can be observed with the finer spatial resolution input data.

A decrease in soil albedo with an increase in soil moisture from AMSR-E data has been demonstrated over a large area. The exponential relationship has been found to be relatively stable for different time and locations. However, more efforts are still needed to improve the quantification of this relationship by using instantaneous (or at least daily) albedo and soil moisture in the future.

Statistics of bare soil broadband albedos were calculated based on soil types and land cover types. The within-class SDEV statistics suggest that: Though both classification schemes could be used as prescribing indicators for soil albedo, land cover type would be a better choice to determine the albedo magnitude for vegetated areas, while soil type is better at characterizing the soil line feature for sparely vegetated areas.

The bare soil broadband albedo could be very useful as one of the key ancillary data for climate models. On the other hand, our proposed method can also generate the bare soil albedo for the spectral bands. The derived bare soil albedo dataset has been demonstrated quite effective in improving the accuracy of satellite LAI and fPAR estimations under low vegetation density conditions [55].

Author Contributions: T.H. and F.G. conceived and designed the experiments; T.H. performed the experiments; S.L., and Y.P. contributed to the data analysis and provided comments and suggestions for the manuscript; T.H. wrote the paper.

Acknowledgments: This work was supported by the National Natural Science Foundation of China grant (41771379), the Key Laboratory of National Geographic State Monitoring of National Administration of Surveying, Mapping and Geoinformation grant (2017NGCMZD02), the USDA project (No. 12451361002825) from the NASA ROSES grant (NNH09ZDA001N) to the University of Maryland. We appreciate the comments from Crystal Schaaf at University of Massachusetts Boston, Martha Anderson at USDA-ARS and the anonymous reviewers to our manuscript. We thank the MODIS albedo team and AMSR-E soil moisture team. Both satellite products were distributed by the NASA Earth Observing System Data and Information System (EOSDIS) and available at http://reverb.echo.nasa.gov. We also thank the USDA NRCS and the Multi-Resolution Land Characteristics Consortium (MRLC) for maintaining and distributing the soil type map and NLCD2006 data used in this study. USDA is an equal opportunity provider and employer.

Conflicts of Interest: The authors declare no conflict of interest. The founding sponsors had no role in the design of the study; in the collection, analyses, or interpretation of data; in the writing of the manuscript, and in the decision to publish the results.

References

1. Liang, S.L.; Wang, K.C.; Zhang, X.T.; Wild, M. Review on estimation of land surface radiation and energy budgets from ground measurement, remote sensing and model simulations. *IEEE J. Sel. Top. Appl. Earth Obs. Remote Sens.* **2010**, *3*, 225–240. [CrossRef]
2. Liang, S.L.; Zhang, X.T.; He, T.; Cheng, J.; Wang, D. Remote sensing of the land surface radiation budget. In *Remote Sensing of Energy Fluxes and Soil Moisture Content*; CRC Press: Boca Raton, FL, USA, 2013; pp. 121–162. [CrossRef]

3. Ganguly, S.; Nemani, R.R.; Zhang, G.; Hashimoto, H.; Milesi, C.; Michaelis, A.; Wang, W.L.; Votava, P.; Samanta, A.; Melton, F.; et al. Generating global Leaf Area Index from Landsat: Algorithm formulation and demonstration. *Remote Sens. Environ.* **2012**, *122*, 185–202. [CrossRef]
4. Myneni, R.B.; Hoffman, S.; Knyazikhin, Y.; Privette, J.L.; Glassy, J.; Tian, Y.; Wang, Y.; Song, X.; Zhang, Y.; Smith, G.R.; et al. Global products of vegetation leaf area and fraction absorbed PAR from year one of MODIS data. *Remote Sens. Environ.* **2002**, *83*, 214–231. [CrossRef]
5. Dickinson, R.E.; Henderson-Sellers, A.; Kennedy, P.J. *Biosphere-Atmosphere Transfer Scheme (BATS) version 1e as coupled to the NCAR Community Climate Model. NCAR Technical Note NCAR/TN-387+STR*; National Center for Atmospheric Research: Boulder, CO, USA, 1993.
6. Bonan, G.B. *A Land Surface Model (LSM Version 1.0) for Ecological, Hydrological, and Atmospheric Studies: Technical Description and User's Guide. NCAR Technical Note NCAR/TN-417+STR*; National Center for Atmospheric Research: Boulder, CO, USA, 1996.
7. Post, D.F.; Fimbres, A.; Matthias, A.D.; Sano, E.E.; Accioly, L.; Batchily, A.K.; Ferreira, L.G. Predicting soil albedo from soil color and spectral reflectance data. *Soil Sci. Soc. Am. J.* **2000**, *64*, 1027–1034. [CrossRef]
8. Liu, H.Z.; Wang, B.M.; Fu, C.B. Relationships between surface albedo, soil thermal parameters and soil moisture in the semi-arid area of Tongyu, Northeastern China. *Adv. Atmos. Sci.* **2008**, *25*, 757–764. [CrossRef]
9. Gascoin, S.; Ducharne, A.; Ribstein, P.; Perroy, E.; Wagnon, P. Sensitivity of bare soil albedo to surface soil moisture on the moraine of the Zongo glacier (Bolivia). *Geophys. Res. Lett.* **2009**, *36*, L02405. [CrossRef]
10. Irons, J.R.; Ranson, K.J.; Daughtry, C.S.T. Estimating big bluestem albedo from directional reflectance measurements. *Remote Sens. Environ.* **1988**, *25*, 185–199. [CrossRef]
11. Matthias, A.D.; Fimbres, A.; Sano, E.E.; Post, D.F.; Accioly, L.; Batchily, A.K.; Ferreira, L.G. Surface roughness effects on soil albedo. *Soil Sci. Soc. Am. J.* **2000**, *64*, 1035–1041. [CrossRef]
12. Zhou, L.; Dickinson, R.E.; Tian, Y.; Zeng, X.; Dai, Y.; Yang, Z.L.; Schaaf, C.B.; Gao, F.; Jin, Y.; Strahler, A.; et al. Comparison of seasonal and spatial variations of albedos from Moderate-Resolution Imaging Spectroradiometer (MODIS) and Common Land Model. *J. Geophys. Res.-Atmos.* **2003**, *108*. [CrossRef]
13. Kala, J.; Evans, J.P.; Pitman, A.J.; Schaaf, C.B.; Decker, M.; Carouge, C.; Mocko, D.; Sun, Q. Implementation of a soil albedo scheme in the CABLEv1.4b land surface model and evaluation against MODIS estimates over Australia. *Geosci. Model Dev.* **2014**, *7*, 2121–2140. [CrossRef]
14. Fensholt, R.; Sandholt, I.; Rasmussen, M.S. Evaluation of MODIS LAI, fAPAR and the relation between fAPAR and NDVI in a semi-arid environment using in situ measurements. *Remote Sens. Environ.* **2004**, *91*, 490–507. [CrossRef]
15. Weiss, M.; Baret, F.; Garrigues, S.; Lacaze, R. LAI and fAPAR CYCLOPES global products derived from VEGETATION. Part 2: validation and comparison with MODIS collection 4 products. *Remote Sens. Environ.* **2007**, *110*, 317–331. [CrossRef]
16. Pisek, J.; Chen, J.M. Mapping forest background reflectivity over North America with Multi-angle Imaging SpectroRadiometer (MISR) data. *Remote Sens. Environ.* **2009**, *113*, 2412–2423. [CrossRef]
17. Pinty, B.; Clerici, M.; Andredakis, I.; Kaminski, T.; Taberner, M.; Verstraete, M.M.; Gobron, N.; Plummer, S.; Widlowski, J.L. Exploiting the MODIS albedos with the Two-stream Inversion Package (JRC-TIP): 2. Fractions of transmitted and absorbed fluxes in the vegetation and soil layers. *J. Geophys. Res.-Atmos.* **2011**, *116*. [CrossRef]
18. Gascoin, S.; Ducharne, A.; Ribstein, P.; Lejeune, Y.; Wagnon, P. Dependence of bare soil albedo on soil moisture on the moraine of the Zongo glacier (Bolivia): Implications for land surface modeling. *J. Geophys. Res.-Atmos.* **2009**, *114*. [CrossRef]
19. Zhou, L.M.; Dickinson, R.E.; Tian, Y.H. Derivation of a soil albedo dataset from MODIS using principal component analysis: Northern Africa and the Arabian Peninsula. *Geophys. Res. Lett.* **2005**, 32. [CrossRef]
20. Liang, X.Z.; Xu, M.; Gao, W.; Kunkel, K.; Slusser, J.; Dai, Y.J.; Min, Q.L.; Houser, P.R.; Rodell, M.; Schaaf, C.B.; et al. Development of land surface albedo parameterization based on Moderate Resolution Imaging Spectroradiometer (MODIS) data. *J. Geophys. Res.-Atmos.* **2005**, *110*. [CrossRef]
21. Houldcroft, C.J.; Grey, W.M.F.; Barnsley, M.; Taylor, C.M.; Los, S.O.; North, P.R.J. New vegetation albedo parameters and global fields of soil background albedo derived from MODIS for use in a climate model. *J. Hydrometeorol.* **2009**, *10*, 183–198. [CrossRef]
22. Jasinski, M.F. Sensitivity of the normalized difference vegetation index to subpixel canopy cover, soil albedo, and pixel scale. *Remote Sens. Environ.* **1990**, *32*, 169–187. [CrossRef]

23. Pettorelli, N.; Vik, J.O.; Mysterud, A.; Gaillard, J.M.; Tucker, C.J.; Stenseth, N.C. Using the satellite-derived NDVI to assess ecological responses to environmental change. *Trends Ecol. Evol.* **2005**, *20*, 503–510. [CrossRef]
24. Kaptue, T.A.T.; Roujean, J.L.; Faroux, S. ECOCLIMAP-II: An ecosystem classification and land surface parameters database of Western Africa at 1 km resolution for the African Monsoon Multidisciplinary Analysis (AMMA) project. *Remote Sens. Environ.* **2010**, *114*, 961–976. [CrossRef]
25. Rechid, D.; Raddatz, T.; Jacob, D. Parameterization of snow-free land surface albedo as a function of vegetation phenology based on MODIS data and applied in climate modelling. *Theor. Appl. Climatol.* **2009**, *95*, 245–255. [CrossRef]
26. Carrer, D.; Meurey, C.; Ceamanos, X.; Roujean, J.-L.; Calvet, J.-C.; Liu, S. Dynamic mapping of snow-free vegetation and bare soil albedos at global 1 km scale from 10-year analysis of MODIS satellite products. *Remote Sens. Environ.* **2014**, *140*, 420–432. [CrossRef]
27. Anderson, M.C.; Kustas, W.P.; Alfieri, J.G.; Gao, F.; Hain, C.; Prueger, J.H.; Evett, S.; Colaizzi, P.; Howell, T.; Chavez, J.L. Mapping daily evapotranspiration at Landsat spatial scales during the BEAREX'08 field campaign. *Adv. Water Resour.* **2012**, *50*, 162–177. [CrossRef]
28. Lobell, D.B.; Asner, G.P. Moisture effects on soil reflectance. *Soil Sci. Soc. Am. J.* **2002**, *66*, 722–727. [CrossRef]
29. Liu, W.D.; Baret, F.; Gu, X.F.; Tong, Q.X.; Zheng, L.F.; Zhang, B. Relating soil surface moisture to reflectance. *Remote Sens. Environ.* **2002**, *81*, 238–246.
30. Wang, K.C.; Wang, P.C.; Liu, J.M.; Sparrow, M.; Haginoya, S.; Zhou, X.J. Variation of surface albedo and soil thermal parameters with soil moisture content at a semi-desert site on the western Tibetan Plateau. *Bound.-Layer Meteorol.* **2005**, *116*, 117–129. [CrossRef]
31. Guan, X.D.; Huang, J.P.; Guo, N.; Bi, J.R.; Wang, G.Y. Variability of soil moisture and its relationship with surface albedo and soil thermal parameters over the Loess Plateau. *Adv. Atmos. Sci.* **2009**, *26*, 692–700. [CrossRef]
32. Liu, S.; Roujean, J.-L.; Kaptue Tchuente, A.T.; Ceamanos, X.; Calvet, J.-C. A parameterization of SEVIRI and MODIS daily surface albedo with soil moisture: Calibration and validation over southwestern France. *Remote Sens. Environ.* **2014**, *144*, 137–151. [CrossRef]
33. Guerschman, J.P.; Scarth, P.F.; McVicar, T.R.; Renzullo, L.J.; Malthus, T.J.; Stewart, J.B.; Rickards, J.E.; Trevithick, R. Assessing the effects of site heterogeneity and soil properties when unmixing photosynthetic vegetation, non-photosynthetic vegetation and bare soil fractions from Landsat and MODIS data. *Remote Sens. Environ.* **2015**, *161*, 12–26. [CrossRef]
34. Baret, F.; Jacquemoud, S.; Hanocq, J.F. About the soil line concept in remote sensing. *Adv. Space Res.* **1993**, *13*, 281–284. [CrossRef]
35. Gitelson, A.A.; Stark, R.; Grits, U.; Rundquist, D.; Kaufman, Y.; Derry, D. Vegetation and soil lines in visible spectral space: a concept and technique for remote estimation of vegetation fraction. *Int. J. Remote Sens.* **2002**, *23*, 2537–2562. [CrossRef]
36. Huete, A.R. A Soil-Adjusted Vegetation Index (SAVI). *Remote Sens. Environ.* **1988**, *25*, 295–309. [CrossRef]
37. Pickup, G.; Chewings, V.H.; Nelson, D.J. Estimating changes in vegetation cover over time in arid rangelands using Landsat MSS data. *Remote Sens. Environ.* **1993**, *43*, 243–263. [CrossRef]
38. Cui, S.; Rajan, N.; Maas, S.J.; Youn, E. An automated soil line identification method using relevance vector machine. *Remote Sens. Lett.* **2014**, *5*, 175–184. [CrossRef]
39. Fox, G.A.; Sabbagh, G.J.; Searcy, S.W.; Yang, C. An automated soil line identification routine for remotely sensed images. *Soil Sci. Soc. Am. J.* **2004**, *68*, 1326–1331. [CrossRef]
40. Schaaf, C.; Wang, Z. *MCD43A1 MODIS/Terra+Aqua BRDF/Albedo Model Parameters Daily L3 Global-500m V006*; NASA EOSDIS Land Processes DAAC: Sioux Falls, SD, USA, 2015. [CrossRef]
41. Gao, B.C. NDWI - A normalized difference water index for remote sensing of vegetation liquid water from space. *Remote Sens. Environ.* **1996**, *58*, 257–266. [CrossRef]
42. Lawrence, P.J.; Chase, T.N. Representing a new MODIS consistent land surface in the Community Land Model (CLM 3.0). *J. Geophys. Res.-Biogeosci.* **2007**, *112*. [CrossRef]
43. USDA. *Soil Taxonomy: A Basic System of Soil Classification for Making and Interpreting Soil Survey*, 2nd ed.; U.S. Department of Agriculture Handbook 436; Soil Survey Staff, Ed.; Natural Resources Conservation Service: Washington, DC, USA, 1999.
44. Baldridge, A.M.; Hook, S.J.; Grove, C.I.; Rivera, G. The ASTER spectral library version 2.0. *Remote Sens. Environ.* **2009**, *113*, 711–715. [CrossRef]

45. Clark, R.N.; Swayze, G.A.; Wise, R.; Livo, E.; Hoefen, T.; Kokaly, R.; Sutley, S.J. USGS digital spectral library splib06a. *U.S. Geol. Surv. Digit. Data Ser. 231* **2007**.
46. Nguy-Robertson, A.; Gitelson, A.; Peng, Y.; Walter-Shea, E.; Leavitt, B.; Arkebauer, T. Continuous monitoring of crop reflectance, vegetation fraction, and identification of developmental stages using a four band radiometer. *Agron. J.* **2013**, *105*, 1769–1779. [CrossRef]
47. Gitelson, A.A.; Gritz, Y.; Merzlyak, M.N. Relationships between leaf chlorophyll content and spectral reflectance and algorithms for non-destructive chlorophyll assessment in higher plant leaves. *J. Plant Physiol.* **2003**, *160*, 271–282. [CrossRef]
48. He, T.; Liang, S.L.; Wang, D.; Shuai, Y.; Yu, Y. Fusion of satellite land surface albedo products across scales using a multiresolution tree method in the north central United States. *IEEE Trans. Geosci. Remote Sens.* **2014**, *52*, 3428–3439. [CrossRef]
49. Masek, J.G.; Vermote, E.F.; Saleous, N.E.; Wolfe, R.; Hall, F.G.; Huemmrich, K.F.; Gao, F.; Kutler, J.; Lim, T.K. A Landsat surface reflectance dataset for North America, 1990–2000. *IEEE Geosci. Remote Sens. Lett.* **2006**, *3*, 68–72. [CrossRef]
50. Fry, J.A.; Xian, G.; Jin, S.M.; Dewitz, J.A.; Homer, C.G.; Yang, L.M.; Barnes, C.A.; Herold, N.D.; Wickham, J.D. Completion of the 2006 National Land Cover Database for the Conterminous United States. *Photogramm. Eng. Remote Sens.* **2011**, *77*, 858–864.
51. Njoku, E.G.; Jackson, T.J.; Lakshmi, V.; Chan, T.K.; Nghiem, S.V. Soil moisture retrieval from AMSR-E. *IEEE Trans. Geosci. Remote Sens.* **2003**, *41*, 215–229. [CrossRef]
52. Njoku, E.G.; Chan, S.K. Vegetation and surface roughness effects on AMSR-E land observations. *Remote Sens. Environ.* **2006**, *100*, 190–199. [CrossRef]
53. Eltahir, E.A.B. A soil moisture rainfall feedback mechanism 1. Theory and observations. *Water Resour. Res.* **1998**, *34*, 765–776. [CrossRef]
54. Liu, Z.; Notaro, M.; Gallimore, R. Indirect vegetation-soil moisture feedback with application to Holocene North Africa climate. *Glob. Chang. Biol.* **2010**, *16*, 1733–1743. [CrossRef]
55. Tao, X.; Liang, S.L.; He, T.; Jin, H.R. Estimation of fraction of absorbed photosynthetically active radiation from multiple satellite data: Model development and validation. *Remote Sens. Environ.* **2016**, *184*, 539–557. [CrossRef]

Article

Intercomparison of Surface Albedo Retrievals from MISR, MODIS, CGLS Using Tower and Upscaled Tower Measurements

Rui Song [1,*], Jan-Peter Muller [1], Said Kharbouche [1] and William Woodgate [2]

[1] Imaging Group, Mullard Space Science Laboratory, University College London, Holmbury St Mary, Dorking, Surrey RH5 6NT, UK; j.muller@ucl.ac.uk (J.-P.M.); s.kharbouche@ucl.ac.uk (S.K.)
[2] Building 801, CSIRO, Black Mountain, Canberra 2601, Australia; William.Woodgate@csiro.au
* Correspondence: rui.song@ucl.ac.uk

Received: 3 February 2019; Accepted: 11 March 2019; Published: 16 March 2019

Abstract: Surface albedo is of crucial interest in land–climate interaction studies, since it is a key parameter that affects the Earth's radiation budget. The temporal and spatial variation of surface albedo can be retrieved from conventional satellite observations after a series of processes, including atmospheric correction to surface spectral bi-directional reflectance factor (BRF), bi-directional reflectance distribution function (BRDF) modelling using these BRFs, and, where required, narrow-to-broadband albedo conversions. This processing chain introduces errors that can be accumulated and then affect the accuracy of the retrieved albedo products. In this study, the albedo products derived from the multi-angle imaging spectroradiometer (MISR), moderate resolution imaging spectroradiometer (MODIS) and the Copernicus Global Land Service (CGLS), based on the VEGETATION and now the PROBA-V sensors, are compared with albedometer and upscaled in situ measurements from 19 tower sites from the FLUXNET network, surface radiation budget network (SURFRAD) and Baseline Surface Radiation Network (BSRN) networks. The MISR sensor onboard the Terra satellite has 9 cameras at different view angles, which allows a near-simultaneous retrieval of surface albedo. Using a 16-day retrieval algorithm, the MODIS generates the daily albedo products (MCD43A) at a 500-m resolution. The CGLS albedo products are derived from the VEGETATION and PROBA-V, and updated every 10 days using a weighted 30-day window. We describe a newly developed method to derive the two types of albedo, which are directional hemispherical reflectance (DHR) and bi-hemispherical reflectance (BHR), directly from three tower-measured variables of shortwave radiation: downwelling, upwelling and diffuse shortwave radiation. In the validation process, the MISR, MODIS and CGLS-derived albedos (DHR and BHR) are first compared with tower measured albedos, using pixel-to-point analysis, between 2012 to 2016. The tower measured point albedos are then upscaled to coarse-resolution albedos, based on atmospherically corrected BRFs from high-resolution Earth observation (HR-EO) data, alongside MODIS BRDF climatology from a larger area. Then a pixel-to-pixel comparison is performed between DHR and BHR retrieved from coarse-resolution satellite observations and DHR and BHR upscaled from accurate tower measurements. The experimental results are presented on exploring the parameter space associated with land cover type, heterogeneous vs. homogeneous and instantaneous vs. time composite retrievals of surface albedo.

Keywords: surface albedo; directional hemispherical reflectance; bi-hemispherical reflectance; tower albedometer; CGLS; MODIS; MISR; upscaling

1. Introduction

Albedo, also known as hemispherical reflectance, is a fundamental radiative parameter for energy partition of incoming solar radiation [1]. Albedo controls the temperature of the Earth's surface in concert with the effect of greenhouse gases. Around 30% of the total incoming radiation is reflected back into space and this is known as the planetary albedo [2]. Most of this radiation is reflected by clouds, snow and ice. Around 4% of the incoming solar irradiance is reflected by the land and ocean surface, which is some 13% of the total radiation reflected in the shortwave [3].

Systematic measurements of albedo have been acquired since the 1940s [4], although the instruments in use today were invented earlier, in the 1920s. Albedometers use a pair of calibrated pyranometers, one looking skywards and the other groundward. In order for an albedometer to cover a large enough region of the Earth's land surface, it is usually mounted at the top of a tower, which can vary in height from 10 m up to over a hundred metres. Such towers were extremely rare until the mid 1990s, so most albedo measurements covered only a small patch of ground and, almost invariably, only snow and ice or grass or concrete or tarmac from a height of a few to 10 m. This meant that such albedo measurements could not be employed to study any long-term trends, as the spatial representativeness of such measurements is very limited. Systematic observations from albedometers from the Baseline Surface Radiation Network (BSRN) network [5] started in 1992 from 10-m high towers. Towers are extremely expensive to construct and maintain and the associated electrical power and/or telecommunications infrastructure requirements makes them fairly rare. As part of the National Oceanic and Atmospheric Administration (NOAA) contribution to BSRN, the surface radiation budget network (SURFRAD) [6] tower-based radiation sensor network was founded in 1995, which now includes seven 10-metre towers. BSRN include measurements of total, direct and diffuse downward and upward radiation, mostly in the shortwave region from 300–3000 nm. FLUXNET is a "global network of regional networks" created by scientists across the world to coordinate regional and global observations from micrometeorological tower sites. These flux tower sites use eddy covariance methods to measure the exchanges of carbon dioxide (CO_2), water vapour and energy between terrestrial ecosystems and the atmosphere. FLUXNET include total albedometer measurements, but only a limited number include measurement of the diffuse component. Most FLUXNET sites are located over forests.

Since the earliest days of satellite radiometer observations, methods to retrieve surface albedo from visible and near-infrared (NIR) geostationary images [7] and polar orbiting images have been developed [8]. These early images typically had pixel resolution around 4–5 km and relied on intercomparison with other satellite data products [9], with little, if any, intercomparison using field-measured albedos. The first example of broadband (shortwave) albedo validation using tower albedometer measurements, upscaled by Landsat 30-m inferred albedos, were made in the early 2000s [10]. After this initial work, effort was focused on finding homogeneous sites to directly compare tower albedometer measurements with satellite-derived albedos from 500 m–3 km [11] without the need for upscaling. Historically, there has been a lack of an appropriate upscaling method for comparing multi-scale albedo measurements. In this work, we develop a general framework based on analysing time series of tower albedometers to retrieve bi-directional reflectance distribution function (BRDF) along with bi-hemispherical reflectance (BHR), called in one particular theoretical case "white sky albedo", with uniform sky irradiance alongside direct hemispherical reflectance (DHR), usually referred to as "black sky albedo" [12]. The materials used in this study include tower measurements derived from the FLUXNET, SURFRAD, and BSRN tower sites, and satellite data products derived by the Copernicus Global Land Service (CGLS) from VEGETATION-2 and Proba-V [13], moderate resolution imaging spectroradiometer (MODIS) [14] and multi-angle imaging spectroradiometer (MISR) [15]. The aim is to develop a new method for comparing ground-level, in situ measurements derived from a tower albedometer against coarse resolution albedos derived from repeat-pass or near-simultaneous multi-angle spaceborne observations. Specific objectives include developing a new method for deriving DHR and BHR from tower albedometer measurements; comparing these derived in situ DHR and BHR tower albedos against CGLS, MODIS and MISR products through a pixel-to-point analysis over a long time-series. A new technique is also developed for upscaling albedo from tower to a coarse resolution based on atmospherically corrected BRFs from high-resolution Earth observation (EO) data, combined with downscaled MODIS BRDF climatology over a larger area. A pixel-to-pixel comparison is presented between DHR and BHR retrieved from CGLS products and DHR and BHR upscaled from in situ measurements using this proposed upscaling technique. The sites are both homogeneous and heterogeneous in land cover and reflectance, and are located on all the continents including Antarctica.

2. Materials and Methods

2.1. Ground Measurements

Measurements between year 2012 and 2016 at 20 tower sites from the FLUXNET, SURFRAD and BSRN networks were used in this study. The sites were located over 5 continents as follows: Europe, North America, South America, Australia and Antarctica, as shown in Figure 1. Table 1 lists the key characteristics of the 20 selected sites, including their associated network, geographic coordinates and land cover type. Shortwave radiation is measured by albedometers at these selected tower sites. A broadband (shortwave) albedometer essentially consists of two pyranometers, which measure the total downward and upwelling radiation. Diffuse radiation is measured by an independent shaded pyranometer using a sun tracker to shield the sensor from direct sunlight [16]. Data from the SURFRAD and BSRN sites were taken between year 2012 and 2016 at 1-min resolution, and data from FLUXNET sites were taken between 10-min and 1-h resolution, depending on the site location. We select for presentation one station from each network including two from the USA and one from Australia representing, tundra with snow/ice, grasslands and evergreen broadleaf respectively.

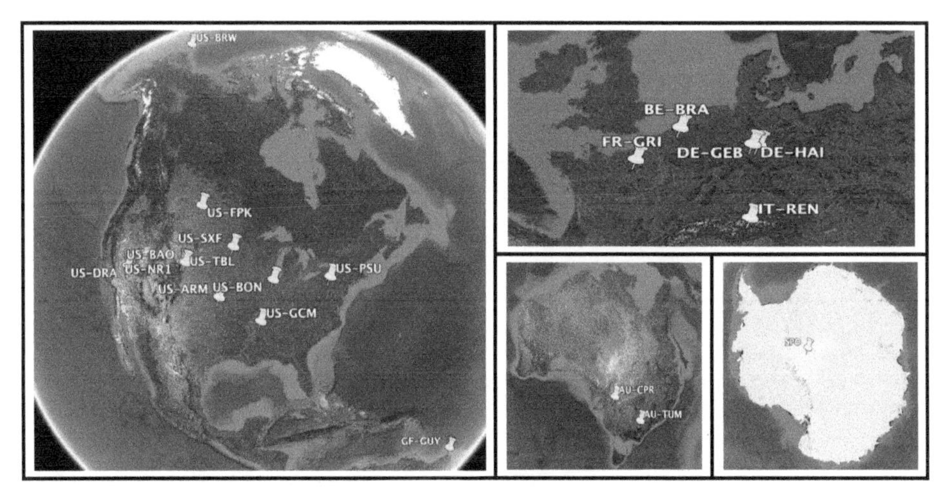

Figure 1. Geographical distribution of selected sites (Google Earth).

2.2. Satellite Albedo

2.2.1. MODIS BRDF/Albedo Products

Based on the three-parameter RossThick–LiSparse–Reciprocal (RTLSR) model, the MODIS bi-directional reflectance distribution function and albedo (BRDF/albedo) products (MCD43A) provide a collection of 500-m, daily resolution data [14]. Clear-sky and atmospherically-corrected surface reflectances from Terra satellite and Aqua satellite, within a 16-day window, were used to retrieve the

BRDF parameters. MCD43A1 provides the 3 BRDF parameters (isotropic, volumetric and geometric) for each of the MODIS bands 1–7, and the visible, near infrared and shortwave bands. The related quality assurance (QA) data are stored in MCD43A2. Based on the retrieved weighting parameters (isotropic, volumetric, and geometric), MCD43A3 derives both DHR (black sky albedo) and BHR (white sky albedo) 500-m data for the corresponding bands. In this study, the 500-m shortwave DHR and BHR from 2012 to 2016, covering the 19 tower sites (except for the SPO), were directly extracted from the MCD43A3 products. For the sites (e.g., the US-BRW and AU-CPR) where there were missing values in the MCD43A3, the DHR and BHR were calculated from the MCD43A1 BRDF parameters. A cross-check was performed to ensure consistency between the two.

2.2.2. Copernicus Global Land Service

The European Union CGLS operates a "a multi-purpose service component" that produces a series of qualified bio-geophysical products on the status and evolution of the land surface at global scale (https://land.copernicus.eu/global/). The land surface parameters produced from the CGLS include the leaf area index (LAI), the fraction of absorbed photosynthetically active radiation (FAPAR) absorbed by the vegetation, the surface albedo, the land surface temperature, the soil moisture, etc. The CGLS albedo products are solely derived from the VEGETATION instrument up until 2014, and since then from the PROBA-V sensors. They are updated every 10 days using a 30-day window. The DHR and BHR are projected onto a regular latitude/longitude grid with a resolution of 1/112° (approx. 1 km at the equator) covering the area from 180°E to 180°W and from 75°N to 60°S. In this study, 1-km CGLS shortwave DHR and BHR products from 2012 to 2016 covering the 19 tower sites were used (no CGLS data over the SPO).

2.2.3. MISR

The multi-angle imaging spectroradiometer (MISR) [17] sensor onboard NASA's Earth Observing System (EOS) Terra satellite provides high accuracy surface albedo products from near simultaneous multi-angular views. The MISR level 2 land/surface albedo products provide land DHR and BHR over four narrow bands: blue (446 ± 21 nm), green (558 ± 15 nm), red (672 ± 11 nm) and near infrared (866 ± 20 nm), at a resolution of 1.1 km. Liang's model [18] is used here to retrieve the total shortwave broadband albedo by linearly combining the spectral albedos as follows:

$$\alpha^{MISR} = 0.126 \cdot \alpha_2 + 0.343 \cdot \alpha_3 + 0.415 \cdot \alpha_4 + 0.0037 \quad (1)$$

where α_2, α_3 and α_4 represent MISR spectral albedos for band 2, 3 and 4, α^{MISR} is the total broadband shortwave albedo. In this study, MISR pixels near each of the tower sites were extracted between 2012 and 2016. It should be noted that, the BHR products from MISR are different to MODIS and CGLS, because they represent the actual blue-sky albedo rather than an idealised white-sky.

2.3. Surface Albedo from Tower Measurements

DHR and BHR [19] are calculated from the ratios of the measured upwelling and downwelling solar radiant fluxes, but they are based on different assumptions about how the atmospheric scattering processes can affect the intensity of downwelling diffuse radiation. If the atmospheric scattering effects are removed, then the illumination can be assumed to originate from a single infinitesimally small point source. In this case, the measured ratio between the upwelling and downwelling radiations becomes the DHR. If the atmospheric scattering effects are included, then the illumination is assumed to be uniform from all angles and this is known as the "white sky". This results in the BHR being calculated from the measured ratio between the upwelling and downwelling radiation. Both DHR and BHR represent extreme cases that rarely exist in the physical "real world". In all previous works of satellite-derived albedo validation using in situ measurements, a compromised value between DHR and BHR has been used for an indirect comparison [20,21]. This compromised value is intended to represent the in situ albedo, which is called the blue-sky (or clear sky) albedo and can be computed as follows:

$$BlueSkyAlbedo = \beta \cdot \mathrm{BHR} + (1 - \beta) \cdot \mathrm{DHR} \quad (2)$$

where β denotes the proportion of diffuse component in downwelling solar radiation. Normally, β is measured at ground-tower sites by a separate pyranometer, which is independent from the pyranometers that measure the total downwelling and upwelling radiations. This independent pyranometer is mounted with a sun tracker that shields the sensor from direct sunlight. If the ground-based pyranometer that measures diffuse radiation is not available, a satellite aerosol product (e.g., MOD04/MYD04 of MODIS or the aerosol optical depth (AOD) retrieved as part of the GlobAlbedo (http://www.globalbedo.org/) product [22]) can be used to estimate β, but this is with high temporal and spatial uncertainties.

Table 1. List of tower sites with key characteristics: acronyms, geographical coordinates, network, footprint (see Equation (12)) and land cover type, defined by International Geosphere-Biosphere Programme (IGBP). Station names in bold are those whose results are shown below.

Station	Acronym	Latitude (°)	Longitude (°)	Network	Footprint	Land Classification (IGBP)
Barrow **	US-BRW	71.323	−156.607	BSRN (http://bsrn.awi.de)	51 m	Snow and Ice
Niwot Ridge #	US-NR1	40.033	−105.546	FLUXNET (https://FLUXNET.ornl.gov)	158 m	Evergreen Needleleaf
Sioux Falls	US-SXF	43.730	−96.620	SURFRAD (https://www.esrl.noaa.gov/gmd/grad/surfrad/)	126 m	Croplands
ARM Southern Great Plains	US-ARM	36.606	−97.489	FLUXNET	25 m	Croplands
Bondville	US-BON	40.052	−88.373	SURFRAD	126 m	Croplands
Boulder atmospheric observatory *	US-BAO	40.050	−105.004	BSRN	3788 m	Cropland Mosaics
Desert Rock *	US-DRA	36.624	−116.019	SURFRAD	126 m	Open Shrublands
Fort Peck *	US-FPK	48.308	−105.102	SURFRAD	126 m	Grasslands
Goodwin Creek	US-GCM	34.255	−89.873	SURFRAD	126 m	Deciduous Broadleaf
Penn State	US-PSU	40.720	−77.931	SURFRAD	126 m	Deciduous Broadleaf
Table Mountain *	US-TBL	40.125	−105.237	SURFRAD	126 m	Bare soil and Rocks
Gebesee *	DE-GEB	51.100	10.914	FLUXNET	76 m	Croplands
Hainich *	DE-HAI	51.070	10.450	FLUXNET	265 m	Mixed Forest
Grignon	FR-GRI	48.844	1.952	FLUXNET	67 m	Croplands
Guyaflux *#	GF-GUY	5.279	−52.925	FLUXNET	290 m	Evergreen Broadleaf
Brasschaat	BE-BRA	51.309	4.521	FLUXNET	240 m	Mixed Forest
Renon	IT-REN	46.587	11.434	FLUXNET	152 m	Evergreen Needleleaf
Tumbarumba *	AU-Tum	−35.657	148.152	FLUXNET	505 m	Evergreen Broadleaf
Calperum #	AU-CPR	−34.003	140.588	FLUXNET	215 m	Closed Shrublands
South Pole *	SPO	−90	59	BSRN	25 m	Snow and ice

Sites marked with * are claimed to be spatially representative, which is sometimes referred to as homogeneous by [11]. ** US-BRW is spatially representative during snow covered periods, but heterogeneous during the snow melt season. N.B. The three sites marked with # do not have diffuse radiation measurements, so the method introduced in Section 2.3.1 is used to estimate diffuse radiation [23].

Blue-sky albedo can be estimated by combining the DHR and BHR data from satellite measurements and the β value from tower measurements. In this way, the estimated albedo value at local solar noon can be used directly for comparison with the ground-based albedo. However, there are two major critical issues in this inter-comparison: (1) β is measured at local solar noon and, often, under cloud-free conditions. The blue-sky albedo in this case is dominated by the DHR because the β value is close to zero under cloud-free conditions. (2) DHR and BHR cannot be assessed separately from the blue-sky albedo using this method. To overcome these issues, a new strategy is proposed in which albedo is derived into the DHR and BHR components separately, solely from in situ tower data. A conceptual flowchart of this processing chain is illustrated in Figure 2, and details are introduced in Sections 2.3.1 and 2.3.2.

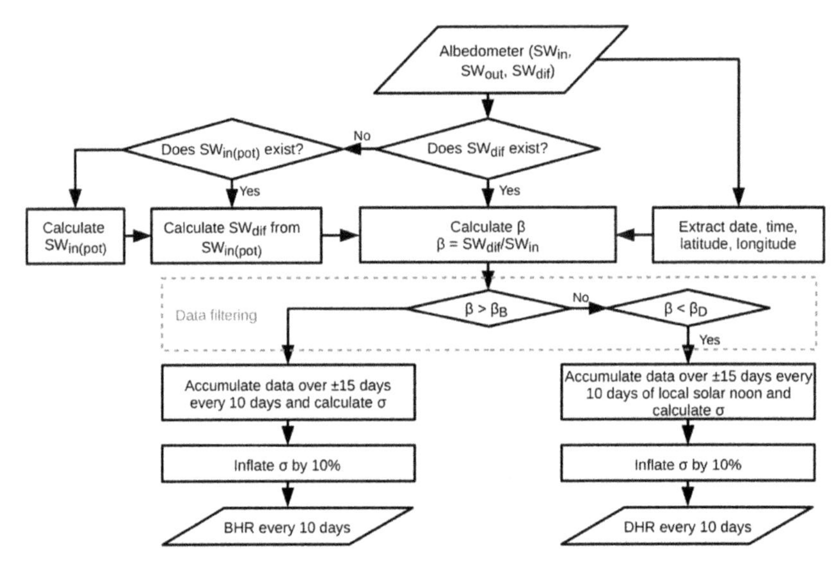

Figure 2. Flowchart illustrating the algorithm for retrieving directional hemispherical reflectance (DHR) and bi-hemispherical reflectance (BHR) from tower measurements. β_B and β_D represent the threshold of the diffuse ratio for filtering BHR and DHR, respectively. σ represents the standard deviation of the calculated BHR and DHR values. SW_{in}, SW_{out} and SW_{dif} represent the downwelling total, upwelling total and downwelling diffuse shortwave radiation, respectively.

2.3.1. Directional Hemispherical Reflectance (DHR)

In Figure 2, the SW_{in}, SW_{out} and SW_{dif} represent the downwelling total, upwelling total, and downwelling diffuse shortwave radiation measured from the tower sites, respectively. The downwelling diffuse radiation was measured at all of the selected SUFRAD and BSRN tower sites, but was unavailable at three FLUXNET tower sites (i.e., the Calperum, Guyaflux and Niwot Ridge sites). To deal with this challenge, we calculated β from the potential top of atmosphere radiance $SW_{in(pot)}$ using Equation (3). The parameter $SW_{in(pot)}$ is provided in the FLUXNET dataset, or, if not, can be estimated based on the geographic location of the tower and the sun-tower geometry [23]. Experimental results demonstrated that diffuse ratios estimated from Equation (3) had a good agreement with real measurements when β was very low or very high.

$$\beta = \left(SW_{in(pot)} - SW_{in}\right) / SW_{in(pot)} \quad (3)$$

According to Equation (2), the DHR can be well approximated by the blue-sky albedo when β tends towards zero. Such low values of β can be reached under completely cloudless conditions with a very low level of aerosol at local solar noon. From empirical heuristic studies we verified that a threshold of $\beta_D = 0.1$ was suitable for screening out undesired tower measurements, because data that meet the condition of $\beta \leq \beta_D$ within ± 1 h local solar noon, for all sites, can be considered sufficient to perform the validation of DHR derived from satellite measurements. A threshold of β_D lower than 0.1

can reduce the amount of remaining data dramatically, because β does not often reach such a low level within the specified ± 1 h at local solar noon, due to broken cloud cover.

Once the ± 1 h local solar noon data are screened by the condition $\beta \leq \beta_D$, the mean and standard deviation (σ) over a sliding time-window of 30 days can be calculated. In this example the time-window for averaging the tower data was set as 30 days for CGLS, so that the in situ DHR and BHR could be compared with these CGLS products. However, a narrower time-window can also be adopted if enough data can be collected for estimating DHRs using this strategy. This method can cause a bias in the derived DHR values because the effect of BHR in the blue-sky albedo is ignored. However, the effect of BHR in the blue-sky albedo will not exceed the predefined threshold of β_D. Therefore, in order to estimate the uncertainty of the DHRs, the σ of the final in situ DHRs is dilated by β_D, so as to represent the effect of BHR on DHR. Figure 3 shows an example of ground-based albedo (SW_{out}/SW_{in}) at different stages of in situ DHR creation from the data of FLUXNET Tumbarumba site in 2015. The main steps of producing in situ DHRs can be summarised as follows: (1) Calculating SW_{out}/SW_{in} for all the data points; (2) filtering data with negative upwelling or downwelling radiation values; (3) filtering data with solar zenith angle larger than 75°, because the linear BRDF model does not work for these zenith angles; (4) retaining data that meet the condition $\beta \leq \beta_D$; (5) retaining data within ± 1 h of local solar noon; (6) applying a weighting function over the time window of 30 days; and (7) including the estimation of uncertainty values.

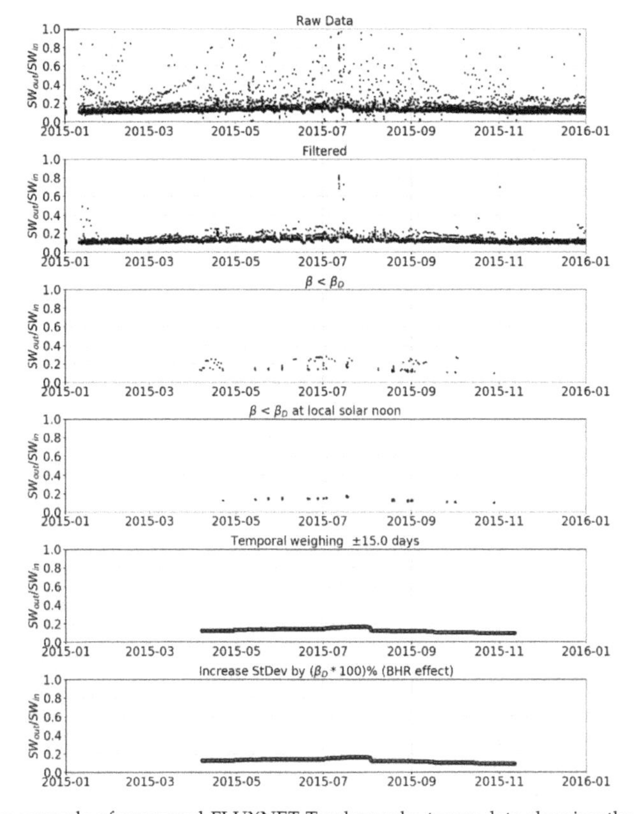

Figure 3. An example of processed FLUXNET Tumbarumba tower data showing the steps in the production of in situ DHRs.

2.3.2. Bi-Directional Hemispherical Reflectance (BHR)

In contrast, the in situ BHR can be approximated by the blue-sky albedo when β tends towards 1. This high diffusion ratio can only be reached under certain conditions, such as under thick unbroken stratus clouds, which are often formed before sunrise and after sundown. Under this assumption, a lower threshold of $\beta_B = 0.9$ was used to filter undesirable measurements from the raw data. Once the data are filtered by this condition ($\beta \geq \beta_B$), the BHR was calculated from the SW_{out}/SW_{in} from the remaining data. Similarly, the mean and the standard deviation over a sliding time-window, which is 30 days in this example, were calculated. The effect of DHR on in situ BHR should be taken into account when assuming the BHR, and can be approximated by the blue-sky albedo. From trial-and-error, the σ of the final in situ BHR need to be increased by $1 - \beta_B$ to represent its uncertainty. Figure 4 shows an example of ground-based albedos at different stages of in situ BHR creation from the data of FLUXNET Tumbarumba site in 2015. The processing and filtering steps were the same as specified above for DHR.

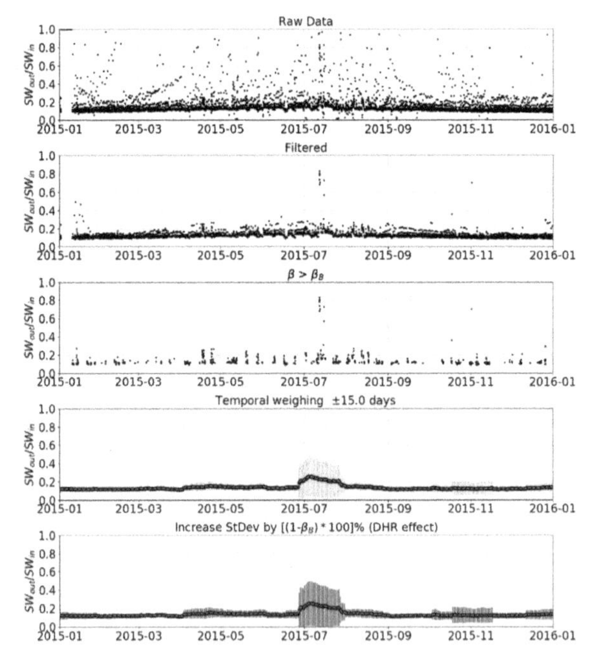

Figure 4. A sample of processed FLUXNET Tumbarumba tower data for producing in situ BHRs.

2.4. *Albedo Upscaling*

According to Liang [10], unless the surface area is large and sufficiently homogeneous or a sufficient number of measurements is obtained during the period of satellite overpass, "point" measurements are not feasible to validate coarse-scale albedo products through a direct comparison. The in situ-based measurements are considered to be spatially representatively at homogenous land surfaces, and; therefore, can be directly used [11]. However, the in situ measurements and coarse-scale albedo are generally at an unmatched scale over heterogeneous land surfaces [24]. In order to validate satellite-derived coarse resolution albedo products from in situ measurements, upscaling from ground "point" data to the coarse resolutions is essential. This upscaling process consists of two steps. First, a "calibration factor" needs to be calculated based on the "point" measurements and the high-resolution EO pixels within the projected field-of-view (FoV) of the tower albedometer. Second, the high-resolution albedo products need to be upscaled to coarse-resolution using the pre-calculated "calibration factor".

High-resolution (<100 m) albedo products are not currently available from EO satellites. However, they can be derived using the methods described in [25], and in [26] for the Chinese HuanJing (HJ) satellite sensor. Their method used early machine learning (neural networks) based on numerous radiative transfer simulations. In this study, atmospherically corrected bi-directional reflectance factors (BRFs), from high-resolution EO, alongside coarse-resolution albedos, predicted from a MODIS BRDF climatology over a larger area, were employed as inputs. Firstly, pure endmembers for the corresponding area of the reference albedometer measurements were extracted based on the high-resolution spectral reflectance data. This was followed by aggregating the derived high-resolution spectral BRFs to coarse resolutions. In the next step, a linear regression was established between the endmember abundance and MODIS BRDF-derived albedo-to-nadir-reflectance ratios. This linear regression was established based on the data at coarse resolutions, but could be used to assign each high-resolution pixel around the tower with a corresponding albedo value. This was based on the assumption that over a larger area, coarse-resolution BRDFs can be represented by high-resolution mosaics of pixel BRDFs, weighted by their coverage proportions. Then, the high-resolution spectral reflectance values were converted to shortwave reflectance values through the use of narrowband to broadband conversion coefficients. In the end, a "calibration factor" was obtained by calculating the ratio of shortwave albedo derived from the albedometer and that from the high-resolution EO. This factor could then be applied to upscale the shortwave albedo from high resolutions to coarser resolution.

2.4.1. Retrieval of High-Resolution Shortwave Albedos

We modified the method proposed by Shuai et al. [27] and used it to retrieve high-resolution shortwave albedos from Landsat-8 high-resolution BRFs and coarse-resolution MODIS BRDF climatology. The idea of Shuai's method is to first classify the spectral features of the land surface through unsupervised classifications, using 6 non-thermal bands of 30-m Landsat data. The Landsat data are then re-projected and aggregated to MODIS resolution, such that the albedo-to-nadir-reflectance ratios can be calculated for the "pure" pixels defined from the classification. Finally, the high-resolution albedo is produced for each of the Landsat pixel using the derived ratios. In Shuai's method, the detection of "pure pixels" from coarse-resolution pixels is an essential requirement for the subsequent calculation of albedo-to-nadir-reflectance ratios. However, this approach is unlikely to find "pure pixels" for each of the classes at a coarse resolution image, especially for heterogeneous land surfaces. In the modified retrieval method, the albedo-to-nadir-reflectance ratios still play a key role in deriving the high-resolution albedos, but the existence of "pure pixels" in coarse resolutions is not necessarily required.

The 1-km MODIS BRDF climatology parameters were produced from the 500-m MCD43A1 products [28]. The surfaces reflectances derived from this kernel-driven BRDF model are described as:

$$R(\lambda,\theta_{in},\theta_{out},\phi) = f_{iso}(\lambda) + f_{vol}(\lambda)k_{vol}(\theta_{in},\theta_{out},\phi) + f_{geo}(\lambda)k_{geo}(\theta_{in},\theta_{out},\phi) \tag{4}$$

where λ is the bandpass of a given spectral channel; θ_{in}, θ_{out} and ϕ are the solar zenith, view zenith and relative azimuth angles, respectively. k is the BRDF RossThick–LiSparse–Reciprocal (RTLSR) kernel and f is the spectrally-dependent kernel weighting, with subscripts *iso*, *vol* and *geo* representing the isotropic, volumetric and geometric-optical components, respectively. Integration of the BRFs over all view angles results in a DHR, and a further integration over all illumination angles results in a BHR:

$$\mathrm{DHR_M}(\theta_{in}(L_8)) = \frac{1}{\pi}\int_0^{2\pi} d\varphi \int_0^1 R_M(\lambda,\ \theta_{in}(L_8),\ \theta_{out},\ \varphi)u_v du_v \tag{5}$$

$$\mathrm{BHR_M} = \frac{1}{\pi}\int_0^{2\pi} d\varphi \int_0^1 \mathrm{DHR_M}(\theta_{in}(L_8))u_s du_s \tag{6}$$

where u_v(=sin θ_{out}) and u_s(=sin θ_{in}) are the variables of integration. The shortwave BRFs at MODIS 1-km resolution for the Landsat-8 solar zenith and view zenith angle are given by

$$R_M(\Omega(L_8)) = R_M(\lambda,\ \theta_{in} = \theta_{in}(L_8),\ \theta_{out} = \theta_{out}(L_8),\ \varphi = \varphi(L_8)) \tag{7}$$

where $\Omega(L_8)$ is the Landsat-8 sun and sensor geometry. Then, the ratios between the shortwave albedo and shortwave reflectance values can be computed for all the pixels at 1-km MODIS resolution:

$$\alpha_D = \frac{DHR_M(\theta_{in}(L_8))}{R_M(\Omega_L)};\ \alpha_B = \frac{BHR_M}{R_M(\Omega_L)} \tag{8}$$

An endmember is defined as a land "type" that is assumed to have a unique spectral signature. Here, the N-FINDR endmember extraction algorithm [29] was adopted to extract the pure endmembers from the 30-m Landsat-8 spectral reflectance data. This was followed by re-projecting the Landsat-8 data from Universal Transverse Mercator (UTM) to MODIS (sinusoidal) projection, and aggregating the pixels from 30-m to 1-km resolution. Then the proportion of each endmember was calculated for each of the aggregated 1-km pixel using a fully constrained least squares (FCLS) linear un-mixing method [30]. Then, the following equation was established:

$$\mathbf{A}\,\mathbf{W} = \mathbf{R} \tag{9}$$

where **A** is a (m, n) matrix with m being the number aggregated pixels, and n being the number of endmembers. Each row of the matrix **A** contains the proportions of derived endmembers. **W** is a (n, 1) matrix containing the weighting parameters. **R** is a (m, 1) matrix with the elements representing the MODIS BRDF climatology albedo-to-nadir-reflectance ratios. The weighting function **W** is solved as,

$$\mathbf{W} = \left(\mathbf{A}^{\mathrm{T}}\,\mathbf{A}\right)^{-1}\mathbf{A}^{\mathrm{T}}\,\mathbf{R} \tag{10}$$

where the superscript T refers to matrix transpose. Given the Landsat-8 spectral reflectance values, the shortwave reflectance values can be calculated through the use of a set of narrowband to broadband conversion coefficients [18]:

$$\alpha^{L_8} = 0.356\alpha_2 + 0.13\alpha_4 + 0.373\alpha_5 + 0.085\alpha_6 + 0.072\alpha_7 - 0.0018 \tag{11}$$

where α_2, α_4, α_5, α_6 and α_7 represent the Landsat-8 blue, red, NIR, SWIR-1 and SWIR-2 narrow bands, respectively. The shortwave reflectance is transformed to shortwave albedo through the following formula:

$$\mathbf{B} = (\mathbf{C}\,\mathbf{W}) \circ \mathbf{L} \tag{12}$$

where **C** is (k, n) matrix with k being the number of 30-m Landsat-8 pixels. Each row of **C** contains the derived endmember proportions of Landsat-8 pixels. **L** is a (k, 1) matrix that contains the shortwave reflectance derived from Equation (11) for each pixel, and $\circ$ is the Hadamard product. The processing chain for generating high-resolution albedo using Landsat-8, as an example, is illustrated in Figure 5.

2.4.2. Upscaling of Albedo from Tower to Coarse Resolutions

The pyranometers that measure downwelling and upwelling radiation are mounted on towers with a fixed height of 10 m at the SURFRAD sites, while the BSRN and FLUXNET sites utilize towers at different heights, usually dependent on the height of the canopy-top. The reference albedo was located by assuming the albedometer measures a circular area from the top of the tower [31]. The diameter of this circular area, which represents the effective projected FoV of the tower albedometer, was estimated as:

$$\mathrm{D} = 2\tan(FoV^{\circ}/2)\cdot(h_{tower} - h_{ToC}) \tag{13}$$

where h_{tower} and h_{ToC} are the height of tower and averaged height of vegetation, respectively. $FoV/2$ is half of the effective field of view in degrees, which is 81° [32]. For a pyranometer mounted on a 10-m tower, the projected FoV on the surface is 126 m. The values for all the sites are given in Table 1. The reference albedo was approximated by averaging the corresponding albedo values of pixels within a high-resolution EO shortwave albedo product. Then, a "calibration factor" could be derived from the ratio between the in situ albedo and the reference albedo. To produce the coarse-resolution albedos, the high-resolution albedo product needed to be aggregated to a coarse resolution first, and then modified with this "calibration factor". The above-mentioned process for producing coarse-resolution albedo product is illustrated in Figure 6.

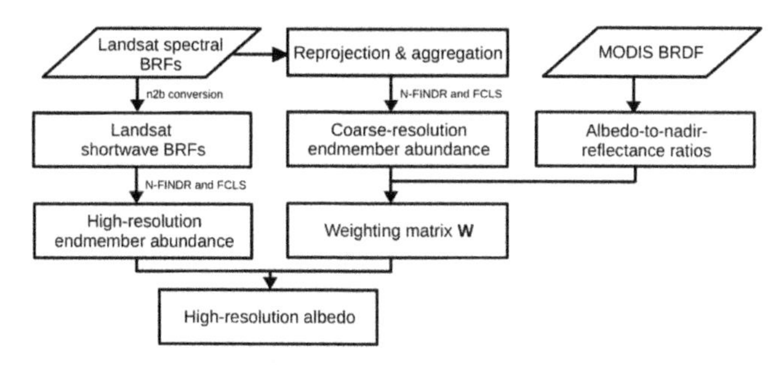

Figure 5. Processing chain for calculating high-resolution albedo from Landsat BRFs and MODIS BRDFs.

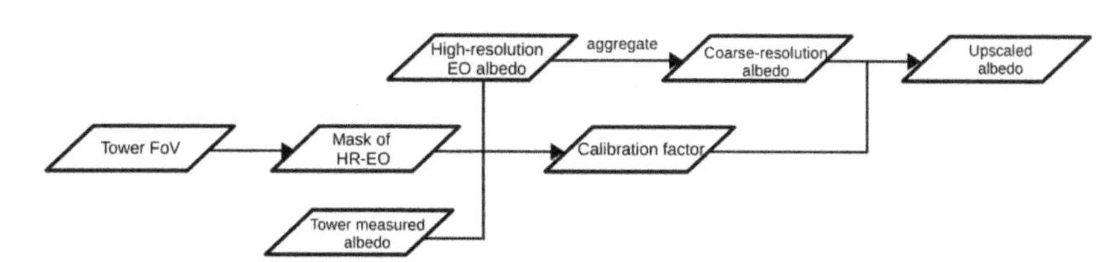

Figure 6. Steps of upscaling high-resolution albedo to coarse-resolution pixel size. The process of calculating high-resolution EO shortwave albedo is illustrated in Figure 5.

3. Results

3.1. Comparison of Surface Albedo between Satellite Products and In Situ Retrievals

Tower measured shortwave radiation data from 19 sites were used to estimate both DHRs and BHRs and evaluate the accuracy of satellite derived values. There were no intercomparison results at the SPO site presented here, due to the lack of higher resolution satellite data covering this region. In Figures 7 and 8, DHRs and BHRs retrieved from tower albedometers were compared with the CGLS, MODIS and MISR products for the following sites: AU-TUM (evergreen broadleaf), US-FPK (grasslands) and US-BRW (snow and ice). The intercomparison of time-series tower and satellite-derived albedo for the other SURFRAD and BSRN sites is provided in the Supplementary Figure S1. Albedo products from satellite observations were produced using different time windows (i.e., 30 days for CGLS, 16 days for MODIS and near simultaneously for MISR (~7 min)). Here, three different time windows (30, 16 and three days) were used in tower albedo retrieval for the corresponding intercomparison with CGLS, MODIS and MISR products, respectively. A three-day rather than a one-day window was used in tower albedo retrievals when comparing with MISR products, because the effective number of measurements acquired from a one-day window was often insufficient to retrieve DHRs or BHRs after data screening.

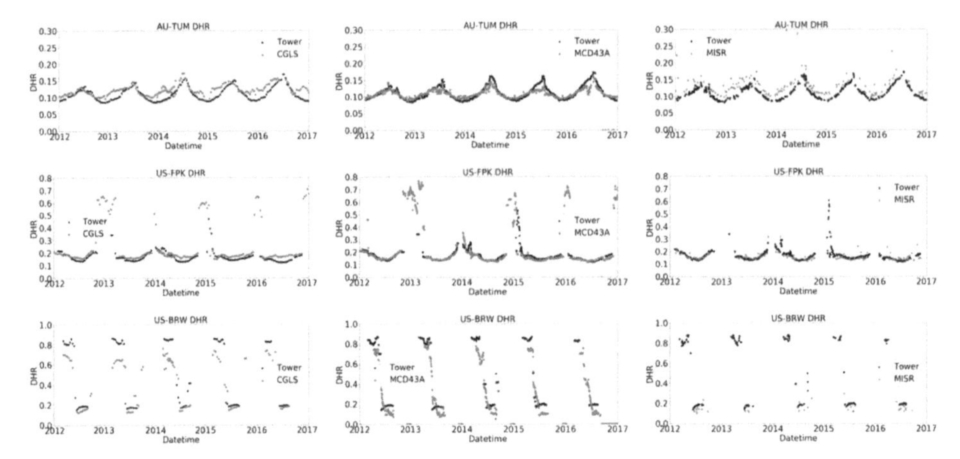

Figure 7. CGLS (column 1), MODIS (column 2) and MISR (column 3) DHR products compared with tower derived DHRs at the Tumbarumba, Fort Peck and Barrow sites.

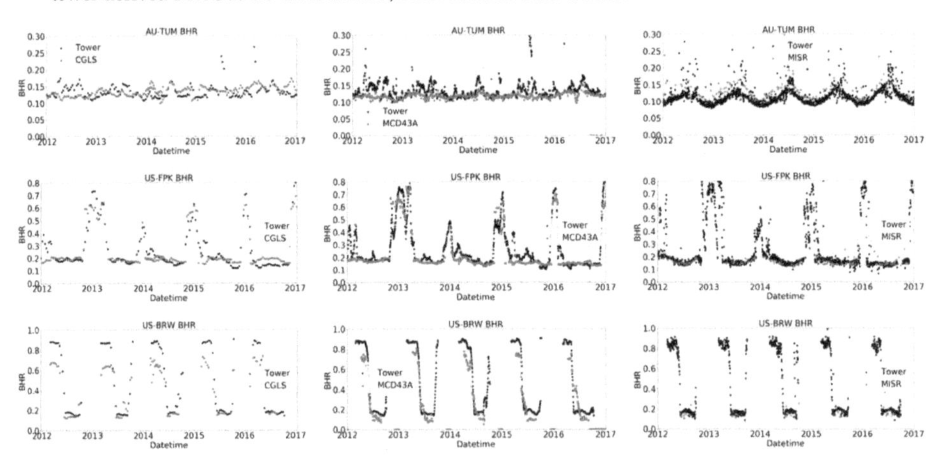

Figure 8. CGLS (column 1), MODIS (column 2) and MISR (column 3) BHR products compared with tower derived BHRs at the Tumbarumba, Fort Peck and Barrow sites.

At some sites (e.g., Tumbarumba and Fort Peck sites), the DHR intercomparison showed a good match in terms of absolute values and seasonal variations. The DHRs at the Barrow site showed a better match during the snow-free season than the snow-covered season. Among the three satellite DHR products, MODIS showed the best agreement with the in situ measurements. At the Tumbarumba site, both CGLS and MISR retrievals showed a systematic overestimation of the DHR, while the MODIS retrievals agreed fairly well with in situ measurements in all time periods. At the Ford Peck site, the MISR DHRs were comparable with MODIS DHRs, whereas the CGLS retrievals were still overestimated during the snow-free season. It is interesting to note that the MODIS retrievals had better performance in picking up the albedo of snow points. At the Barrow site, the MISR retrievals were closer to the in situ measurements during the snow-free season than the CGLS and MODIS retrievals.

The intercomparison of BHR measurements at the four sites discussed above are displayed in Figure 8, and results for the other sites can be found in Supplementary Figure S2. It should be noted that the BHR products provided in MISR were a very close approximation to the blue-sky albedo, rather than the white-sky albedo. Therefore, the tower albedos were directly retrieved from the ratio between the upwelling and downwelling radiation for this specific comparison with MISR retrievals.

The tower data for the purpose of MISR BHR comparison were screened over a ±1-h window at local solar noon during one day. The variation of surface albedo was dependent on the solar zenith angles, please see [33] in more details for an explanation of why solar noon is employed.

Generally speaking, the DHR retrievals showed better agreement between satellite and in situ measurements than the BHR retrievals. In our method for BHR retrievals, the illumination was assumed to be uniform from all angles when the diffuse ratio was larger than β_B. However, not all the tower data screened for BHR retrievals could meet this condition. This was the error source that may reduce the accuracy of BHR retrievals. Similarly, the MODIS DHR retrievals showed the best agreement with tower measurements, followed by the MISR retrievals, and then followed by the CGLS retrievals.

The albedo values derived from satellite products and tower retrieval are summarised in the 2D scatterplots shown in Figure 9. for DHRs for three selected sites. At the Tumbarumba site, all the satellite products were well-correlated with the in situ retrievals, with a bias value less than 0.025. The Fort Peck site also showed a good correlation, except for some points which were incorrectly identified as snow in CGLS and MODIS. The MISR products showed a better performance at the Barrow site during the snow-free season, while the CGLS and MODIS were better in picking up snow, although the snow-covered DHRs were often underestimated.

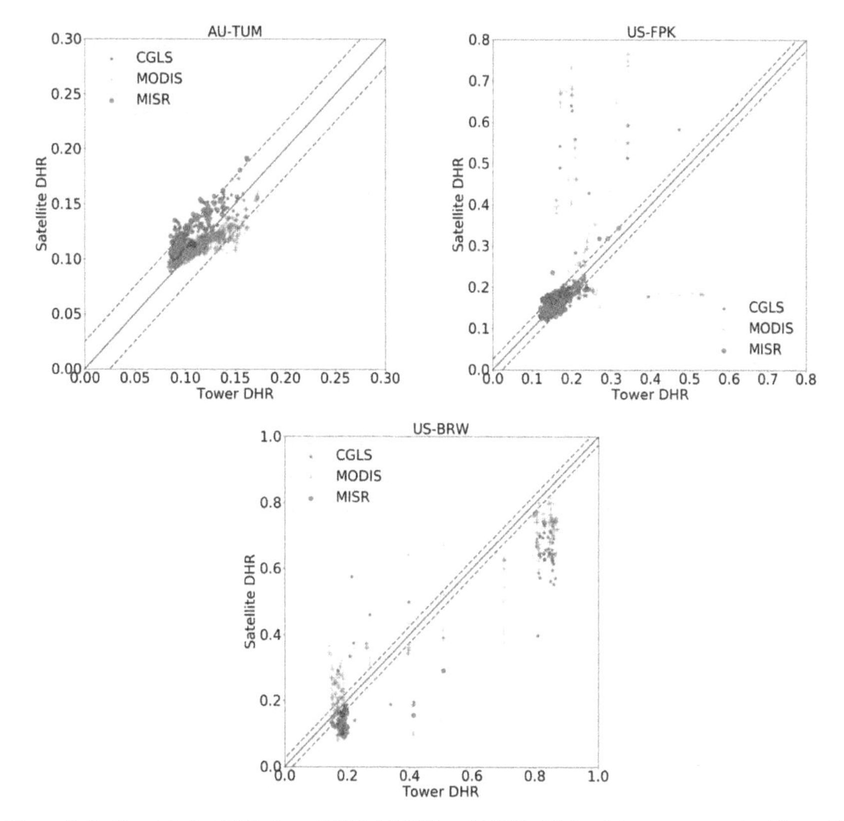

Figure 9. Scatterplots for DHRs from CGLS, MODIS and MISR. All the data are summarised from all the results from 2012-01-01 to 2016-12-31 for all four selected sites. The blue, green and red lines indicate CGLS, MODIS and MISR DHR products. The central solid lines are 1:1 lines (perfect correlation), and the outer dashed lines are 0.025 offset dashed lines.

The albedo values derived from satellite products against corresponding tower retrievals are summarised in Figure 10 for BHRs. Again, the satellite products and tower retrievals showed a better agreement of DHR values than BHR values. Large biases occured during the snow-covered season, which could be observed at the Fort Peck and Barrow sites.

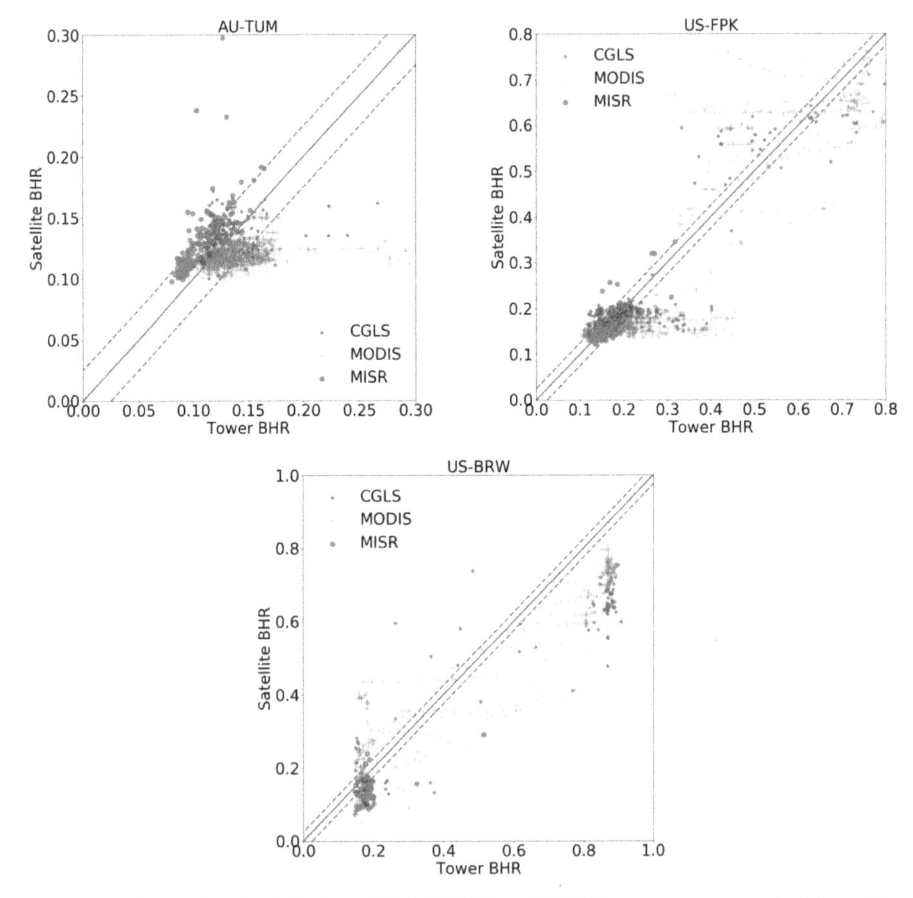

Figure 10. Scatterplots for BHRs from CGLS, MODIS, and MISR. The data are summarised from all the results from 2012-01-01 to 2016-12-31. The blue, green and red lines indicate CGLS, MODIS and MISR BHR products. The central solid lines are 1:1 lines, and the outer dashed lines are 0.025 offset dashed lines.

3.2. Comparison of Surface Albedo between Coarse-Resolution Satellite Products and Upscaled Tower Values

DHRs and BHRs retrieved from tower data were upscaled to 1-km resolution and compared with CGLS products to assess the performance of this upscaling strategy. Coincident Landsat-8 30-m albedo data, which are used as a bridge to fill gaps between the small footprint tower measurements and the coarse-resolution measurements, were produced using the method introduced in Section 2.4.1. Scatter plots between the upscaled albedo and CGLS 1-km albedo are displayed in Figure 11 for DHR comparisons and Figure 12 for BHR comparisons, respectively. Comparisons for other sites are given in the Supplementary Figure S5. MODIS BRDF climatology data were used as input in the upscaling process; therefore, here the upscaled values were not directly compared with the MODIS albedo products. The sparsity of MISR albedo products severely increased the difficulty of finding cloudless Landsat-8 data.

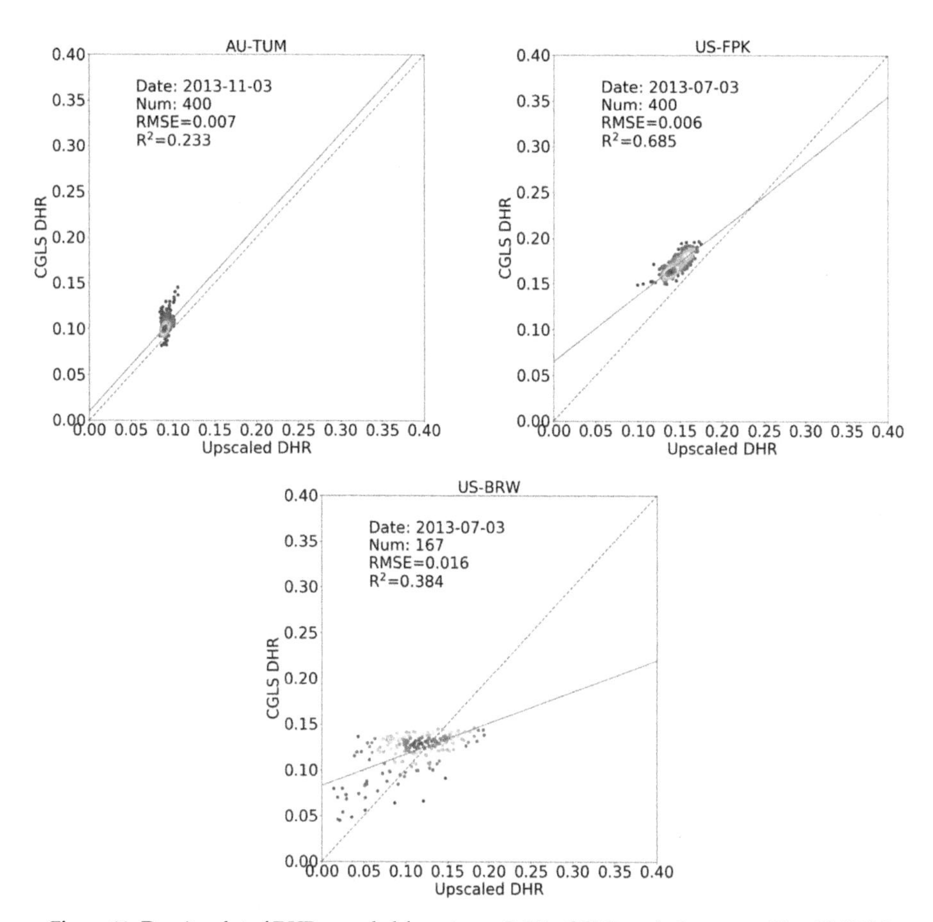

Figure 11. Density plot of DHR upscaled from tower FoV to CGLS resolution over a 20 × 20 CGLS pixel region.

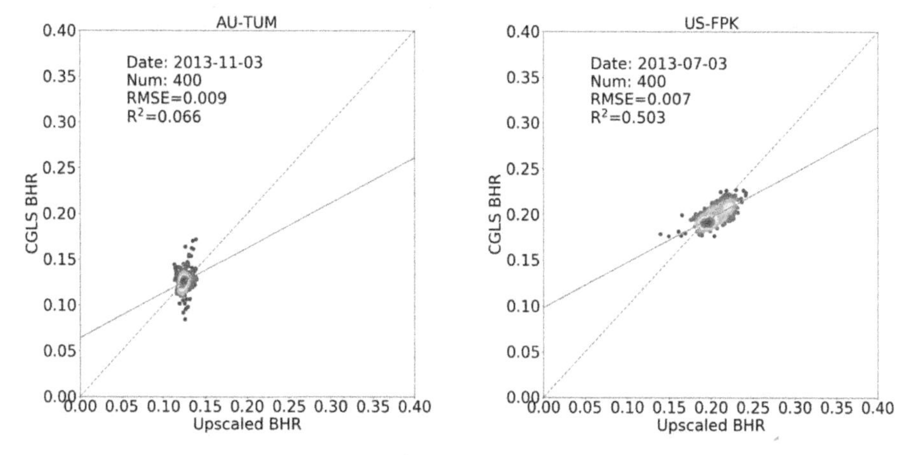

Figure 12. *Cont.*

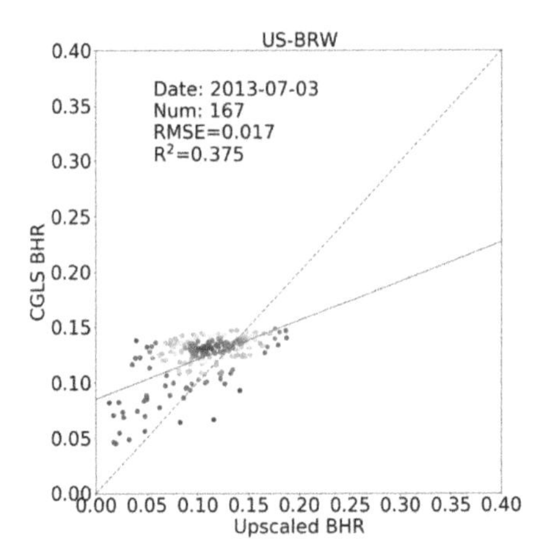

Figure 12. Density plot of BHR upscaled from tower FoV to CGLS resolution over a 20 × 20 CGLS pixel region.

At the Fort Peck site, the upscaled DHRs were well correlated with the CGLS DHRs, with a R-squared ($\boldsymbol{R}^2$) of 0.685 and root-mean-square-error (RMSE) of 0.006. The Barrow site had fewer pixels upscaled to coarse resolution than the other sites, and the upscaled DHRs showed larger differences with CGLS for the pixels with smaller albedo values, due to melt-ponds and tundra in this region. At the Tumbarumba site, most pixels were clustered around the 1:1 line, because of the good agreement in "point-to-pixel" time series analysis. But the upscaled DHRs had a relatively small value of $\boldsymbol{R}^2$ (0.233) when compared with CGLS DHRs, which suggests that the upscaling coefficient was not suitable for upscaling to a region covering 20 × 20 pixels around the tower. The BHR upscaling results were close to the DHR upscaling results in terms of $\boldsymbol{R}^2$ and RMSE.

4. Discussion

The DHR and BHR values retrieved from the tower albedometer data were first directly compared with satellite values derived from the pixel near the tower location. Generally speaking, the homogeneous sites had a better agreement between tower and satellite retrievals than the heterogenous sites. For the homogeneous sites, except for the US-BAO site that appears to have anomalous tower data since the year 2016, all the other sites (AU-TUM, US-TBL, US-FPK, US-DRA, US-BRW) showed good agreement with the satellite retrievals during the snow-free season. Among the heterogenous sites, the US-BON, US-GCM and US-PSU had large differences between the tower and satellite retrievals, while the US-SXF showed good agreement.

MODIS products showed the best agreement with tower retrievals, followed by MISR products, and then followed by CGLS products. The MODIS products appeared to have a good performance in picking up snow-covered points, which can be seen from the time-series analysis at the US-FPK, US-SXF and US-TBL sites. The MISR products were comparable with MODIS products at most of the sites studied in this work, and better than MODIS products at sites like US-BRW during the snow-free season. However, the MISR products were produced using a near-simultaneous retrieval, and compared with tower DHRs generated over a three-day window and tower BHRs generated in one day. In this case, the agreement between the MISR and tower retrieval suggests that the MISR products were closer to the actual surface albedo values.

The albedo values retrieved from both homogeneous and heterogenous sites were upscaled to coarse resolutions through the use of Landsat-8 spectral reflectance and MODIS BRDF climatology data. There was no obvious difference in the agreement between upscaled albedos and coarse-resolution albedos over homogeneous and heterogenous sites. But relatively better correlations could be still be found at homogeneous sites, such as US-DRA with a R^2 of 0.81 for DHR comparisons. There are several sources that can affect the accuracy of upscaled albedo values. First of all, the accuracy of the generation of the high-resolution albedo values plays an important role in the upscaling process. Secondly, the upscaling coefficient calculated from the tower FoV is accurate for a local area because of the high coherence. If it is applied to a larger area, errors are more likely to be introduced at pixels further away from the tower. This can explain why at some homogeneous sites (e.g., AU-TUM with a R^2 of 0.233 on DHRs) the upscaled albedos appeared to have a poor correlation with coarse-resolution albedos. However, for heterogenous sites, such as US-GCM, a R^2 value larger than 0.5 could be found between the upscaled albedos and coarse-resolution albedos. This suggests that an optimal sample size for maximising this upscaling can be determined, and that this upscaling method can be applied to both homogeneous and heterogenous surfaces.

5. Conclusions

In this study, a new method is introduced which allows the derivation of DHR and BHR values from tower albedometer measurements. This method was applied to derive DHRs and BHRs over 20 tower sites, including both homogeneous and heterogenous land surfaces from the FLUXNET, SURFRAD and BSRN networks between the years 2012 and 2016. The retrieved DHR and BHR values were directly compared with the satellite albedo values, including CGLS, MODIS and MISR retrievals. The MODIS 16-day albedo products show the best agreement with in situ retrievals, whilst the MISR near-simultaneous measurements show a similar good agreement with in situ retrievals for a smaller time-window. The CGLS 30-day products have larger biases than MODIS and MISR products. Overall, the direct intercomparison with tower albedometer derived values shows a better match over the homogeneous sites than the heterogenous sites. The agreement between tower and satellite retrieved DHR values are better than the BHR values. This is because DHRs are only measured at local solar noon, whereas BHRs are derived from measurements at all possible solar zenith angles.

A surface albedo upscaling method, for tower FoV albedos to coarse resolutions, is described. This method employs atmospherically corrected BRFs from high-resolution EO alongside coarse-resolution albedos predicted from a MODIS BRDF climatology over a larger area as inputs. The high-resolution albedo values are retrieved from the MODIS BRDF derived albedo-to-nadir-reflectance ratios. This method was applied to upscale tower measured DHR and BHR values to 1-km resolutions and compared with the CGLS products. These results imply that this surface albedo upscaling strategy can be applied to both homogeneous and heterogenous surfaces if the optimal sample size for optimising this upscaling is known. For example, one of the sites, where 3×3 km were compared, shows that the pixel to the north-west of where the tower is located yields a better correlation than the pixel containing the tower, which appears to be associated with the observation that the land cover changes in the south-east. This will be explored in more detail in future.

Supplementary Materials: The following are available online at http://www.mdpi.com/2072-4292/11/6/644/s1, Figure S1: CGLS, MODIS and MISR DHR products compared with tower derived DHRs. Figure S2: CGLS, MODIS and MISR BHR compared with tower derived BHRs. Figure S3: Scatter plot of DHR values retrieved from CGLS, MODIS and MISR between 2012 and 2016. Figure S4: Scatter plot of BHR values retrieved from CGLS, MODIS and MISR between 2012 and 2016. Figure S5: Density plot of DHR and BHR upscaled from tower FoV to CGLS resolution over a 20*20 pixel region.

Author Contributions: Conceptualization, J.-P.M. (Jan-Peter Muller); methodology, R.S. (Rui Song), S.K. (Said Kharbouche); software, R.S., S.K.; validation, R.S.; formal analysis, R.S.; investigation, R.S.; resources, J.-P.M.; data curation, J.-P.M.; writing—original draft preparation, R.S.; writing—review and editing, J.-P.M., S.K., and W.W. (William Woodgate); visualization, R.S.; supervision, J.-P.M.; project administration, J.-P.M.; funding acquisition, J.-P.M.

Funding: This research was funded by the European Commission Joint Research Centre grant number [FWC932059], part of the Global Component of the European Union's Copernicus Land Monitoring Service.

Acknowledgments: This work used JASMIN, the UK's collaborative data analysis environment http://jasmin.ac.uk. This work has been undertaken using data from the Global Component of the European Union's Copernicus Land Monitoring Service, European Commission Joint Research Centre FWC [932059]. We would like to thank Nadine Gobron and Christian Lanconelli of JRC Ispra for fruitful discussions. We would like to thank NOAA for access to their datasets through SURFRAD (http://www.esrl.noaa.gov/gmd/grad/surfrad/) and the BSRN (http://bsrn.awi.de) [5] for access to their datasets. This work also used tower albedometer data acquired by the FLUXNET community and in particular by the following networks: AmeriFlux (U.S. Department of Energy, Biological and Environmental Research, Terrestrial Carbon Program (DE-FG02-04ER63917 and DE-FG02-04ER63911), CarboEuropeIP, CarboItaly, CarboMont, OzFlux, USCCC. This study has been undertaken using data from GBOV "Ground Based Observation for Validation" (https://land.copernicus.eu/global/gbov), part of the Global Component of the European Union's Copernicus Land Monitoring Service. GBOV product developments are managed by ACRI-ST from the research work of University College London, University of Leicester, University of Southampton, University of Valencia and Informus GmbH. We acknowledge the financial support to the tower albedometer data harmonization provided by CarboEuropeIP, FAO-GTOS-TCO, iLEAPS, Max Planck Institute for Biogeochemistry, National Science Foundation, University of Tuscia, Université Laval and Environment Canada and US Department of Energy and the database development and technical support from Berkeley Water Center, Lawrence Berkeley National Laboratory, Microsoft Research eScience, Oak Ridge National Laboratory, University of California-Berkeley, University of Virginia.

Conflicts of Interest: The authors declare no conflict of interest.

Abbreviations

The following abbreviations are used in this manuscript:

AOD	Aerosol Optical Depth
BHR	Bi-Hemispherical Reflectance
BRF	Bi-Directional Reflectance Factor
BRDF	Reflectance Distribution Function
BSRN	Baseline Surface Radiation Network
CGLS	Copernicus Global Land Service
DHR	Directional Hemispherical Reflectance
EOS	Earth Observing System
FCLS	Fully Constrained Least Squares
FoV	Field-of-View
FAPAR	Fraction of Absorbed Photosynthetically Active Radiation
HR-EO	High-Resolution Earth Observation
HJ	HuanJing
IGBP	International Geosphere-Biosphere Programme
LAI	Leaf Area Index
MODIS	Moderate Resolution Imaging Spectroradiometer
MISR	Multi-Angle Imaging Spectroradiometer
NIR	Near-Infrared
NOAA	National Oceanic and Atmospheric Administration
QA	Quality Assurance
RTLSR	RossThick–LiSparse–Reciprocal
RMSE	Root-Mean-Square-Error
SURFRAD	Surface Radiation Budget Network
UTM	Universal Transverse Mercator

References

1. Dickinson, R.E. Land surface processes and climate-surface albedos and energy balance. *Adv. Geophys.* **1983**, *25*, 305–353. [CrossRef]
2. Harrison, E.F.; Minnis, P.; Barkstrom, B.R.; Gibson, G. Radiation budget at the top of the atmosphere. In *Atlas of Satellite Observations Related to Global Change*; Gurney, R.J., Foster, J.L., Parkinson, C.L., Eds.; Cambridge University Press: London, UK, 1993; pp. 19–38.
3. Myneni, R.B.; Asrar, G.; Tanre, D.; Choudhury, B.J. Remote sensing of solar radiation absorbed and reflected by vegetated land surfaces. *IEEE Trans. Geosci. Remote Sens.* **1992**, *30*, 302–314. [CrossRef]

4. Fritz, S. The Albedo of the Ground and Atmosphere. *Bull. Am. Meteorol. Soc.* **1948**, *29*, 303–312. [CrossRef]
5. Driemel, A.; Augustine, J.; Behrens, K.; Colle, S.; Cox, C.; Cuevas-Agulló, E.; Denn, F.M.; Duprat, T.; Fukuda, M.; Grobe, H.; et al. Baseline Surface Radiation Network (BSRN): Structure and data description (1992–2017). *Earth Syst. Sci. Data* **2018**, *10*, 1491–1501. [CrossRef]
6. Augustine, J.A.; Hodges, G.B.; Cornwall, C.R.; Michalsky, J.J.; Medina, C.I.; Augustine, J.A.; Hodges, G.B.; Cornwall, C.R.; Michalsky, J.J.; Medina, C.I. An update on SURFRAD–the GCOS Surface Radiation Budget Network for the Continental United States. *J. Atmos. Ocean. Technol.* **2005**, *22*, 1460–1472. [CrossRef]
7. Pinty, B.; Szejwach, G. A New Technique for Inferring Surface Albedo from Satellite Observations. *J. Appl. Meteorol. Climatol.* **1985**, *24*, 741–750. [CrossRef]
8. Gutman, G. A Simple Method for Estimating Monthly Mean Albedo of Land Surfaces from AVHRR Data. *J. Appl. Meteorol.* **1988**, *27*, 973–988. [CrossRef]
9. Csiszar, I.; Gutman, G. Mapping global land surface albedo from NOAA AVHRR. *J. Geophys. Res. Solid Earth* **1999**, *104*, 6215–6228. [CrossRef]
10. Liang, S.; Fang, H.; Chen, M.; Shuey, C.J.; Walthall, C.; Daughtry, C.; Morisette, F.; Schaaf, C.; Strahler, A. Validating MODIS land surface reflectance and albedo products: Methods and preliminary results. *Remote Sens. Environ.* **2002**, *83*, 149–162. [CrossRef]
11. Cescatti, A.; Marcolla, B.; Vannan, S.K.S.; Pan, J.Y.; Román, M.O.; Yang, X.; Ciais, P.; Cook, R.B.; Law, B.E.; Matteucci, G.; et al. Intercomparison of MODIS albedo retrievals and in situ measurements across the global FLUXNET network. *Remote Sens. Environ.* **2012**, *121*, 323–334. [CrossRef]
12. Strahler, A.; Muller, J.-P.; Lucht, W.; Schaaf, C.; Tsang, T.; Gao, F.; Xiaowen, L.; Lewis, P.; Barnsley, M.J. MODIS BRDF Albedo Product Algorithm Theoretical Basis Document Version 5.0. Available online: https://modis.gsfc.nasa.gov/data/atbd/atbd_mod09.pdf (accessed on 12 January 2019).
13. Lacaze, R.; Smets, B.; Trigo, I.; Calvet, J.C.; Jann, A.; Camacho, F.; Baret, F.; Kidd, R.; Defourny, P.; Tansey, K.; et al. The Copernicus Global Land Service: Present and future. In Proceedings of the EGU General Assembly, Vienna, Austria, 7–12 April 2013.
14. Schaaf, C.B.; Gao, F.; Strahler, A.H.; Lucht, W.; Li, X.W.; Tsang, T.; Strugnell, N.C.; Zhang, X.Y.; Jin, Y.F.; Muller, J.P.; et al. First operational BRDF, albedo nadir reflectance products from MODIS. *Remote Sens. Environ.* **2002**, *83*, 135–148. [CrossRef]
15. Martonchik, J.V.; Diner, D.J.; Pinty, B.; Verstraete, M.M.; Myneni, R.B.; Knyazikhin, Y.; Gordon, H.R. Determination of land and ocean reflective, radiative, and biophysical properties using multiangle imaging. *IEEE Trans. Geosci. Remote Sens.* **1998**, *36*, 1266–1281. [CrossRef]
16. NOAA. SURFRAD Overview: Surface Radiation Budget Monitoring. 2017. Available online: https://www.esrl.noaa.gov/gmd/grad/surfrad/overview.html (accessed on 1 September 2018).
17. Diner, D.J.; Beckert, J.C.; Reilly, T.H.; Bruegge, C.J.; Conel, J.E.; Kahn, R.A.; Martonchik, J.V.; Ackerman, T.P.; Davies, R.; Gerstl, S.A.; et al. Multi-angle Imaging SpectroRadiometer (MISR) instrument description and experiment overview. *IEEE Trans. Geosci. Remote Sens.* **1998**, *36*, 1072–1087. [CrossRef]
18. Liang, S. Narrowband to broadband conversions of land surface albedo: I Algorithms. *Remote Sens. Environ.* **2000**, *76*, 213–238. [CrossRef]
19. Schaepman-Strub, G.; Schaepman, M.E.; Painter, T.H.; Dangel, S.; Martonchik, J.V. Reflectance quantities in optical remote sensing—definitions and case studies. *Remote Sens. Environ.* **2006**, *103*, 27–42. [CrossRef]
20. Lucht, W.; Schaaf, C.B.; Strahler, A.H. An algorithm for the retrieval of albedo from space using semiempirical BRDF models. *IEEE Trans. Geosci. Remote Sens.* **2000**, *38*, 977–998. [CrossRef]
21. Román, M.O.; Gatebe, C.K.; Shuai, Y.M.; Wang, Z.S.; Gao, F.; Masek, J.G.; He, T.; Liang, S.L.; Schaaf, C.B. Use of in situ and airborne multiangle data to assess MODIS- and Landsat-based estimates of directional reflectance and albedo. *IEEE Trans. Geosci. Remote Sens.* **2013**, *51*, 1393–1404. [CrossRef]
22. Muller, J.-P. GlobAlbedo Final Product Validation Report 2012. Available online: http://www.globalbedo.org/docs/GlobAlbedo_FVR_V1_2_web.pdf (accessed on 10 January 2019).
23. Mousavi Maleki, S.A.; Hizam, H.; Gomes, C. Estimation of Hourly, Daily and Monthly Global Solar Radiation on Inclined Surfaces: Models Re-Visited. *Energies* **2017**, *10*, 134. [CrossRef]
24. Wu, X.D.; Wen, J.G.; Xiao, Q.; Liu, Q.; Peng, J.J.; Dou, B.C.; Li, X.H.; You, D.Q.; Tang, Y.; Liu, Q.H. Coarse scale in situ albedo observations over heterogeneous snow-free land surfaces and validation strategy: A case of MODIS albedo products preliminary validation over northern China. *Remote Sens. Environ.* **2016**, *184*, 25–39. [CrossRef]

25. Liang, S. A direct algorithm for estimating land surface broadband albedos from MODIS imagery. *IEEE Trans. Geosci. Remote Sens.* **2003**, *41*, 136–145. [CrossRef]
26. Qu, Y.; Liang, S.; Liu, Q.; He, T.; Liu, S.; Li, X. Mapping surface broadband albedo from satellite observations: A review of literatures on algorithms and products. *Remote Sens.* **2015**, *7*, 990–1020. [CrossRef]
27. Shuai, Y.; Masek, J.G.; Gao, F.; Schaaf, C.B. An algorithm for the retrieval of 30-m snow-free albedo from Landsat surface reflectance and MODIS BRDF. *Remote Sens. Environ.* **2011**, *115*, 2204–2216. [CrossRef]
28. Muller, J.P.; Lopez-Saldana, G.; Kharbouche, S.; Danne, O.; Lattanzio, A.; Schulz, J.; Lewis, P. Optimal estimation for the retrieval of traceable and validated albedo: Lessons learnt from the ESA-GlobAlbedo and EU-QA4ECV projects. 2019; in preparation.
29. Winter, M.E. N-FINDR: An algorithm for fast autonomous spectral end-member determination in hyperspectral data. In Proceedings of the SPIE 3753, Imaging Spectrometry V, Denver, CO, USA, 27 October 1999. [CrossRef]
30. Heinz, D.; Chang, C.-I.; Althouse, M.L.G. Fully constrained least-squares based linear unmixing hyperspectral image classification. In Proceedings of the IEEE 1999 International Geoscience and Remote Sensing Symposium, Hamburg, Germany, 28 June–2 July 1999; pp. 1401–1403. [CrossRef]
31. Adams, J.; Gobron, N.; Widlowski, J.-L.; Mio, C. A model-based framework for the quality assessment of surface albedo in situ measurement protocols. *J. Quant. Spectrosc. Radiat. Transf.* **2016**, *180*, 126–146. [CrossRef]
32. Michalsky, J.J.; Harrison, L.C.; Berkheiser, W.E. Cosine response characteristics of some radiometric and photometric sensors. *Sol. Energy* **1995**, *54*, 397–402. [CrossRef]
33. Román, M.O.; Schaaf, C.B.; Lewis, P.; Gao, F.; Anderson, G.P.; Privette, J.L.; Strahler, A.H.; Woodcock, C.E.; Barnsley, M. Assessing the coupling between surface albedo derived from MODIS and the fraction of diffuse skylight over spatially-characterized landscapes. *Remote Sens. Environ.* **2010**, *114*, 738–760. [CrossRef]

Article

A Method for Landsat and Sentinel 2 (HLS) BRDF Normalization

Belen Franch [1,2,*], Eric Vermote [2], Sergii Skakun [1,2], Jean-Claude Roger [1,2], Jeffrey Masek [2], Junchang Ju [2,3], Jose Luis Villaescusa-Nadal [1,2] and Andres Santamaria-Artigas [1,2]

1 Department of Geographical Sciences, University of Maryland, College Park, MD 20742, USA; skakun@umd.edu (S.S.); roger63@umd.edu (J.-C.R.); jvillaes@terpmail.umd.edu (J.L.V.-N.); asantam@umd.edu (A.S.-A.)

2 NASA Goddard Space Flight Center, Greenbelt, MD 20771, USA; eric.f.vermote@nasa.gov (E.V.); jeffrey.g.masek@nasa.gov (J.M.); junchang.ju@nasa.gov (J.J.)

3 Earth System Science Interdisciplinary Center, University of Maryland, College Park, MD 20740, USA

* Correspondence: befranch@umd.edu

Received: 24 January 2019; Accepted: 11 March 2019; Published: 15 March 2019

Abstract: The Harmonized Landsat/Sentinel-2 (HLS) project aims to generate a seamless surface reflectance product by combining observations from USGS/NASA Landsat-8 and ESA Sentinel-2 remote sensing satellites. These satellites' sampling characteristics provide nearly constant observation geometry and low illumination variation through the scene. However, the illumination variation throughout the year impacts the surface reflectance by producing higher values for low solar zenith angles and lower reflectance for large zenith angles. In this work, we present a model to derive the bidirectional reflectance distribution function (BRDF) normalization and apply it to the HLS product at 30 m spatial resolution. It is based on the BRDF parameters estimated from the MODerate Resolution Imaging Spectroradiometer (MODIS) surface reflectance product (M{O,Y}D09) at 1 km spatial resolution using the VJB method (Vermote et al., 2009). Unsupervised classification (segmentation) of HLS images is used to disaggregate the BRDF parameters to the HLS spatial resolution and to build a BRDF parameters database at HLS scale. We first test the proposed BRDF normalization for different solar zenith angles over two homogeneous sites, in particular one desert and one Peruvian Amazon forest. The proposed method reduces both the correlation with the solar zenith angle and the coefficient of variation (CV) of the reflectance time series in the red and near infrared bands to 4% in forest and keeps a low CV of 3% to 4% for the deserts. Additionally, we assess the impact of the view zenith angle (VZA) in an area of the Brazilian Amazon forest close to the equator, where impact of the angular variation is stronger because it occurs in the principal plane. The directional reflectance shows a strong dependency with the VZA. The current HLS BRDF correction reduces this dependency but still shows an under-correction, especially in the near infrared, while the proposed method shows no dependency with the view angles. We also evaluate the BRDF parameters using field surface albedo measurements as a reference over seven different sites of the US surface radiation budget observing network (SURFRAD) and five sites of the Australian OzFlux network.

Keywords: HLS; Landsat; Sentinel 2; BRDF; albedo; SURFRAD; OzFlux

---

1. Introduction

The Harmonized Landsat/Sentinel-2 (HLS) project [1] provides a surface reflectance product that combines observations from USGS/NASA's Landsat 8 and ESA's Sentinel-2 satellites at moderate spatial resolution (30 m). The main goal is to provide a unique dataset based on both satellites' data to improve the revisit time to 3–5 days depending on the latitude [1]. Along with a common atmospheric correction algorithm [2], geometric resampling to 30 m spatial resolution and geographic registration [1], the product is also corrected for BRDF effects and band pass adjustment.

The purpose of the band pass adjustment is to correct discrepancies between data coming from different sensors due differences their Relative Spectral Response (RSR) functions. In the case of the near infrared (NIR) spectral band, the spectral difference between Sentinel-2 and Landsat-8 is ~0.03%, due to the almost identical spectral functions of the analogous bands. In the red band, however, these differences are of ~3% [3] and need to be corrected. To correct the differences in the red band, Villaescusa-Nadal [3] employed a Spectral Band Adjustment Factor (SBAF) exponential model dependent on the NDVI which reduced the discrepancy between the sensors by ~55%.

Regarding the BRDF correction, Landsat and Sentinel-2 sampling characteristics provide nearly constant observation geometry and low illumination variation within each scene. However, when a surface reflectance time series combining measurements from both sensors is created, variations due to differences in view geometry between them arise. In extreme cases, the differences in the view zenith angle for a ground target can reach 20° from adjacent orbits only a few days apart. Additionally, variations in the seasonal illumination also impact the surface reflectance value. Gao [4] concluded that for Landsat-like narrow swath sensors, the major BRDF effect arises from the day of year effect, and can cause variations of 0.04–0.06 reflectance compared to mid-summer observations. Such angular effects can be corrected using a Bidirectional Reflectance Distribution Function (BRDF) model. However, the narrow angular sampling of moderate resolution sensors such as Landsat and Sentinel 2 complicates the BRDF parameters retrieval.

The official HLS product (version 1.4) provides a BRDF normalized product based on the BRDF normalization proposed by Roy [5]. In this product, the view angle is set to nadir and the solar zenith angle is fixed through time but varies for each tile based on the latitude. However, this correction is not optimal because the considered coefficients are constant and do not depend on the surface type or condition. Previous studies have worked towards BRDF estimation at moderate resolution for correcting the BRDF effects or for generating an albedo product of medium spatial resolution sensors. Shuai [6] used the MODIS Bidirectional Reflectance Distribution Function (BRDF) MDC43 product at 500 m [7] to generate a Landsat surface albedo product that later was implemented back to the 80s to the "pre-MODIS era" [8]. Their method achieves a Root Mean Square Error (RMSE) generally lower than 0.03 (12%) when compared with in situ data. However, this method does not provide a BRDF product at Landsat scale. Flood [9] used constant BRDF parameters to adjust surface reflectance from Landsat (TM and ETM+) and SPOT 5 images to a standard angular configuration. Gao [4] proposed a method to correct BRDF effects on Landsat and the advanced wide field sensor (AWiFS) based on detecting homogeneous areas with USDA cropland data layer (CDL) map to build a BRDF Look-Up Map (LUM) based on MODIS BRDF data. Van Doninck [10] assessed the directional effects on Landsat TM/ETM+ imagery over Amazonian forests and evaluated five different normalization techniques, including MODIS-based and Landsat-based BRDF. They concluded that the two MODIS-based methods analyzed only removed half of the angular effect in the infrared bands while the best results were obtained parametrizing the BRDF model using pairs of Landsat images. Franch [11] presented an alternative algorithm to calculate the Landsat (TM and ETM+) surface albedo based on the BRDF parameters estimated from the MODerate Resolution Imaging Spectroradiometer (MODIS) Climate Modeling Grid (CMG) surface reflectance product (M{O,Y}D09) using the VJB method [12,13]. The validation of the results against field measurements showed an RMSE of 0.015 (7%). Franch [14] tested this method to correct the BRDF effects of the HLS data over the Amazon forest in Peru. Their results show a good performance of the model when applied to HLS data with an error of 0.01 in the surface albedo retrieval.

In this paper, we describe the adaptation of the method of Franch [11,14] for HLS BRDF normalization, apply the proposed method to the HLS product from 2013 to 2017 and validate its applicability to Landsat 8 and Sentinel 2 data, test it on three homogeneous sites to assess its capability to remove BRDF artifacts, and finally evaluate it through the comparison to ground albedo measurements over SURFRAD and OzFlux networks.

2. Materials and Methods

2.1. Materials

2.1.1. HLS Data

The harmonization process of the HLS product is based on a set of algorithms for atmospheric correction, cloud and cloud-shadow screening, and co-registration of data from the Operational Land Imager (OLI) onboard Landsat-8 and the Multi-Spectral Instrument (MSI) onboard Sentinel-2. The HLS product is also corrected for bandpass difference and BRDF effects. However, in this work, we considered uncorrected data to explore the feasibility of the proposed algorithm for BRDF normalization. We used all HLS data (version 1.4) available from January 2013 to December 2017.

2.1.2. MODIS Data

We processed MODIS surface reflectance Collection 6 data at 1 km spatial resolution (M{OY}D09GA, Vermote, 2015) over the same areas and period, to derive the BRDF parameters using the VJB method [12,13].

2.1.3. Homogeneous Sites

Three homogeneous sites were selected for this study. First, we studied the impact of the solar zenith angle (SZA) variation through the year over two homogeneous sites. The first site is located on a homogeneous area of dense tropical forest in the Tambopata region of the Peruvian Amazon (12.818S, 69.281W, HLS tile 19LDI) where an albedometer is retrieving continuous measurements [15]. In this work, we have not included the albedo validation of this site, where Franch [14] reported an error of 0.01. This area is located in the southwest Amazon moist forest ecoregion of dense forests with a canopy height between 30 to 50 m high (https://www.worldwildlife.org/ecoregions/nt0173). The second site is located on a homogeneous desert area in Yuma, Arizona (32.499N, 114.498W, HLS tile 11SQS). This area is a lower-elevation section of the Sonoran desert in the southwestern United States and northwest of Mexico. We also analyzed the impact of the view zenith angle (VZA) variation. Because in the tropics the Landsat scan direction is closely aligned with the solar principal plane and the angular effects can be as strong as 20% in reflectance [16], we selected a site located close to the equator on the Brazilian Amazon basin (2.144S, 58.999W, HLS tile 21MTT). This area is located in the Uatuma-Trombetas moist forests ecoregion (Figure 1). The forests are characterized by a dense vegetation with a high number of small-diameter (less than 200 mm) to medium-diameter (200 to 600 mm) stems and a canopy height varying from 20 to 30 m, with emergent trees to 40 m (https://www.worldwildlife.org/ecoregions/nt0173).

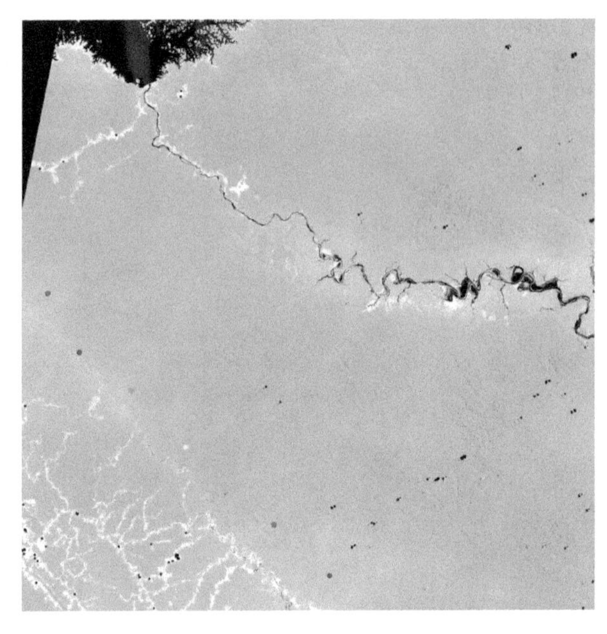

Figure 1. Harmonized Landsat/Sentinel-2 (HLS) image RGB composite on August 21th of 2016 in the Brazilian Amazon area. The seven dots next to the river show the location and size of the averaged area to analyze the transect covering different VZA (see Section 2.2.3).

2.1.4. SURFRAD Data

The global downwelling and upwelling solar radiations are continuously recorded by skyward facing and downward facing Eppley model 8–48 pyranometers mounted on a meteorological tower. Table 1 specifies the height of each tower above the target.

Table 1. Geolocation and brief description of the US surface radiation budget observing network (SURFRAD) sites considered in this study.

Station Name	Network	Location Latitude, Longitude	Land cover Type	HLS Tile	Tower Height above Target
Desert Rock Station	SURFRAD	36.6232N, 116.0196W	Sparse vegetation	11SNA	10 m
Table Mountain	SURFRAD	40.1256N, 105.2378W	Sandy with exposed rocks, sparse grasses and shrubs	13TDE	10 m
Bondville	SURFRAD	40.0516N, 88.3733W	Agriculture	16TCK	10 m
Goodwin creek	SURFRAD	34.2547N, 89.8729W	Pasture grass and sparsely distributed deciduous trees	16SBD	10 m
Penn state university	SURFRAD	40.7203N, 77.9310W	Agriculture Research field	18TTL	10 m
Fort Peck	SURFRAD	48.3078N, 105.1017W	Sparse vegetation	13UDP	10 m
Sioux Falls	SURFRAD	43.7340N, 96.62328W	Prairie grasses	14TPP	10 m

2.1.5. OzFlux Data

TERN OzFlux is a national ecosystem research network set up to provide the Australian and global ecosystem modelling communities with nationally consistent observations of energy, carbon, and water exchange between the atmosphere and key Australian ecosystems. TERN OzFlux is part of an international network (FluxNet) of over 500 flux stations that is designed to provide continuous, long-term micrometeorological measurements to monitor the state of ecosystems globally [17].

From the 40 sites in this network, we selected 5 sites for validation purposes according to data availability during the period analyzed (2013 to 2017). In these sites, a pair of pyranometers from Kipp & Zonen CNR4 measure energy balance between incoming and reflected short-wave radiation at different heights (Table 2). The surface albedo is estimated in the spectral range 300–2800 nm.

Table 2. Geolocation and brief description of the OzFlux sites considered in this study.

Station Name	Network	Location Latitude, Longitude	Land cover Type	HLS Tile	Tower Height above Target
Calperum	OzFlux	34.0027S, 140.5877E	Sand dunes with trees and shrubs	54HVH	20 m
Cumberland Plain	OzFlux	33.6152S, 150.7236E	Dry sclerophyll forest	56HKH	29 m (~6 m above canopy)
Whroo	OzFlux	36.6732S, 145.0294E	Eucalyptus forest	55HCV	36 m (~10 m above canopy)
Wombat	OzFlux	37.4222S, 144.0944E	Eucalyptus forest	55HBU	30 m (~5 m above canopy)
Yanco	OzFlux	34.9893S, 146.2907E	Grassland	55HDB	2 m

2.2. Methods

2.2.1. Current HLS BRDF Normalization

The HLS product (version 1.4) is normalized for BRDF effects using the method proposed by Roy [5]. This method uses fixed BRDF coefficients for each spectral band (i.e., a constant BRDF shape), derived from a large number of pixels in the MODIS 500 m BRDF product (MCD43) that are globally and temporally distributed.

$$\rho^N(\theta_s^{out}, 0, 0; \lambda) = \rho(\theta_s, \theta_v, \phi; \lambda) * c(\lambda) \tag{1}$$

$$c(\lambda) = \frac{k_0(\lambda) + k_1(\lambda)F_1(\theta_s^{out}, 0, 0) + k_2(\lambda)F_2(\theta_s^{out}, 0, 0)}{k_0(\lambda) + k_1(\lambda)F_1(\theta_s, \theta_v, \phi) + k_2(\lambda)F_2(\theta_s, \theta_v, \phi)} \tag{2}$$

where θ_s is the sun zenith angle, θ_v is the view zenith angle, ϕ is the relative azimuth angle, F_1 is the volume scattering kernel, based on the Rossthick function derived by Roujean [18], and F_2 is the geometric kernel, based on the Li-sparse model [19] but considering the reciprocal form given by Lucht [20] and corrected for the Hot-Spot process proposed by Maignan [21]. θ_s^{out} is the sun zenith angle of the normalized data, which is set to constant value depending on the tile central latitude [1]. The BRDF coefficients of the model (k_0, k_1, and k_2) are provided in [1,5].

2.2.2. Proposed BRDF Normalization Method

For our proposed method, we first subset a 9 × 9 km area centered over the albedometer tower locations. All images are then masked for clouds and cloud shadow effects using the product's cloud mask. Next, we run an unsupervised classification over every scene using the Iterative Self-Organizing Data Analysis Technique Algorithm (ISODATA) developed by Ball, G.H. and Hall, D.J. (1965). This classification is used to unmix the BRDF parameters retrieved from MODIS (at 1 km spatial resolution) to HLS spatial resolution (30 m). We run the ISODATA algorithm for maximum three iterations with the maximum number of clusters setup at 15.

Following the Vermote et al. (2009) notations, the surface reflectance (ρ) is written as:

$$\rho(\theta_s, \theta_v, \phi) = k_0\left[1 + \frac{k_1}{k_0}F_1(\theta_s, \theta_v, \phi) + \frac{k_2}{k_0}F_2(\theta_s, \theta_v, \phi)\right] \tag{3}$$

where F_1 and F_2 are fixed functions of the observation geometry, and k_0, k_1, and k_2 are free parameters. From this notation, we define V as k_1/k_0 and R for k_2/k_0. In order to invert the MODIS BRDF parameters (V and R) we use the VJB method [12,13]. This method assumes that the difference between the successive observations in time is mainly attributed to directional effects while the variation of k_0 is supposed small. Additionally, it assumes that R and V are represented by a linear function of the NDVI.

$$V = V_0 + V_1 \text{ NDVI} \tag{4}$$

$$R = R_0 + R_1 \text{ NDVI} \tag{5}$$

We unmix the BRDF parameters to the HLS 30 m spatial resolution, by assuming that the surface reflectance of a MODIS pixel can be written as a weighted sum of n Landsat classes:

$$\begin{aligned} k_0^{1km_x} + k_1^{1km_x} F_1 & (\theta_s, \theta_v, \varnothing) + k_2^{1km_x} F_2(\theta_s, \theta_v, \varnothing) \\ & = A_x \left(k_0^{C_1} + k_1^{C_1} F_1(\theta_s, \theta_v, \varnothing) + k_2^{C_1} F_2(\theta_s, \theta_v, \varnothing) \right) \\ & = B_x \left(k_0^{C_2} + k_1^{C_2} F_1(\theta_s, \theta_v, \varnothing) + k_2^{C_2} F_2(\theta_s, \theta_v, \varnothing) \right) + \cdots \\ & = N_x \left(k_0^{C_n} + k_1^{C_n} F_1(\theta_s, \theta_v, \varnothing) + k_2^{C_n} F_2(\theta_s, \theta_v, \varnothing) \right) \end{aligned} \tag{6}$$

where A_x, B_x ..., N_x represent proportions of each the n classes within the x MODIS pixel. BRDF parameters for each class, which are unknowns in Equation (6), can be derived through matrix inversion. We only invert the k_1 and k_2 parameters since they describe the directional shape, while k_0 describes the reflectance magnitude. Considering the HLS surface reflectance and the classification image, we derive k_0^{HLS} as shown in Equation (7).

$$k_0^{HLS} = \rho^{HLS}(\theta_s, \theta_v, \phi) - k_1^{HLS} F_1(\theta_s, \theta_v, \phi) - k_2^{HLS} F_2(\theta_s, \theta_v, \phi) \tag{7}$$

Figure 2 shows the red band BRDF parameters inverted using this method on an HLS image on June 20th of 2017 over the Yuma desert in Arizona (US).

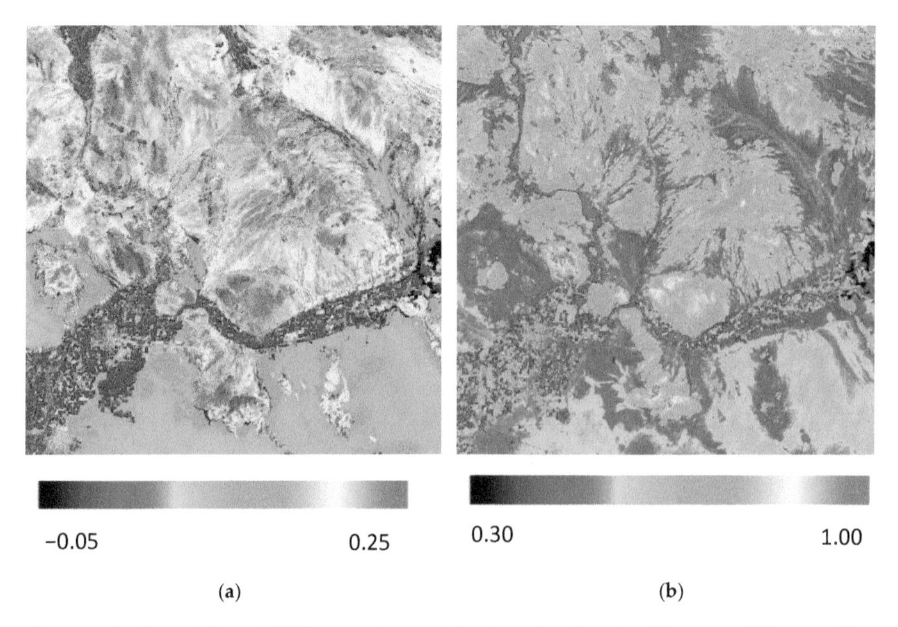

Figure 2. Red band (**a**) R and (**b**) V parameters applying inverting Equation (6) over an HLS image (tile 11SQS) on June 20th of 2017 in Yuma, Arizona (US).

The BRDF inversion is applied to each image. However, this may generate noise in the BRDF inversion caused mainly by clouds. The presence of clouds reduces the number of cloud-free pixels considered in the BRDF inversion, limiting the number of observations available for inversion (note that just images with a minimum of 30% cloud free pixels can be processed to have enough information to invert the model). Additionally, the inaccuracy of the cloud mask [22] may introduce noise in the BRDF unmixing of a given class.

Based on the individual BRDF unmixing of each image in the HLS database (starting from 2013 for Landsat 8, 2015 for Sentinel 2A and 2017 for Sentinel 2B), our goal is to reproduce the BRDF parameters V_0, V_1, R_0, and R_1 in Equations (4) and (5) at HLS spatial resolution. Next, we apply a linear regression for the V and R parameters versus the NDVI to derive the V_0, V_1, R_0, and R_1 parameters at HLS pixel level.

Finally, with the BRDF parameters at HLS resolution, we apply Equation (8) to derive the normalized surface reflectance (ρ^N) at 45 degrees of solar zenith angle and nadir observation (Vermote et al., 2009):

$$\rho^N(45,0,0) = \rho(\theta_s,\theta_v,\phi) * \frac{1 + VF_1(45,0,0) + RF_2(45,0,0)}{1 + VF_1(\theta_s,\theta_v,\phi) + RF_2(\theta_s,\theta_v,\phi)} \tag{8}$$

2.2.3. Temporal Evaluation of Homogeneous Sites

We test the performance of the BRDF normalization by applying the proposed method to two homogeneous sites: forest and desert. To evaluate the correction and assuming that these surfaces remain stationary, we estimate the coefficient of variation (CV), for the reflectance time series. The CV is defined as the ratio of the standard deviation to the mean (Equation (9)).

$$\mathrm{CV} = \frac{stddev(\rho_0,\rho_1,\ldots\rho_n)}{average(\rho_0,\rho_1,\ldots\rho_n)} \tag{9}$$

where ρ_i is the surface reflectance of day i.

2.2.4. Spatial Evaluation of an Equatorial Region

We evaluate the spatial impact of the view zenith angle variation in a dense Amazonian forest close to the equator. To do this, we extract a transect consisting of seven circular homogeneous regions of 800 pixels next to the Urubu river, which cover different values of the VZA. The condition for homogeneity is that the standard deviation is lower than 5% in any band. From these we can estimate the average and standard deviation for each band. The results from using the directional reflectance, the HLS BRDF normalized, and the proposed BRDF normalization are then compared. A p-value for each case was calculated which provides a test of the null hypothesis that a coefficient for each regression is equal to zero, i.e., the regressor has no effect on the dependent variable (VZA in our case). A small p-value indicates that the null hypothesis can be rejected, meaning that there is a statistically significant relationship between the regressor and dependent variable. We used this statistic index to assess how significant the reflectance dependency with the VZA was.

2.2.5. Albedo Validation

Using the BRDF parameters at HLS resolution, we derive the surface albedo which is validated against field measurements. The surface albedo is estimated following the same methodology as the official MCD43 MODIS product [7] by integrals of the BRDF model through the black-sky albedo and the white-sky albedo [7]. Surface albedo is typically estimated in the spectral range 280–2800 nm and is therefore comparable with the broadband MODIS albedo (300–5000 nm). Therefore, we estimate the broadband albedo using the narrow to broadband equation proposed by Liang [23].

For validation purposes, assuming that every pyranometer has an effective field of view of 81°, the downward pyranometer covers an area equal to tan(81°)*height. According to this equation and considering that the tower height in every SURFRAD is 10 m, the area covered by the pyranometer is 63 m in radius (3 by 3 HLS pixels). Then, the weighted average of the nine HLS pixels is estimated considering the angle of the instrument and the center of each pixel. The same approach is applied to OzFlux sites to account for the field of view as function of the height of the instrument.

SURFRAD instruments provide radiation measurements every minute, and we just consider the observations that match the time of Landsat 8 and Sentinel 2 overpass time. On the other hand, since OzFlux instruments provide radiation measurements every 30 min, a linear interpolation is done to estimate the value at the Landsat 8 and Sentinel 2 overpass time. A careful analysis of cloud-free conditions around the tower was performed to discard observations that may include residual effects.

3. Results

3.1. Temporal Evaluation of Homogeneous Sites

First, focusing on the Peruvian Amazon forest, Figure 3 shows the directional (red), BRDF-normalized using Roy [5] (green), and BRDF-normalized using the proposed BRDF normalization (black) surface reflectance versus solar zenith angle for each day of the year (DOY) considered in the time series (2013–2017). Note that we have represented differently Landsat 8 (dots) and Sentinel 2 (triangle) data. Results are shown for (a) the red band, (b) the near infrared (NIR) band, (c) the Normalized Difference Vegetation Index (NDVI), and (d) the SZA for each observation. Given the high frequency of cloud cover in this area, a total of 23 scenes were used in this analysis through the entire period 2013–2017.

All parameters show good consistency for both satellites' products. The directional surface reflectance (red) shows a clear dependency with the SZA in the RED (r^2 = 0.56) but mostly in the NIR band (r^2 = 0.82), showing higher values for low solar angles and lower values for higher angles. This effect is not fully corrected with the current BRDF-normalization of the HLS product (green), which after the correction still shows a dependency on the SZA in both bands. However, it is corrected on the BRDF-normalized RED reflectance and minimized in the NIR reflectance by decreasing the correlation coefficient and providing a slope closer to zero when using the proposed algorithm (black), which normalize all observations to SZA = 45° and nadir observation. The NDVI barely shows any dependency on the SZA for any product. Figure 4 shows a subset of the NIR band directional (left) and BRDF-normalized using the proposed algorithm (right) surface reflectance of the HLS image that is mostly affected by the SZA effect according to Figure 3. The directional reflectance image shows higher values than the BRDF-normalized one over the scene.

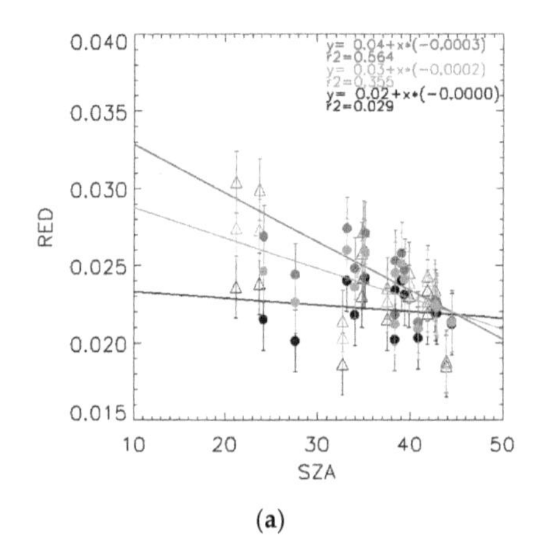

(a)

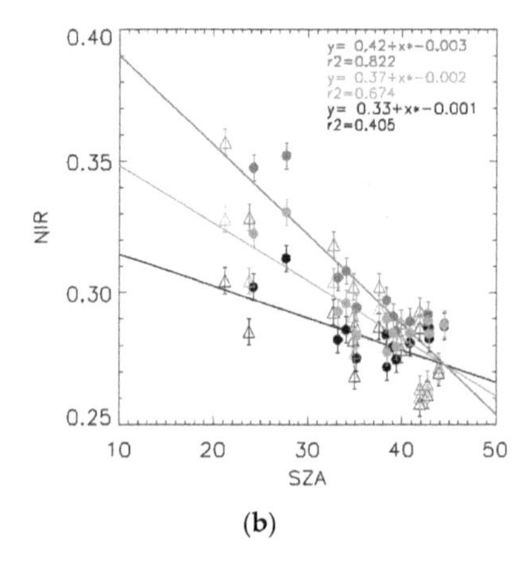

(b)

Figure 3. *Cont.*

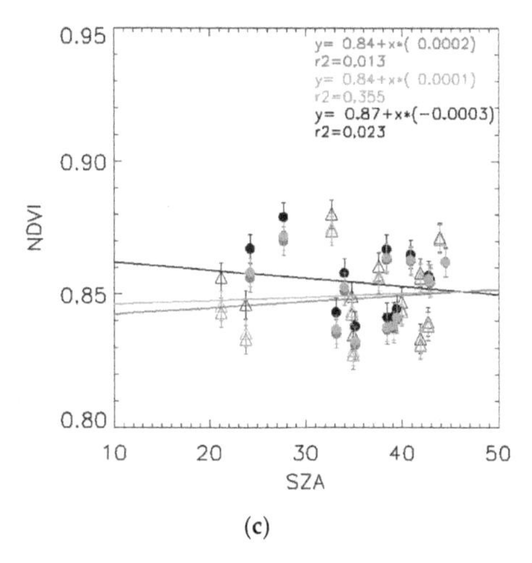

(c)

Figure 3. Peruvian Amazon pixel Landsat 8 (dots) and Sentinel 2 (triangles) surface reflectance in the (**a**) RED, (**b**) Near infrared (NIR) and (**c**) Normalized Difference Vegetation Index (NDVI), with no normalization (red color), HLS BRDF normalization (green color) and the proposed BRDF-normalization (black color) from 2013 to 2017 versus the Solar Zenith Angle (SZA). The error bars displayed represent the uncertainty of the Landsat 8 surface reflectance product [2], assuming the same error for Sentinel 2. Adapted from Franch [14].

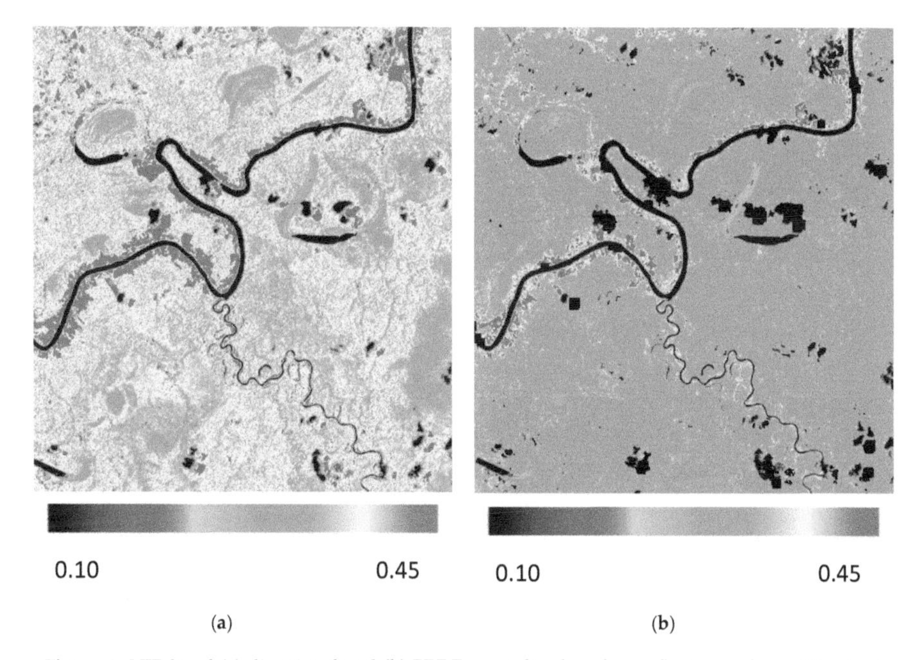

(a) (b)

Figure 4. NIR band (**a**) directional and (**b**) BRDF-normalized surface reflectance of an HLS subset centered on the Peruvian Amazon tower on December 12th of 2015.

Similar to Figure 3, Figure 5 shows the directional (red), BRDF-normalized using Roy [5] (green), and BRDF-normalized using the proposed BRDF normalization (black) surface reflectance versus the SZA in the desert site in Yuma (Arizona, US). The SZA of this site shows a greater variation during the year (from 20 to 60 degrees) compared to the previous site (from 20 to 45 degrees). Despite this, the directional reflectance barely shows any dependency with the SZA. However, the HLS BRDF-normalized data shows a high dependency (green) with higher values for high SZA and lower values for low SZA. This is not the case when applying the proposed algorithm (black), that shows more stable surface reflectance values.

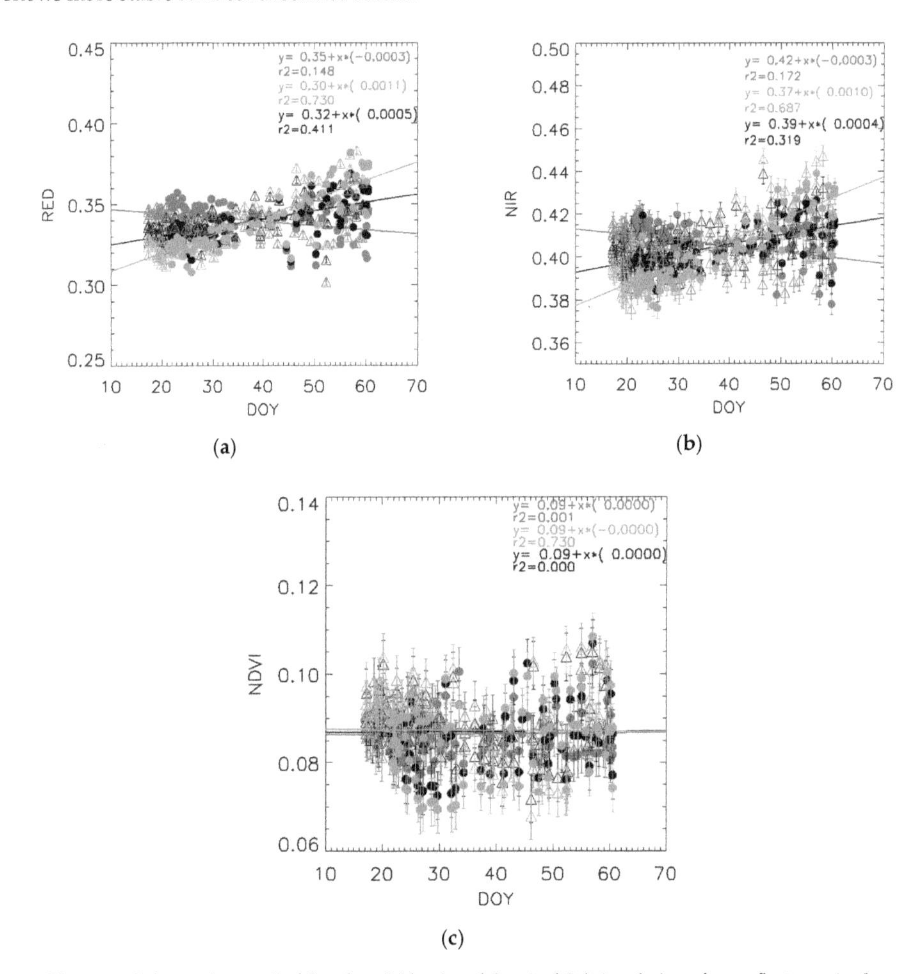

Figure 5. Arizona desert pixel Landsat 8 (dots) and Sentinel 2 (triangles) surface reflectance in the (**a**) RED, (**b**) NIR, and (**c**) NDVI with no normalization (red), HLS BRDF normalization (green), and the proposed BRDF-normalization (black) from 2013 to 2017 versus the SZA. The error bars displayed represent the uncertainty of the Landsat 8 surface reflectance product [2], assuming the same error for Sentinel 2.

Tables 3 and 4 show the coefficient of variation of the two homogeneous sites considered. The forest shows a high CV in the red (11%) and the NIR (8%) directional surface reflectance. The CV reduction in both red and NIR bands is around 2% when applying the current HLS correction and 4% when applying the proposed algorithm. However, the NDVI CV remains low and constant for all surface reflectances. The desert's CV, on the other hand, is much lower and does not show any significant change for any band or the NDVI when applying the proposed algorithm, but it does show an increase when applying the current BRDF correction of about 2% to 3%.

Table 3. Coefficient of variation (CV) of the Peruvian Amazon pixel.

CV (%)	RED	NIR	NDVI
Directional reflectance	11.4	8.3	1.6
Current HLS BRDF normalization	9.3	6.0	1.6
Proposed BRDF normalization	7.6	4.5	1.6

Table 4. Coefficient of variation (CV) of a desert site in Arizona (USA).

CV (%)	RED	NIR	NDVI
Directional reflectance	2.8	2.3	6.8
Current HLS BRDF normalization	5.4	4.2	10.1
Proposed BRDF normalization	3.3	2.5	8.2

3.2. Spatial Evaluation of an Equatorial Region

Figure 6 shows the HLS NIR surface reflectance image. The directional reflectance (Figure 6a) shows higher values along the western part (compared to the eastern) where the VZA is larger (Figure 6c). Note that the solar azimuth angle image shows nearly constant values of 63 degrees. This means that the western part is observed in the backscattering direction, which explains the higher values observed. The current HLS correction (Figure 6b) reduces this angular effect. However, the west area still shows higher values. Finally, when applying the proposed normalization (Figure 6c) this illumination effect is minimized.

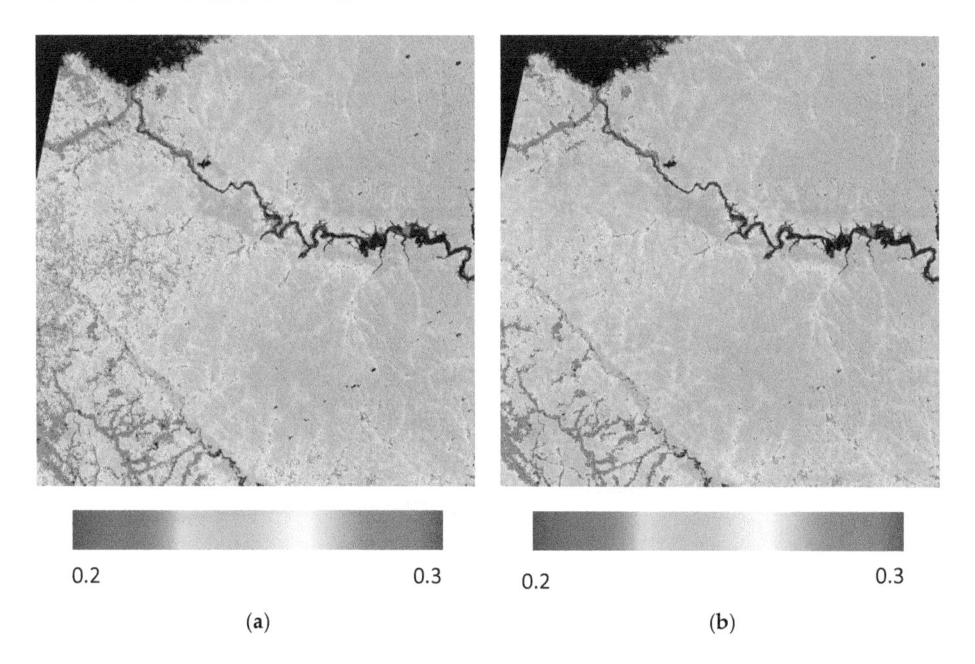

(a) (b)

Figure 6. *Cont.*

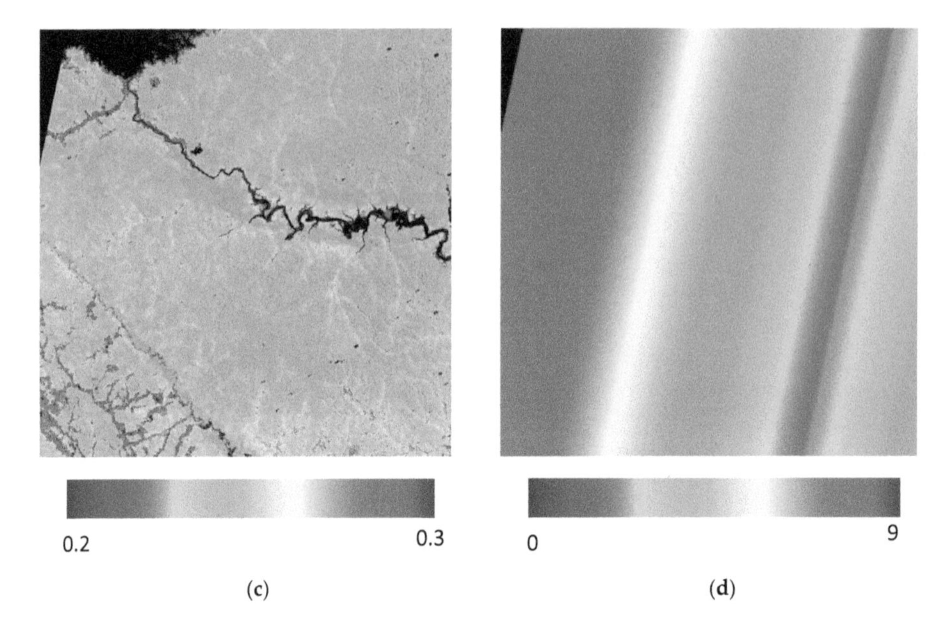

Figure 6. HLS NIR surface reflectance: (**a**) directional, (**b**) using the current BRDF normalization and (**c**) using the proposed normalization. (**d**) The view zenith angle of each pixel. Image on August 21th of 2016 in the Brazilian Amazon area.

Figure 7 shows the transect surface reflectance versus the view zenith angle. Though there is variability through the different samples, we have added a trendline to show the angular dependency of each dataset. The directional reflectance shows the highest values and the largest slope with a p-value of 0.02, and it is followed by the current HLS BRDF normalization with a p-value of 0.08. The proposed method shows the lowest and more stable values with a slope closer to zero and a p-value of 0.31, meaning that there is no significant relationship between the corrected reflectance and the VZA. Additionally, the similar intercepts of the linear regressions show that the three datasets converge to a similar value at zero degrees observation. The analysis of the red band and NDVI (Appendix A) show similar conclusions but the directional effect is lower.

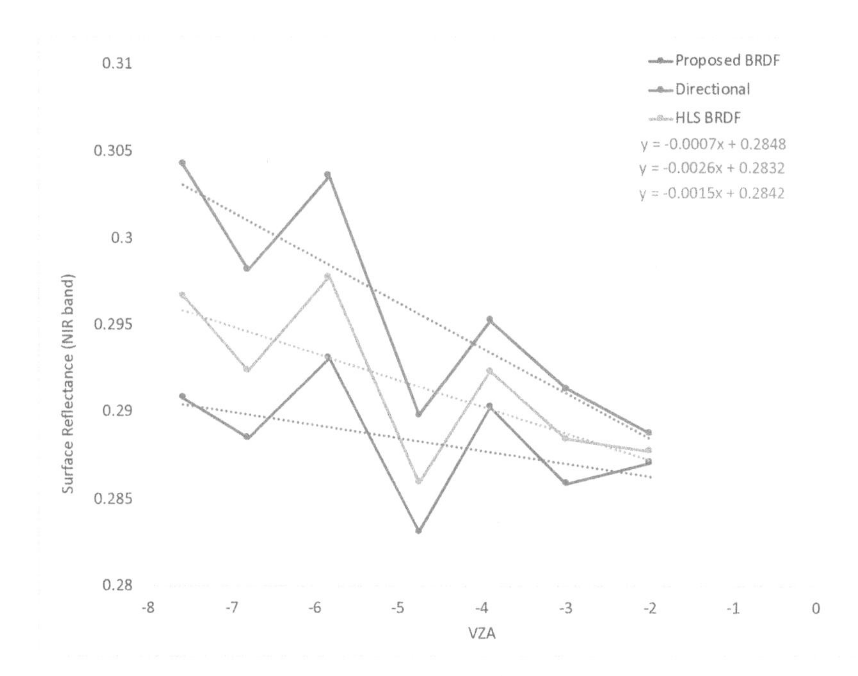

Figure 7. HLS NIR surface reflectance transect values versus the view zenith angle (VZA).

3.3. *Surface Albedo Validation*

We also evaluate the BRDF normalization by using the BRDF parameters to derive the surface albedo, which can be validated with field measurements. Figure 8 shows the validation of the surface albedo over (a) the SURFRAD network, (b) the OzFlux network, and (c) both networks. Each marker in the plot represents an individual Landsat 8 or Sentinel 2 overpass. OzFlux validation shows lower errors (0.015–0.016) than SURFRAD (0.020). Sentinel 2 albedo validation in SURFRAD shows an overestimation for albedos higher than 0.2. However, this is observed in a total of 5 dates in Desert Rock and may be caused by an inaccurate atmospheric correction during those dates. Nevertheless, the overestimation is included within the RMSE of the sites. Combining both networks, the surface albedo validation shows a RMSE of 0.019 for Landsat 8 and 0.018 for Sentinel 2. Figure 8d shows the result of applying the narrow to broadband equation to the directional reflectance. In this case, the errors increase to 0.028 for Landsat 8 and 0.030 for Sentinel 2. Note that an equivalent evaluation cannot be included for the current HLS BRDF normalization method because in Roy [5] it is explicitly stated that "this approach is not applicable for generation of Landsat surface albedo, which requires a full characterization of the surface BRDF".

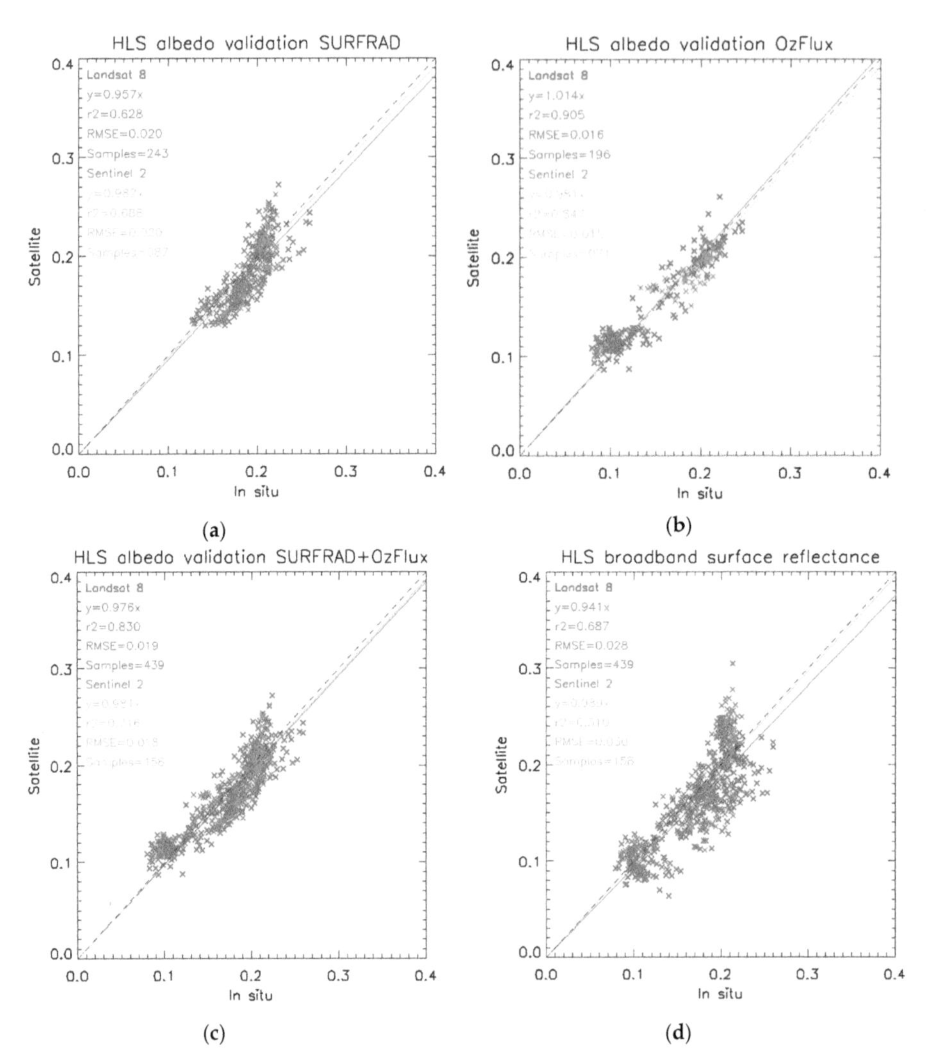

Figure 8. Broadband blue sky surface albedo validation of all the (**a**) SURFRAD, (**b**) OzFlux sites, and (**c**) combining both sites considered from 2013 to 2017. (**d**) The broadband directional surface reflectance comparison with surface albedo measurements.

4. Discussion and Conclusions

Quality of atmospheric and BRDF correction is essential for any application of the HLS dataset. In this work, we present a new method to improve the BRDF normalization of the HLS data. It is based on the Franch [11] method, which was originally designed and evaluated to derive the surface albedo of Landsat TM and ETM+ data. In this manuscript we improve it, apply it to HLS data, explore its feasibility to correct the BRDF effects of the dataset, and validate it over two homogeneous sites and several land cover types of the SURFRAD and OzFlux networks.

The proposed improvement is based on reproducing the BRDF parameters' NDVI dependency [12,13] at moderate spatial resolution (HLS). In the operational context, this means that we create an HLS pixel-based database of the BRDF parameters (V_0, V_1, R_0, and R_1) that is regularly updated to account for any land cover change. In this way, the effect of outliers and poor-quality pixels on the BRDF inversion is minimized, resulting in a more stable and robust BRDF model. We acknowledge that the assumption of a BRDF model being a function of the NDVI has limitations. For example, on sparse forests where the NDVI is not a good descriptor of canopy structure [24]. However, this simple model might be used for a rough correction of BRDF effects in reflectance time series. Although a full inversion of the BRDF model will give better results, some applications, such as real time processing, may want to trade accuracy for simplicity [25].

When applying the proposed method, the results show a decrease in the surface reflectance timeseries CV of 4% and a decrease of the correlation coefficient with the SZA for the forest site, and little to no dependency on the desert site (which is barely affected by angular effects). In contrast, the current HLS BRDF normalization algorithm under-corrects the BRDF effects on the forests site, and increases the CV on the desert site. It is known that vegetation in the Amazon has a phenological cycle as a result of small variations in temperature and precipitation throughout the year [26]. However, directional effects on surface reflectance can be larger than the differences in reflectance between floristically different vegetation types [27,28].

The evaluation of the spatial variability of the view zenith angle showed a clear dependency in the backscatter direction, leading to higher directional reflectance for larger view angles. The current BRDF method reduces this dependency, but still shows an under-correction of the signal in the dense forest area analyzed. This under correction was already reported by Roy [5]. Finally, the proposed method shows no dependency with the view zenith angle.

The evaluation against surface albedo field measurements show an RMSE error of 0.019 for Landsat 8 and 0.018 for Sentinel-2. The OzFlux network error is similar to Franch [11] (RMSE of 0.016), while SURFRAD sites show slightly larger errors. Note that Franch [11] evaluated the method over five SURFRAD sites from 2003 to 2006. The higher error in the SURFRAD network compared to OzFlux might be caused by higher errors in the atmospheric correction algorithm since the atmosphere over Australia exhibit lower aerosol content than over the US. Compared to previous studies that provide albedo validation, the proposed method shows lower errors than Shuai [6] (RMSE 0.024). These results are further evidence of the good performance of the surface reflectance product of Landsat 8 and Sentinel 2.

Author Contributions: Conceptualization and methodology, B.F. and E.V.; data curation: S.S., J.J. and A.S.-A.; supervision: J.-C.R. and J.M.; validation B.F.; writing: B.F., J.L.V.-N. and A.S.-A.

Funding: This work was supported by the NASA grant "Support for the HLS (Harmonized Landsat-Sentinel-2) Project" (no. NNX16AN88G).

Acknowledgments: This work was supported by the NASA grant "Support for the HLS (Harmonized Landsat-Sentinel-2) Project" (no. NNX16AN88G).

Conflicts of Interest: The authors declare no conflict of interest.

Appendix A

In this appendix, we include the transect estimations of the red band and the NDVI and compare their view zenith angle dependency of the directional data compared to each BRDF normalization method.

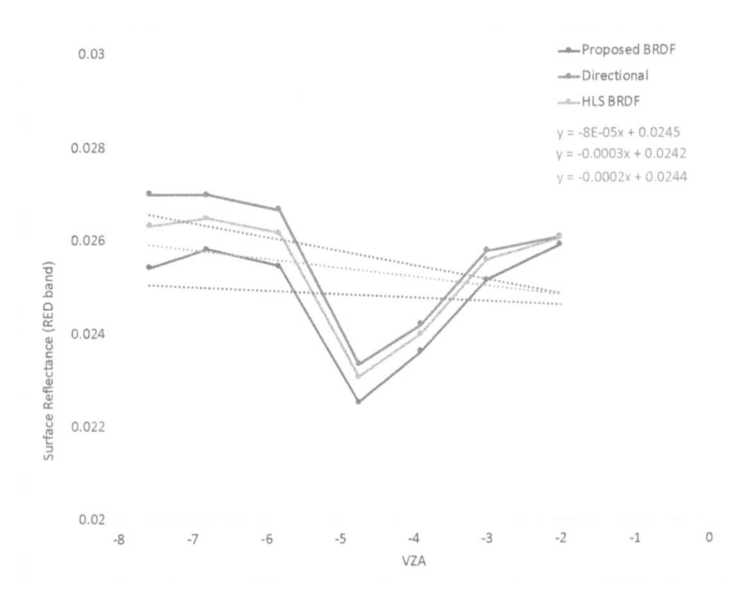

Figure A1. HLS red surface reflectance transect values versus the view zenith angle (VZA).

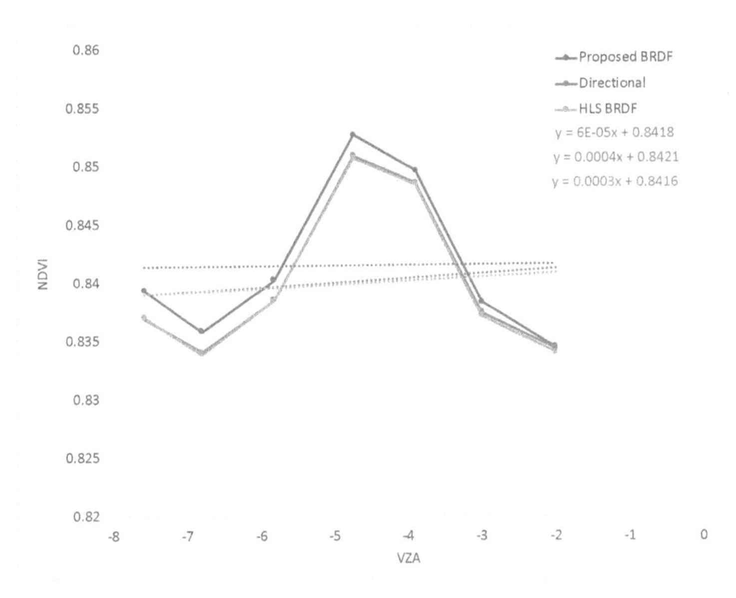

Figure A2. HLS NDVI transect values versus the view zenith angle (VZA).

References

1. Claverie, M.; Ju, J.; Masek, J.G.; Dungan, J.L.; Vermote, E.F.; Roger, J.-C.; Skakun, S.V.; Justice, C. The Harmonized Landsat and Sentinel-2 surface reflectance data set. *Remote Sens. Environ.* **2018**, *219*, 145–161. [CrossRef]
2. Vermote, E.; Justice, C.; Claverie, M.; Franch, B. Preliminary analysis of the performance of the Landsat 8/OLI land surface reflectance product. *Remote Sens. Environ.* **2016**, *185*, 46–56. [CrossRef]

3. Villaescusa-Nadal, J.L.; Franch, B.; Roger, J.; Vermote, E.F.; Skakun, S.; Justice, C. Spectral Adjustment Model's Analysis and Application to Remote Sensing Data. *IEEE J. Sel. Top. Appl. Earth Obs. Remote Sens.* **2019**, 1–12. [CrossRef]
4. Gao, F.; He, T.; Masek, J.G.; Shuai, Y.; Schaaf, C.B.; Wang, Z. Angular Effects and Correction for Medium Resolution Sensors to Support Crop Monitoring. *IEEE J. Sel. Top. Appl. Earth Obs. Remote Sens.* **2014**, *7*, 4480–4489. [CrossRef]
5. Roy, D.P.; Zhang, H.K.; Ju, J.; Gomez-Dans, J.L.; Lewis, P.E.; Schaaf, C.B.; Sun, Q.; Li, J.; Huang, H.; Kovalskyy, V. A general method to normalize Landsat reflectance data to nadir BRDF adjusted reflectance. *Remote Sens. Environ.* **2016**, *176*, 255–271. [CrossRef]
6. Shuai, Y.; Masek, J.G.; Gao, F.; Schaaf, C.B. An algorithm for the retrieval of 30-m snow-free albedo from Landsat surface reflectance and MODIS BRDF. *Remote Sens. Environ.* **2011**, *115*, 2204–2216. [CrossRef]
7. Schaaf, C.B.; Gao, F.; Strahler, A.H.; Lucht, W.; Li, X.; Tsang, T.; Strugnell, N.C.; Zhang, X.; Jin, Y.; Muller, J.-P.; et al. First operational BRDF, albedo nadir reflectance products from MODIS. *Remote Sens. Environ.* **2002**, *83*, 135–148. [CrossRef]
8. Shuai, Y.; Masek, J.G.; Gao, F.; Schaaf, C.B.; He, T. An approach for the long-term 30-m land surface snow-free albedo retrieval from historic Landsat surface reflectance and MODIS-based a priori anisotropy knowledge. *Remote Sens. Environ.* **2014**, *152*, 467–479. [CrossRef]
9. Flood, N.; Danaher, T.; Gill, T.; Gillingham, S. An Operational Scheme for Deriving Standardised Surface Reflectance from Landsat TM/ETM+ and SPOT HRG Imagery for Eastern Australia. *Remote Sens.* **2013**, *5*, 83–109. [CrossRef]
10. Van doninck, J.; Tuomisto, H. Evaluation of directional normalization methods for Landsat TM/ETM+ over primary Amazonian lowland forests. *Int. J. Appl. Earth Obs. Geoinf.* **2017**, *58*, 249–263. [CrossRef]
11. Franch, B.; Vermote, E.F.; Claverie, M. Intercomparison of Landsat albedo retrieval techniques and evaluation against in situ measurements across the US SURFRAD network. *Remote Sens. Environ.* **2014**, *152*. [CrossRef]
12. Vermote, E.; Justice, C.O.; Breon, F.M. Towards a Generalized Approach for Correction of the BRDF Effect in MODIS Directional Reflectances. *IEEE Trans. Geosci. Remote Sens.* **2009**, *47*, 898–908. [CrossRef]
13. Franch, B.; Vermote, E.F.; Sobrino, J.A.; Julien, Y. Retrieval of Surface Albedo on a Daily Basis: Application to MODIS Data. *IEEE Trans. Geosci. Remote Sens.* **2014**, *52*, 7549–7558. [CrossRef]
14. Franch, B.; Vermote, E.; Skakun, S.; Roger, J.-C.; Santamaria-Artigas, A.; Villaescusa-Nadal, J.L.; Masek, J. Toward Landsat and Sentinel-2 BRDF Normalization and Albedo Estimation: A Case Study in the Peruvian Amazon Forest. *Front. Earth Sci.* **2018**, *6*, 185. [CrossRef]
15. Vihermaa, L.E.; Waldron, S.; Domingues, T.; Grace, J.; Cosio, E.G.; Limonchi, F.; Hopkinson, C.; Rocha, H.R.; Gloor, E. Fluvial carbon export from a lowland Amazonian rainforest in relation to atmospheric fluxes. *J. Geophys. Res. Biogeosciences* **2016**, *121*, 3001–3018. [CrossRef]
16. Nagol, J.R.; Sexton, J.O.; Kim, D.-H.; Anand, A.; Morton, D.; Vermote, E.; Townshend, J.R. Bidirectional effects in Landsat reflectance estimates: Is there a problem to solve? *ISPRS J. Photogramm. Remote Sens.* **2015**, *103*, 129–135. [CrossRef]
17. Baldocchi, D.; Falge, E.; Gu, L.; Olson, R.; Hollinger, D.; Running, S.; Anthoni, P.; Bernhofer, C.; Davis, K.; Evans, R.; et al. FLUXNET: A new tool to study the temporal and spatial variability of ecosystem-scale carbon dioxide, water vapor, and energy flux densities. *Bull. Am. Meteorol. Soc.* **2001**, *82*, 2415–2434. [CrossRef]
18. Roujean, J.-L.; Leroy, M.; Deschamps, P.-Y. A bidirectional reflectance model of the Earth's surface for the correction of remote sensing data. *J. Geophys. Res. Atmos.* **1992**, *97*, 20455–20468. [CrossRef]
19. Li, X.; Strahler, A.H. Geometric-optical bidirectional reflectance modeling of a conifer forest canopy. *IEEE Trans. Geosci. Remote Sens.* **1986**, *GE-24*, 906–919. [CrossRef]
20. Lucht, W. Expected retrieval accuracies of bidirectional reflectance and albedo from EOS-MODIS and MISR angular sampling. *J. Geophys. Res. Atmos.* **1998**, *103*, 8763–8778. [CrossRef]
21. Maignan, F.; Bréon, F.-M.; Lacaze, R. Bidirectional reflectance of Earth targets: Evaluation of analytical models using a large set of spaceborne measurements with emphasis on the Hot Spot. *Remote Sens. Environ.* **2004**, *90*, 210–220. [CrossRef]
22. Skakun, S.; Vermote, E.F.; Roger, J.-C.; Justice, C.O.; Masek, J.G. Validation of the LaSRC cloud detection algorithm for Landsat 8 images. *IEEE J. Sel. Top. Appl. Earth Obs. Remote Sens.* **2019**, *12*. [CrossRef]
23. Liang, S. Narrowband to broadband conversions of land surface albedo I: Algorithms. *Remote Sens. Environ.* **2001**, *76*, 213–238. [CrossRef]

24. Stenberg, P.; Rautiainen, M.; Manninen, T.; Voipio, P.; Smolander, H. Reduced simple ratio better than NDVI for estimating LAI in Finnish pine and spruce stands. *Silva Fennica* **2004**, *38*, 431. [CrossRef]
25. Bréon, F.-M.; Vermote, E. Correction of MODIS surface reflectance time series for BRDF effects. *Remote Sens. Environ.* **2012**, *125*, 1–9. [CrossRef]
26. Muro, J.; Tuomisto, H.; Higgins, M.A.; Moulatlet, G.M.; Ruokolainen, K. Floristic composition and across-track reflectance gradient in Landsat images over Amazonian forests. *ISPRS J. Photogramm. Remote Sens.* **2016**, *119*, 361–372. [CrossRef]
27. Higgins, M.A.; Asner, G.P.; Perez, E.; Elespuru, N.; Tuomisto, H.; Ruokolainen, K.; Alonso, A. Use of Landsat and SRTM data to detect broad-scale biodiversity patterns in Northwestern Amazonia. *Remote Sens.* **2012**, *4*, 2401–2418. [CrossRef]
28. Van doninck, J.; Tuomisto, H. A Landsat composite covering all Amazonia for applications in ecology and conservation. *Remote Sens. Ecol. Conserv.* **2018**, *4*, 197–210. [CrossRef]

Article

Evaluating the Spatial Representativeness of the MODerate Resolution Image Spectroradiometer Albedo Product (MCD43) at AmeriFlux Sites

Hongmin Zhou [1,2,*], Shunlin Liang [3,4], Tao He [4], Jindi Wang [1,2], Yanchen Bo [1,2] and Dongdong Wang [3]

1. The State Key Laboratory of Remote Sensing Science, Institute of Remote Sensing Science and Engineering, Faculty of Geographical Science, Beijing Normal University, Beijing 100875, China; wangjd@bnu.edu.cn (J.W.); boyc@bnu.edu.cn (Y.B.)
2. Beijing Engineering Research Center for Global Land Remote Sensing Products, Beijing Normal University, Beijing 100875, China
3. Department of Geographical Sciences, University of Maryland, College Park, MD 20742, USA; sliang@umd.edu (S.L.); ddwang@umd.edu (D.W.)
4. School of Remote Sensing and Information Engineering, Wuhan University, Wuhan 430079, China; taohers@whu.edu.cn

* Correspondence: zhouhm@bnu.edu.cn; Tel.: +86-10-5880-6011

Received: 11 February 2019; Accepted: 28 February 2019; Published: 6 March 2019

Abstract: Land surface albedo is a key parameter in regulating surface radiation budgets. The gridded remote sensing albedo product often represents information concerning an area larger than the nominal spatial resolution because of the large viewing angles of the observations. It is essential to quantify the spatial representativeness of remote sensing products to better guide the sampling strategy in field experiments and match products from different sources. This study quantifies the spatial representativeness of the MODerate Resolution Image Spectroradiometer (MODIS) (collection V006) 500 m daily albedo product (MCD43A3) using the high-resolution product as intermediate data for different land cover types. A total of 1820 paired high-resolution Landsat Thematic Mapper (TM) and coarse-resolution (MODIS) albedo data from five land cover types were used. The TM albedo data was used as the spatial-complete high resolution data to evaluate the spatial representativeness of the MODIS albedo product. Semivarioagrams were estimated from 30 m Landsat data at different spatial scales. Surface heterogeneity was evaluated with sill value and relative coefficient of variation. The 30 m Landsat albedo data was aggregated to 450 m–1800 m using two different methods and compared with MODIS albedo product. The spatial representativeness of MODIS albedo product was determined according to the surface heterogeneity and the consistency of MODIS data and the aggregated TM value. Results indicated that for evergreen broadleaf forests, deciduous broadleaf forests, open shrub lands, woody savannas and grasslands, the MODIS 500 m daily albedo product represents a spatial scale of approximately 630 m. For mixed forests and croplands, the representative spatial scale was approximately 690 m. The difference obtained was primarily because of the complexity of the landscape structure. For mixed forests and croplands, the structure of the landscape was relatively complex due to the presence of different forest and plant types in the pixel area, whereas the other landscape structures were considerably simpler.

Keywords: albedo; spatial representativeness; semivariogram; MODIS; Landsat

---

1. Introduction

Land surface albedo describes the fraction of incoming solar radiation reflected by the surface of the land. It influences the surface energy budget [1], and it is essential for global and regional estimation of energy and mass exchanges between the Earth's surface and the atmosphere [2–6]. An accuracy of 0.02–0.05 is required by climate models for global surface albedo [7,8]. To monitor the spatio-temporal changes in land surface albedo, albedo products are routinely generated from satellite data, such as the Polarization and Directionality of the Earth's Reflectance (POLDER) [9–12], the Medium Resolution Imaging Spectrometer (MERIS) [13], the Clouds and the Earth's Radiant Energy System (CERES) [11,12], the Visible Infrared Imaging Radiometer Suite (VIIRS) [14,15], and the Airborne Visible Infrared Imaging Spectrometer (AVIRIS) [16]. The Moderate Resolution Imaging Spectroradiometer (MODIS) onboard the Earth Observation System (EOS) Terra and Aqua satellites routinely provided data to derive the land surface shortwave and visible albedos [17–19] used to calibrate and improve albedo parameterizations for land, weather, and climate models [20–26]. Assessment of the accuracy of these products is important because it is critical to the scientific community for various applications. Feedback from this activity will help improve the generation of these products [27].

Direct comparison with the ground-based observations of albedo values is the commonly used method to assess the accuracy of remote sensing albedo products [28–30]. The tower observation is compared directly with the satellite product [31–33] based on the assumption that the satellite and tower observation have the same footprint or the landscape is homogeneous. Researchers subsequently recognized that the land surface albedo varies strongly in space and across seasons, because of which land surface homogeneity is now examined prior to the "point-to-pixel" comparison. Only sites whose representativeness is adequate for satellite pixel scales are selected in the direct validation and "heterogeneous" sites are excluded [34]. The method most suited to a comparison for all kinds of landscapes is using higher spatial resolution data as intermediate data [27,35,36]. Burakowski, et al. [37] used airborne hyperspectral imagery to validate MODIS albedo product in snow-covered areas; Mira, et al. [10] used the convolved albedo onboard the Formosat-2 Taiwanese satellite as a reference to evaluate the newly released MCD43D product. However, since no global high-resolution albedo product (at a level of tens of meters) is available, validation using intermediate data has been conducted at a limited number of locations.

Prior to evaluating the remote sensing product, identifying the spatial representativeness of the products is essential. It can help ground sampling point settings and match remote sensing data from multiple sources. MODIS albedo products are retrieved using observations covering a large area that depends on the view zenith angles. Although observations are weighted by angular coverage before albedo retrieval, the actual coverage of the pixel is always larger than the nominal spatial resolution [38,39]. Efforts have been made to characterize the effective resolution of the MODIS gridded product [40]. Mira, et al. [10] used Formost-2 data at a resolution of 8 m to characterize the equivalent point spread function of MODIS albedo at a 1 km pixel. The Full Width at Half Maximum (FWHM), recognized as the effective resolution, has been confirmed to represent the footprint of MODIS data for accurate validation. Campagnolo et al. [41] used extensive time series data (2003–2014) at a large size of the linear natural target in the Netherlands to analyze the effective spatial resolution of the MODIS albedo product (the spatial scale the pixel represents). They verified that the spatial representativeness of the MODIS daily albedo product approximately varied from 606 m to 843 m. Campagnolo and Montano [40] estimated the point spread function (PSF) of nominal 250 m MODIS gridded surface reflectance products, and discovered that the spatial representativeness varied from 344 m to 835 m along the rows, and between 292 m and 523 m along the columns. Their work helps users understand the spatial properties of the satellite product, but their work was based only on a single area, and a general adaptable result that is based on different kinds of land surfaces is needed.

The MODIS BRDF/albedo product was derived with a semi-empirical, kernel-driven BRDF model. Data for 16-day, multi-angular, cloud-free, atmosphere-corrected surface reflectance was compiled to apply the retrieval procedure. To better characterize the rapid change of the land surface, the daily albedo product was retrieved using the same semi-empirical algorithm, but with the single day of interest emphasized by being weighted more heavily [39,42] in the 16-day composite period.

The spatial representativeness and accuracy of the newly released daily albedo product (MCD43A3, V006) have not been tested for different types of land cover for a long time series.

The high-resolution (30–80 m) satellite albedo product is essential in understanding the climatic consequences of land cover change and medium-to-fine scale applications [43]. It is also a key bridge to the assessment of coarse-resolution products. He et al. [44] developed a method to estimate both snow-covered and snow-free albedo from the Chinese environment and disaster monitoring and forecasting small satellite constellation (HJ) satellite data. Zhou, et al. [15] then derived 30 m albedo from Landsat 7 and Landsat 8 using a similar algorithm. He et al. [45] evaluated 30 m albedo product estimated from Landsat Multispectral Scanner (MSS), Thematic Mapper (TM), Enhanced Thematic Mapper Plus (ETM+), and Operational Land Imager (OLI) at Surface Radiation (SURFRAD), AmeriFlux, Baseline Surface Radiation Network (BSRN), and Greenland Climate Network (GC-Net) sites, with results indicating that the direct estimation approach can generate reliable albedo estimates with accuracy of 0.022 to 0.034 in terms of the root mean square error (RMSE). The derivation of global land surface albedo product using Landsat sensors makes it possible to better understand energy transfer between the land surface and the atmosphere at global and regional scales. Furthermore, it makes the assessment of the coarse-resolution albedo product possible, as well as scale transformations for different land cover types and landscapes.

2. Datasets

Three types of datasets were used to quantitatively determine the spatial representativeness of the MODIS daily albedo product: (1) Tower-measured surface albedo datasets from AmeriFlux sites were used as field truth to calibrate 30 m albedo data. (2) The MODIS daily albedo product MOD43A3 (V006, 500 m, daily) was assessed, and annual land cover data were downloaded to identify different land cover types. (3) Landsat TM level-one data were used to estimate the 30 m spatial resolution of the albedo, which was then used as the spatial complete high resolution data to evaluate the spatial representativeness of the MODIS 500 m spatial resolution albedo product.

2.1. AmeriFlux Site Data

The AmeriFlux network is a community of sites and scientists measuring the amounts of carbon, water, energy fluxes, and related environmental variables in ecosystems across the Americas [46]. It supplies a long period of field observations and features wide coverage of forests, grasslands, croplands, shrub lands, wetlands, savannas, and other geographies (e.g., urban). A total of 166 sites are distributed in North and South America, of which 109 feature continuous radiation measurement. Level-2 data of downward and upward radiation from these 109 sites were downloaded from the Carbon Dioxide Information Analysis Center (CDIAC) at Oak Ridge National Laboratory (ORNL). The instruments mounted on the tower including Kipp and Zonen (CNR1, CM-3, OR CM-6), and the Eppley-PSP albedometer/pyranometer. The instruments were fitted with domes to collect radiation fluxes in the broadband shortwave domain (0.3–2.8 μm). Data received from each site were reviewed and incorporated into the network of the AmeriFlux database. The data review process includes checking for consistent units, naming conventions, reporting intervals, and reforming to ensure consistency with the larger network-wide database [47]. Radiation data were collected every half hour during the years indicated in Table 1 for each site. The overall range of years recorded at one site or more is 1995 to 2013. A region of 10 × 10 km around each site was selected as the research area. Information pertaining to the sites used in this study—site name, latitude, longitude, land cover type, and the years when data are available—is listed in Table 1.

Table 1. Flux sites used in the analysis.

SITE_NAME	LOCATION_LAT	LOCATION_LONG	IGBP	TOWER_BEGAN	TOWER_END	SITE_NAME	LOCATION_LAT	LOCATION_LONG	IGBP	TOWER_BEGAN	TOWER_END
ARM Southern Great Plains site—Lamont	36.61	−97.49	CRO	2002	2013	Mary's River (Fir) site	44.65	−123.55	ENF	2005	2013
Bondville	40.01	−88.29	CRO	1996	2013	Metolius Young Pine Burn	44.32	−121.61	ENF	2010	2013
Bondville (companion site)	40.01	−88.29	CRO	2004	2008	Flagstaff—Unmanaged Forest	35.09	−111.76	ENF	2006	2010
Brooks Field Site 10—Ames	41.69	−93.69	CRO	2001	2013	Duke Forest—loblolly pine	35.98	−79.09	ENF	2001	2008
Brooks Field Site 11—Ames	41.97	−93.69	CRO	2001	2013	Howland Forest (west tower)	45.21	−68.75	ENF	1999	2013
Curtice Walter—Berger cropland	41.63	−83.35	CRO	2011	2013	Metolius—second young aged pine	44.32	−121.61	ENF	2004	2009
Fermi National Accelerator Laboratory—Batavia (Agricultural site)	41.86	−88.22	CRO	2005	2013	Metolius—intermediate aged ponderosa pine	44.45	−121.56	ENF	2002	2013
Mead—irrigated continuous maize site	41.17	−96.48	CRO	2001	2013	Howland Forest (main tower)	45.20	−68.74	ENF	1996	2013
Mead—irrigated maize-soybean rotation site	41.16	−96.47	CRO	2001	2013	Poker Flat Research Range Black Spruce Forest	65.12	−147.49	ENF	2011	2013
Mead—rainfed maize-soybean rotation site	41.18	−96.44	CRO	2001	2013	Saskatchewan—Western Boreal, forest burned in 1977.	54.49	−105.82	ENF	2004	2006
Ponca City	36.77	−97.13	CRO	1997	2001	Flagstaff—Managed Forest	35.14	−111.73	ENF	2006	2010
Rosemount—G21	44.43	−93.05	CRO	2002	2013	Niwot Ridge Forest (LTER NWT1)	40.03	−105.55	ENF	1998	2013
Sioux Falls Portable	43.24	−96.90	CRO	2007	2013	Howland Forest (harvest site)	45.21	−68.73	ENF	2000	2013
Twitchell Corn	38.10	−121.64	CRO	2012	2013	UCI-1930 burn site	55.91	−98.52	ENF	2001	2005
Twitchell Alfalfa	38.12	−121.65	CRO	2013	2013	UCI-1964 burn site wet	55.91	−98.38	ENF	2002	2004
Bartlett Experimental Forest	44.06	−71.29	DBF	2004	2013	Quebec—Eastern Boreal	49.27	−74.04	ENF	2001	2013
Morgan Monroe State Forest	39.32	−86.41	DBF	1999	2013	NC_Clearcut	35.81	−76.71	ENF	2005	2013
Duke Forest-hardwoods	35.97	−79.10	DBF	2003	2013	GLEES	41.37	−106.24	ENF	2004	2013
Willow Creek	45.81	−90.08	DBF	1999	2013	Wind River Crane Site	45.82	−121.95	ENF	1998	2013
Chestnut Ridge	35.93	−84.33	DBF	2005	2013	UCI-1850 burn site	55.88	−98.48	ENF	2002	2005
Silas Little—New Jersey	39.91	−74.60	DBF	2004	2013	UCI-1964 burn site	55.91	−98.38	ENF	2001	2005
Univ. of Mich. Biological Station	45.56	−84.71	DBF	1999	2013	UCI-1981 burn site	55.86	−98.49	ENF	2001	2005
Walker Branch Watershed	35.96	−84.29	DBF	1995	1999	Black Hills	44.16	−103.65	ENF	2003	2008
Missouri Ozark Site	38.74	−92.20	DBF	2004	2013	Chimney Park	41.07	−106.12	ENF	2009	2013
Oak Openings	41.55	−83.84	DBF	2004	2013	NC_Loblolly Plantation	35.80	−76.67	ENF	2005	2013
UMBS Disturbance	45.56	−84.70	DBF	2007	2013	Quebec—Eastern Boreal, Mature Black Spruce.	49.69	−74.34	ENF	2003	2013
Fermi National Accelerator Laboratory—Batavia	41.84	−88.24	GRA	2004	2013	Valles Caldera National Preserve (Mixed Conifer)	35.89	−106.53	ENF	2007	2013
ARM USDA UNL OSU Woodward Switchgrass 1	36.43	−99.42	GRA	2009	2013	Valles Caldera National Preserve (Ponderosa pine)	35.86	−106.60	ENF	2007	2013
ARM USDA UNL OSU Woodward Switchgrass 2	36.64	−99.60	GRA	2009	2013	Ontario—Turkey Point 1939 Plantation White Pine	42.71	−80.36	ENF	2003	2013
Konza Prairie LTER (KNZ)	39.08	−96.56	GRA	2006	2013	Bonanza Creek	63.92	−145.74	OSH	2003	2013
Walnut Gulch Kendall Grasslands	31.74	−109.94	GRA	2004	2013	UCI-1989 burn site	55.92	−98.96	OSH	2001	2005
Audubon Research Ranch	31.59	−110.51	GRA	2002	2013	Saskatchewan—Western Boreal	54.09	−106.01	OSH	2001	2006

Table 1. *Cont.*

SITE_NAME	LOCATION_LAT	LOCATION_LONG	IGBP	TOWER_BEGAN	TOWER_END	SITE_NAME	LOCATION_LAT	LOCATION_LONG	IGBP	TOWER_BEGAN	TOWER_END
Duke Forest-open field	35.97	−79.09	GRA	2001	2013	Anaktuvuk River Severe Burn	68.99	−150.28	OSH	2008	2013
Diablo	37.68	−121.53	GRA	2010	2013	Anaktuvuk River Moderate Burn	68.95	−150.21	OSH	2008	2013
Santa Rita Grassland	31.79	−110.83	GRA	2008	2013	Anaktuvuk River Unburned	68.93	−150.27	OSH	2008	2013
Canaan Valley	39.06	−79.42	GRA	2004	2013	Walden	40.78	−106.26	OSH	2006	2008
Cottonwood	43.95	−101.85	GRA	2007	2009	UCI-1998 burn site	56.64	−99.95	OSH	2002	2005
Fort Peck	48.31	−105.10	GRA	2000	2013	Sevilleta (LTER desert shrubland)	34.33	−106.74	OSH	2007	2013
Brookings	44.35	−96.84	GRA	2004	2013	Santa Rita Creosote	31.91	−110.84	OSH	2008	2013
Walnut River Watershed (Smileyburg)	37.52	−96.86	GRA	2001	2004	Juniper savanna site (Willard)	34.43	−105.86	OSH	2007	2013
Sevilleta (LTER desert grassland)	34.36	−106.70	GRA	2007	2013	Walnut Gulch Lucky Hills Shrub	31.74	−110.05	OSH	2007	2013
Flagstaff—Wildfire	35.45	−111.77	GRA	2006	2010	Imnavait Creek Watershed Tussock Tundra	68.61	−149.30	OSH	2007	2013
KUOM Turfgrass Field	45.00	−93.19	GRA	2005	2009	Imnavait Creek Watershed Heath Tundra	68.61	−149.30	OSH	2007	2013
Kansas Field Station	39.06	−95.19	GRA	2007	2013	Pinon-juniper site (Mountainair)	34.44	−106.24	OSH	2007	2013
Corral Pocket	38.09	−109.39	GRA	2001	2007	Everglades (long hydroperiod marsh)	25.55	−80.78	WET	2007	2013
Shidler—Oklahoma	36.93	−96.68	GRA	1997	2001	Everglades (short hydroperiod marsh)	25.44	−80.59	WET	2007	2013
Vaira Ranch—Ione	38.41	−120.95	GRA	2000	2013	Twitchell East End Wetland	38.10	−121.64	WET	2013	2013
Goodwin Creek	34.25	−89.87	GRA	2002	2006	Winous Point North Marsh	41.46	−83.00	WET	2011	2013
Santarem-Km83-Logged Forest	-3.02	−54.97	EBF	2000	2003	Twitchell Wetland West Pond	38.11	−121.65	WET	2012	2013
Shark River Slough (Tower SRS-6) Everglades	25.36	−81.08	EBF	2004	2013	Imnavait Creek Watershed Wet Sedge Tundra	68.61	−149.31	WET	2007	2013
Ontario—Groundhog River, Boreal Mixedwood Forest.	48.22	−82.16	MF	2003	2013	Olentangy River Wetland Research Park	40.02	−83.02	WET	2011	2013
Santa Rita Mesquite	31.82	−110.87	WSA	2004	2013	Ivotuk	68.49	−155.75	WET	2003	2013
Freeman Ranch—Mesquite Juniper	29.95	−98.00	WSA	2004	2013	Lost Creek	46.08	−89.98	WET	2001	2013
Sylvania Wilderness Area	46.24	−89.35	MF	2001	2013	Freeman Ranch—Woodland	29.94	−97.99	CSH	2004	2013
Fort Dix	39.97	−74.43	MF	2005	2008						

Note: * SITE_NAME is the full name of the selected AmeriFlux sites. *IGBP* is the International Geosphere Biosphere Program.CRO is cropland, DBF is deciduous broadleaf forest, ENF is evergreen needleleaf forest, GRA is grassland, MF is mixed forest, OSH is open shrubs, WET is permanent wetlands, WSA is woody savannas.

2.2. MODIS Data

2.2.1. MCD43 BRDF/Albedo Product

The MODIS Albedo product (MCD43A3 V06) provides daily albedo data on the Earth's surface for each pixel at a grid resolution of 500 m. A 16-day period of cloud-free surface reflectance from both the Terra and the Aqua is used to derive the daily data, with weight as a function of quality, the observation coverage, and the temporal distance from the day of interest. The newly broadcasted albedo product has been utilized for regional applications (e.g., forest, agriculture, and disturbance monitoring), and is now downloadable from the Land Processes Distributed Active Archive Center (LP DAAC). The 16-day composed daily albedo product has been validated over typical agricultural landscapes [10,41], grasslands, and agriculture and forest surfaces [39], but has not been extensively validated for other land cover types using massive amounts of data.

2.2.2. MCD12Q1 Land Cover

Land cover plays a major role in the climate and biogeochemistry of the Earth's ecosystem. The MODIS land cover type product provides a suite of land cover types by using spectral and temporal information derived from the MODIS. It characterizes five global land cover classification systems—the International Geosphere Biosphere Program (*IGBP*), university of Maryland (UMD), Leaf Area Index/Fraction of Photosynthetically Active Radiation absorbed by vegetation (LAI/fPAR), New Patriotic Party (NPP), and Plant Functional Types (PFT)—at annual time steps and a 500 m spatial resolution. Land cover type assessment and quality control information are also included. Given that the land cover type may change over time, yearly MODIS land cover type data (MCD12Q1) are downloaded, and the IGBP land cover type is used.

2.3. Landsat-Retrieved Albedo

The TM onboard the Landsat 4 and 5 satellites with seven spectral bands covered the shortwave range at a resolution of 30 m from 1984 to 2011. It is an optimal data source for regional and global land surface change research because of its long period of operation [48–52]. The high-spatial-resolution (30 m), long-term albedo product based on Landsat data has been derived [43,45,53]. In this study, we adopt the algorithm of He et al. [45] to derive the global long-term land surface albedo product from Landsat TM data. L1T data that were temporally consistent with AmeriFlux site observations were downloaded from the U.S. Geological Survey (USGS) website. To minimize the influence of cloud coverage, only the highest data quality (quality flag is 9) and a maximum cloud cover of 10% were used. In this paper, 2034 scenes of TM data were downloaded, covering all land cover types and seasons, where 1820 scenes corresponded to high-quality MODIS data. TM data were first rectified to sinusoidal projections and resampled to 30 m resolution using the nearest neighbor algorithm.

The accuracy of meso-scale data needs to be verified and calibrated before they are used to bridge field measurements and coarse-resolution products. Landsat TM albedo data was first calibrated with the field data. The Landsat TM overpassed at local time 10:00–11:00 AM. To guarantee the temporal consistency between TM data and field observation, field observations 15 min before and after the TM passing over were averaged. The field observation covers an area larger than 30 m, so we assume that the land surface at 30 m scale is homogeneous and field observation can compare directly with TM data. However, for each AmeriFlux site, the tower-located TM pixel was picked out according to the tower location. Sometimes, the tower was not located at the center of the pixel and possible geometric correction error may also exist, so albedo of 3×3 TM pixels around the tower were averaged and compared with the field observations.

Figure 1 shows a scatter plot of the field measurements and the TM albedo. The correction coefficient was derived from least squares regression between TM data and ground measurement. The determination coefficient was 0.877, RMSE and bias were 0.033 and 0.00035, respectively. The calibrated

TM data with the determined correlation were then used as a bridge for comparison with the MODIS albedo product as well as for further analysis to determine the spatial representativeness of the later.

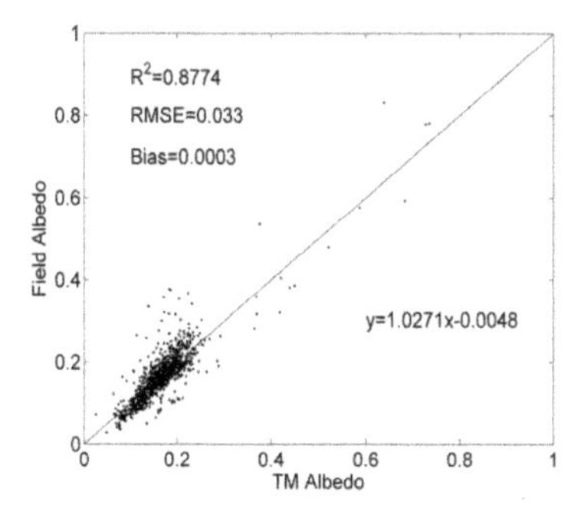

Figure 1. Field data comparison with Landsat Thematic Mapper (TM) data, which is the average of 3 × 3 pixels. The fitting function is then used for TM albedo data calibration.

3. Methods

To quantitatively evaluate the spatial representativeness of the MODIS daily albedo product, 30 m albedo data were derived from Landsat TM data and then calibrated using field observation data. Semivariagrams are calculated with calibrated 30 m albedo data for different scales, in which sill value is considered as the key index to measure the magnitude of field homogeneity, i.e., sill values of adjacent scales differ from each other significantly, indicating that the field homogeneity changed significantly between these two scales. Relative coefficient of variation (R_{CV}) is also employed to determine the land surface homogeneity: R_{CV} tends to be 0, which indicates that the adjacent scales have similar homogeneity, the MODIS pixel represents the larger scale. The last index used is the determination coefficient, 30 m albedo data is aggregated to different scales and compared with MODIS data. The scale in which it has the highest determination coefficient indicates the MODIS pixel represents the spatial scale best.

3.1. Variogram Model Parameters from TM Data

When using tower observation to validate the remote sensing albedo product, the spatial representativeness of the observation footprint was investigated on semivariogram models [37,39,47,54]. In this study, the method of deriving variogram functions to analyze surface albedo with TM data was used [36]. The variogram estimator $\gamma(h)$ was used to obtain the half-average squared difference between albedo values within a certain distance. According to Román et al. [47], the isotropic spherical variogram model [55] was used to fit the variogram model parameters—the range (a), the sill (c), and the nugget effect (c_0), as below:

$$\gamma_{sph}(h) = \begin{cases} c_0 + c\left(1.5\frac{h}{a} - 0.5\left(\frac{h}{a}\right)^3\right) for\ 0 \le h \le a \\ c_0 + c\ \ for\ h > a \end{cases} \quad (1)$$

The range is the distance from a point beyond which there is no further correlation of the albedo associated with that point. It is the average patch size of the landscape in landscape ecology, which represents a region that differs from its surroundings, but is not necessarily inter-homogeneous. The sill is the maximum semivariance, and is the ordinate value of the range at which the variogram levels off to an asymptote. The non-zero value of the variogram when h = 0 is called the nugget effect. It depends on the variance associated with small-scale variation, measurement errors, or their combination [56]. The range, the sill, and the nugget effect all reveal the spatial variation of the land surface and the scale effect associated with remote sensing data [47,57]. It has been suggested that the land surface is homogeneous (representative) when the sill value is less than 0.001 [37]. In this paper, the semivariogram model was used as well. The 30 m Landsat albedo was first re-projected to a sinusoidal projection and the Landsat pixel located at the center of the MODIS 500 m pixel was determined. The semivariogram was calculated from Landsat data. The model parameters were fitted according to the spherical model.

3.2. Relative Coefficient of Variation

The indices deduced from the parameters of the semivariogram model and the statistical values were also used in this study. The relative coefficient of variation (R_{CV}), the scale requirement index, the relative proportion of structural variation, and the relative strength of spatial correlation, derived from the semivariogram model, were first used by Román et al. [47] to depict spatial variation. When the measurement site was spatially representative, the overall variation between the internal components of the measurement site (scale 1) and the adjacent landscape (scale 2) should have been similar in magnitude, and the R_{CV} should have approached zero. The R_{CV} was also calculated to check the spatial variation in the landscape. To calculate it, the coefficient of variation (cv) was first calculated as the ratio of the standard deviation to the mean. The R_{CV} is given below:

$$R_{CV} = \frac{CV_{scale2} - CV_{scale1}}{CV_{scale1}} \tag{2}$$

3.3. Evaluation Strategies

To determine the spatial representativeness of the 500 m product used in this study, variogram estimation was performed at nine scales for each site. When estimating the variogram, the common spatial step was one MODIS pixel, and according to the result of [41], three scales were added between 1 × 1 and 2 × 2 MODIS pixels; that is, 21 × 21, 23 × 23, and 29 × 29 TM pixels. The 2.5 × 2.5 MODIS pixels were also added to keep an intensive estimation. The estimating scales used in this study are shown in Table 2.

Table 2. Variogram estimating scales selected in this study.

Scale	1	2	3	4	5	6	7	8	9
TM pixels	15 × 15	21 × 21	23 × 23	29 × 29	31 × 31	39 × 39	47 × 47	61 × 61	75 × 75
Scale in meters	450	630	690	870	930	1170	1410	1830	2250
MODIS pixels	1 × 1	-	-	-	2 × 2	2.5 × 2.5	3 × 3	4 × 4	5 × 5

To make the analysis representative, for each site, not only was the tower-located MODIS pixel analyzed, the research area was enlarged to 9 × 9 pixels with the tower-located MODIS pixel as the central pixel. In this research area, for every MODIS pixel, the semivariogram parameters (nugget, sill, range) and the statistical value, including mean and standard deviation, were calculated at the nine scales illustrated in Table 2.

The spatial representativeness was evaluated according to the calculated parameters and values. The sill value represents the magnitude of spatial variability. In this study, the sill values of 9 × 9 MODIS pixel in every research area were compiled (for every scale, sill values were compiled as a group; thus, we obtained nine groups of data), and the paired t-test was implemented every adjacent scale data pair (e.g., 15 × 15 and 21 × 21) to find whether the land surface at adjacent scales was significantly different. Statistical significance was determined by the p-value. If the p-value was zero, it indicated that the difference of the land surface heterogeneity at adjacent scales was not significant; the central MODIS pixel was able to represent an area determined by the larger scale.

R_{CV} depicts spatial variation in the landscape. In this study, R_{CV} was calculated between each pair of scales in Table 2, then all the data from 109 sites and 81 pixels were compiled and the histogram was plotted. The scale in which the R_{CV} value is smallest indicates that the central pixel has the similar spatial representativeness in these adjacent scales.

Taking the calibrated TM data as the albedo reference, we aggregated the TM value at different scales and compared the aggregated 30 m albedo data directly with the 500 m albedo product, and two aggregation methods were used. One was simple average, and the other considered the point spread function. Campagnolo and Montano [40] and Mira et al. [10] used the convolution of a Gaussian function to characterize the optical PSF of the MODIS instrument, assuming that the central area of the pixel made a greater contribution to the signal. In this paper, we adopt the Gaussian function, as was done by Campagnolo and Montano [40] and Mira et al. [10], but set an asymmetric Gaussian point spread function. The PSF was defined below.

$$\mathrm{PSF}(x,y) = \frac{1}{2\pi a^2} exp^{-0.5(x^2/a^2+y^2/a^2)} \tag{3}$$

where

$$\mathrm{a} = \frac{\mathrm{FWHM}}{2.355} \tag{4}$$

The FWHM is the full width at half maximum of the PSF. In this study, we set it as the sensor spatial resolution as Campagnolo and Montano [40], which also represents the spatial representativeness of the pixel.

The determination coefficient was used to judge the satisfactoriness of the consistency between the two datasets. The scale where the highest determination coefficient appeared was considered that the MODIS data has the best spatial representativeness.

4. Results

4.1. Spatial Representativeness Determined by Sill Value

A paired t-test was performed on the sill value for each set of paired TM and MODIS data. The spatial variation was determined according to the criterion that if the p-value was zero, it confirmed the null hypothesis—land surface variation at adjacent scales was not different, and the pixel can represent the larger scale scape. For all data pairs, the histogram of spatial variation is shown in Figure 2. The median value was also used as the indicator of effective spatial resolution, as in Campagnolo et al. [41]. In this situation, the median value was 2, which suggests that the variation in the land surface was subtle within the 21 × 21 TM scale, and the proper aggregation scale for TM albedo data was 21 × 21 pixels. The 21 × 21 aggregation scale represented a 630 m spatial scale, which is consistent with the result of [41].

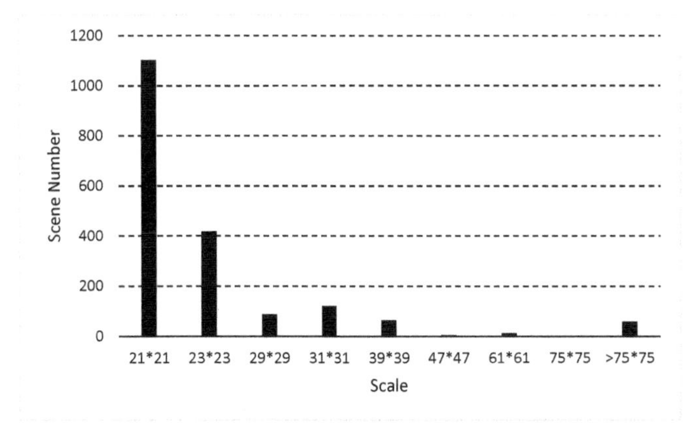

Figure 2. The proper aggregation scale histogram according to the sill value of each scene. The scale is defined with TM pixels. The scene number is the total number of the scenes at each scale where the MODerate Resolution Image Spectroradiometer (MODIS) pixel represents the spatial scale well.

Table 3 shows the proper aggregation scale for every land cover type of AmeriFlux sites. For most land cover types, the land surface was stable within 21 × 21 TM pixels, and for mixed forest and croplands, the proposed effective spatial resolution was 690 m (approximately 23 × 23 TM pixels).

Table 3. Spatial representativeness of MODIS 500 m albedo product for different land cover types.

Land Cover Type	Suggested Spatial Scale	Suggested Spatial Representativeness in Meters
Evergreen Broadleaf forest	2	630
Deciduous Broadleaf forest	2	630
Mixed forest	3	690
Open shrublands	2	630
Woody savannas	2	630
Grasslands	2	630
Croplands	3	690

4.2. Spatial Representativeness Determined by Relative Coefficient of Variation

Figure 3 shows the histogram of the R_{CV} at each scale. To depict the spatial variation at adjacent scales, the R_{CV} was computed from the CV values at each adjacent scale. The median value was calculated as in Figure 3 to describe the integral spatial variation at each scale. From Figure 3, we can see that R_{CV1}, which represents the R_{CV} of the first scale, was the largest for all scales. This means that the spatial variation between scale 1 (15 × 15 TM pixels) and scale 2 (21 × 21 TM pixels) was significant. The median values of R_{CV2} and R_{CV4} were small, which means that the effective spatial resolution should have been 630 m or 870 m. The conclusion was partly consistent with that of the t-test on the sill values. However, the step sizes of the scales were different: for scales 1 and 2, the step was 6 TM pixels, but for scales 2 and 3, it was only 2. Hence, the low level of the variation in the sill value and the small value of the R_{CV} could have been deduced from the small step size. To verify this conclusion, we then reduced the analysis step size and aggregated the TM data at different scales to explore the proper representative scale.

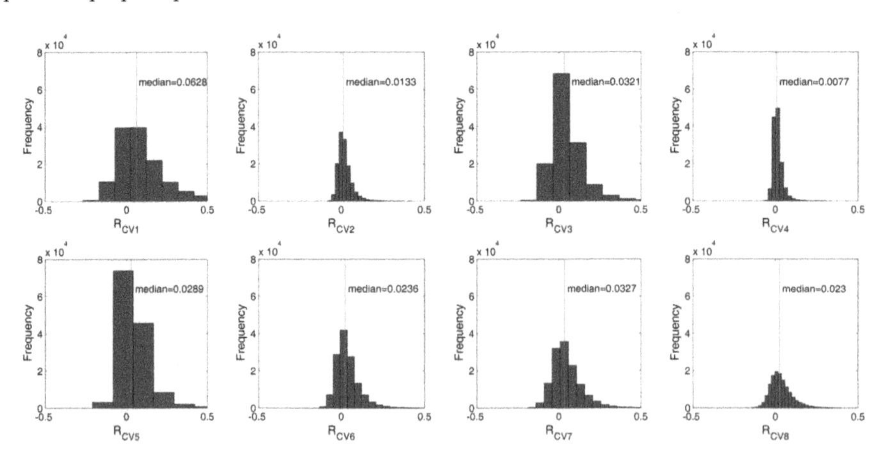

Figure 3. Histogram of R_{CV} at (**a**)–(**h**): scale 1–scale 8. The median is the median value of R_{CV} at each scale.

4.3. Spatial Representativeness Determined by Direct Comparison

We refined the step size to find the highest correlation coefficient between TM and MODIS to determine the proper aggregation scale and use it as the representative spatial scale of MODIS data. The step size was set to 2 TM pixels. Figure 4 shows that the coefficient varied with aggregation scale. TM data were aggregated from 15 × 15 to 61 × 61 TM pixels. From Figure 4 we can see that when

23 × 23 TM pixels were aggregated, the correlation coefficient was the highest and the root mean squared error was the lowest compared with MODIS data. This corresponds to the results in Table 3 in some degree.

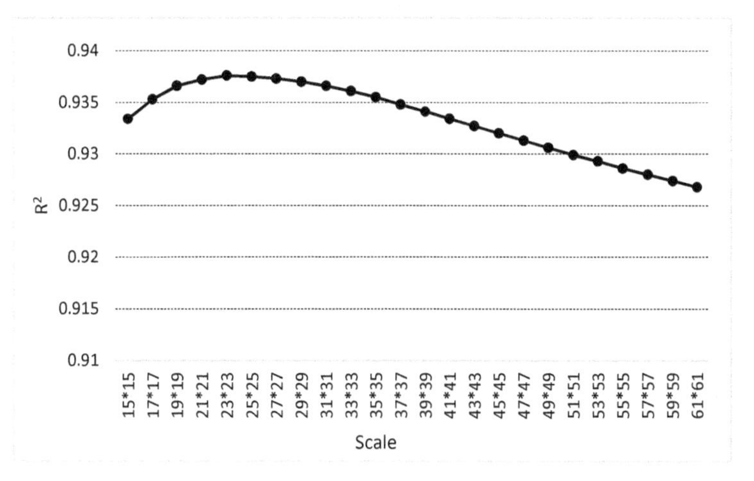

Figure 4. R^2 variation with aggregation scale.

We then checked the comparative accuracy of different land cover types. Table 4 shows the accuracy of comparison in different land cover types at the suggested aggregation scale according to Table 3. For all the land cover types, the RMSE was less than 0.03 and bias less than 0.018. The R^2 of evergreen broadleaf forests was the lowest, mainly because the land surface of this type was snow free, and all albedo values clustered together. For all other land cover types, the R^2 value was higher than 0.86, and for croplands, it reaches 0.965.

Table 4. Comparison in terms of accuracy between TM and MODIS for different land types.

LC	Land Cover Type	R^2	RMSE	BIAS
2	Evergreen Broadleaf forest	0.688	0.016	0.014
4	Deciduous Broadleaf forest	0.863	0.018	0.013
5	Mixed forest	0.926	0.026	0.014
7	Open shrublands	0.947	0.018	0.014
8	Woody savannas	0.944	0.030	0.017
10	Grasslands	0.941	0.021	0.009
12	Croplands	0.965	0.023	0.003

5. Discussion

5.1. Errors Induced by Landsat Albedo Estimation and Spatial Discrepancy

The Landsat albedo estimation algorithm has been validated on a variety of sensors and land cover types [15,16,44,58]. In this study, it was first calibrated using field observations. The errors induced by the estimation algorithm were hence eliminated.

In most cases, the sites were not located at the center of a MODIS pixel. Comparing the MODIS product with field observations or high-resolution data at the site tends to induce errors due to spatial discrepancy. In this study, high-resolution data were first calibrated with the field observations. When evaluating the spatial representativeness of the MODIS product, Landsat pixels located at the center of the MODIS pixels were selected, so the analysis focused on this area could guarantee the spatial agreement to the greatest extent.

5.2. *Difference Deduced by Upscaling Methods*

When upscaling finer-resolution data to a coarser resolution, the simple average method is generally used. Mira et al. [10] assessed an equivalent PSF based on image correlation analysis using an aggregated albedo convolved with PSF over an agricultural landscape. The results indicated that convolving the PSF can reduce uncertainty by up to 0.02 (10%). We checked the difference deduced by the upscaling method.

5.2.1. Simple Average Method

The TM value was first averaged then compared directly with the MODIS data. Figure 5 shows a comparison between the MODIS daily albedo data and the TM data. The results indicate that these datasets agreed well with RMSE less than 0.03 and R^2 greater than 0.92. Although the recommended aggregation scales were scale 2 and 3 (21 × 21 and 23 × 23 TM pixels), the difference in accuracy between the scales was not significant (R^2 ranged from 0.9341 to 0.9442; RMSE ranged from 0.0239 to 0.0249; the biases were nearly identical).

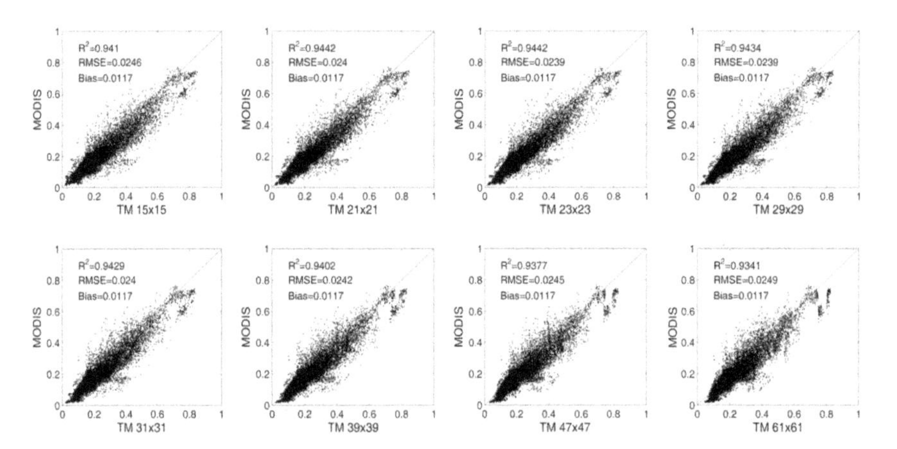

Figure 5. Comparison between MODIS albedo data with TM data aggregated at (**a**) 15 × 15, (**b**) 21 × 21, (**c**) 23 × 23, (**d**) 29 × 29, (**e**) 31 × 31, (**f**) 39 × 39, (**g**) 47 × 47, and (**h**) 61 × 61 scales.

5.2.2. Aggregation with PSF

MODIS gridded products are the outputs of a sampled image system that combines an imaging system with a sampling procedure. The imaging system used was characterized by the sensor PSF, which is considered as the convolution of line spread functions in the along-scan and the along-track directions [10,40]. The FWHM is an important parameter in characterizing pixel resolution. We implemented the PSF function to TM data upscaling. The FWHM was set as the first eight scale values in Table 2, and a weight smaller than 20% was neglected [10].

Table 5 shows a comparison between MODIS data and the TM albedo aggregated with the PSF. The first aggregation scale (15 × 15 TM pixels) has a high R^2 value (0.944), while the second and third aggregation scales (21 × 21 and 23 × 23 TM pixels) has the lowest RMSE (0.0239).

Comparing the results of aggregation with the simple average and those obtained by convolving them with the PSF, the highest R^2 value appeared when the aggregation scale was 23 × 23 TM pixels. The RMSE of the simple average and those obtained through the convolution of the PSF were nearly identical (0.0239 and 0.0238, respectively), indicating that for all datasets, the difference deduced by the aggregation method was not significant.

Table 5. Comparative accuracy of MODIS albedo and TM data aggregated with PSF.

Scale	1	2	3	4	5	6	7	8
TM pixels	15 × 15	21 × 21	23 × 23	29 × 29	31 × 31	39 × 39	47 × 47	61 × 61
R^2	0.944	0.9437	0.943	0.9404	0.9395	0.9354	0.9309	0.9207
RMSE	0.0239	0.0238	0.0238	0.024	0.0241	0.0243	0.0245	0.0248
BIAS	0.0117	0.0117	0.0116	0.0115	0.0114	0.011	0.0103	0.0081

5.3. *Effect of Land Surface Heterogeneity*

We then focused the analysis on tower located pixel for each site. For each scene, the semivariogram was calculated, from 15 × 15 to 77 × 77 TM pixels at a step of 2 TM pixels. Figure 6 shows the heterogeneous number of scenes at each step (sill value larger than 0.001). On average, 257 scenes were heterogeneous on 42 sites.

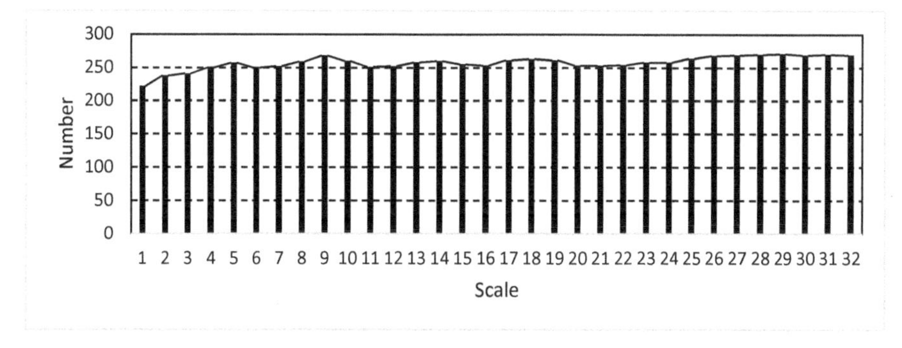

Figure 6. Number of heterogeneous scenes at each scale from 15 × 15 to 77 × 77 TM pixels at a step of 2 TM pixels.

We divided the heterogeneous landscape into three types. Figure 7 shows the general land surfaces and their spatial variations. Figure 7A shows a cropland from US-Ne3 site, located at the center of a rain-fed maize–soybean rotation field. The crop had been harvested, and only bare soil was explored in the scene. The trend in the variation in sill value was not significant with increasing scale. This meant that although the landscape was heterogeneous, the magnitude of heterogeneity at the scales was stable. The difference between the MODIS and the TM values decreased with increasing scale. In this case, with the scale enlarged, the influence of land surface heterogeneity diminished. The MODIS pixel represents an area much larger than its nominal resolution.

Figure 7B shows a US-Fmf site. This was an evergreen needleleaf forest site. The land surface heterogeneity was mainly due to the snow in the upper-right corner. With increasing scale, the land surface became more heterogeneous and the sill value increased from 0.0034 to 0.0059. The discrepancy between MODIS and aggregated TM pixels increased accordingly. For this situation, we may conclude that the more heterogeneous the land surface is, the greater the discrepancy is.

Figure 7C is CA-Fuf site. It is an evergreen needleleaf forest site not far from the US-Fmf site. The land surface heterogeneity was much higher than in the former two scenes, mainly due to the irregular surface and snow. The discrepancy between MODIS and TM albedo was correspondingly significant. In this case, we can hardly determine the effective spatial representativeness of the pixel.

Land surface heterogeneity is a key factor affecting effective spatial representativeness. In general, the more heterogeneous the land surface is, the larger the effective spatial representativeness is. When the magnitudes of land surface heterogeneity at different scales are similar, with the scale increased, land surface homogeneity increases.

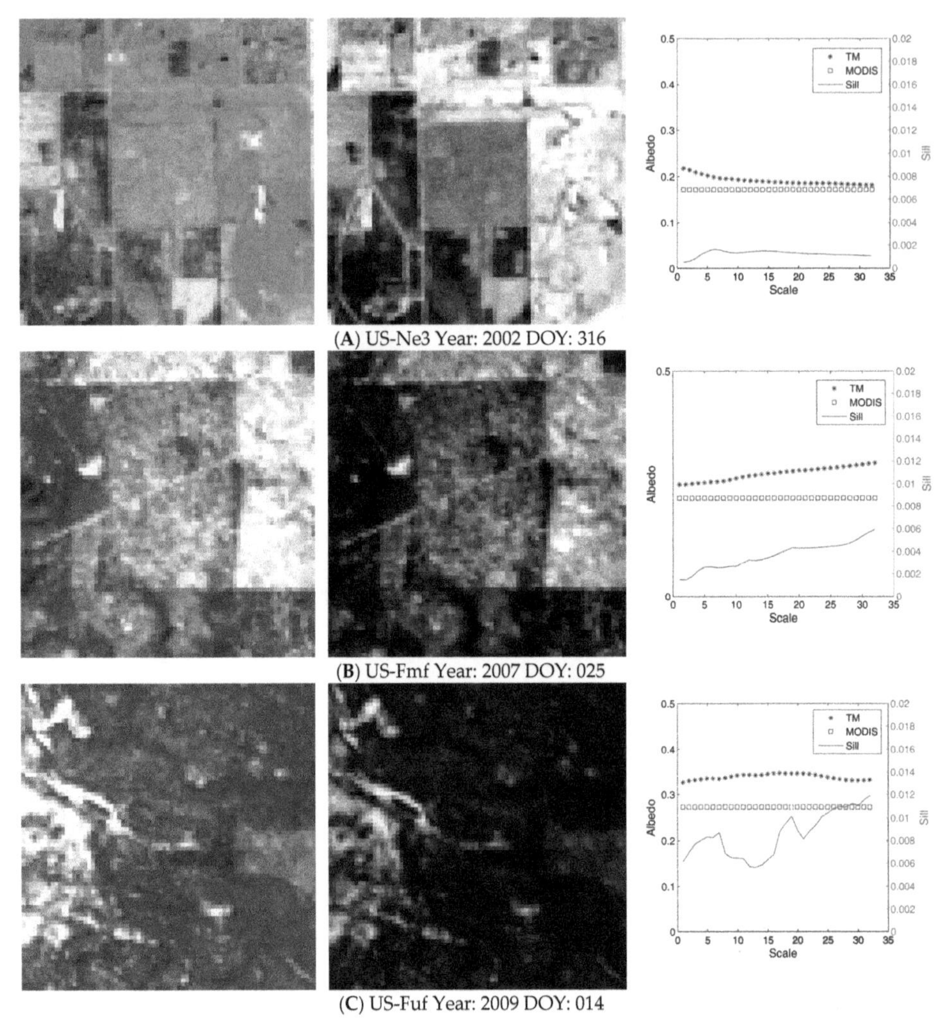

Figure 7. Three heterogeneous land surface types at (**A**) US-Ne3 cropland site, (**B**) US-Fmf evergreen needleleaf forest site and (**C**) CA-Fuf evergreen needleleaf forest site. For each site, the left figure is a false color composition of TM data, the middle figure is albedo from TM data, the right figure shows the aggregated TM albedo values at each scale, MODIS albedo, and sill value from scales 1 (15 × 15 TM pixels) to 32 (77 × 77 TM pixels) at a step of 2 TM pixels.

6. Conclusions

Land surface albedo is an important component of surface energy budgets. The validation of the satellite product is a precondition for its scientific use. Prior to evaluating the remote sensing products, identifying the spatial representativeness of the products is essential. It can help ground sampling point settings and match remote sensing data from multiple sources but it is difficult to determine the effective spatial representativeness of satellite albedo product from a physical perspective, as multi-temporal data are composed to derive the product.

In this study, we evaluated the spatial representativeness of MODIS daily albedo product (MCD43A3) at AmeriFlux sites. Among 1820 paired high-resolution (TM) and coarse-resolution (MODIS) albedo over different land cover types used in this study, around 74.5% pixels were found to be heterogeneous pixels. In order to derive the most reliable spatial representativeness of MODIS albedo product, the land surface heterogeneity was first assessed by the field-calibrated TM albedo; semivarioagrams were then calculated from 30 m Landsat data at different spatial scales. Sill value and relative coefficient of variation were employed as key indices to determine the land surface heterogeneity. The 30 m Landsat albedo data was aggregated to 450 m–1800 m using direct average method and convolved with PSF method. The aggregated data was then compared with MODIS albedo product. The spatial representativeness of MODIS albedo product was determined according to the surface heterogeneity and the consistency of MODIS data and the aggregated TM value.

The results indicate that for most MODIS pixels their spatial representativeness tend to be larger than the 500 m nominal resolution. More specifically, for evergreen broadleaf forests, deciduous broadleaf forests, open shrublands, woody savannas, and grasslands, the effective spatial representativeness of the MODIS albedo was about 630 m; for mixed forest and croplands, the effective spatial resolution was about 690 m. The accuracy of the MODIS 500 m albedo product was high, with a correlation coefficient of 0.94 and RMSE 0.024 when compared with the calibrated TM albedo estimates. The choice of spatial aggregation method between simple spatial averaging and PSF-weighted averaging did not result in any significant difference in determining the spatial representativeness of MODIS albedo. It is also found that the spatial representativeness was difficult to determine at the sites where surface heterogeneity was very high (e.g., covered with evergreen needleleaf forest or partial snow).

In this study, long time period and large space data sets are used for spatial representativeness evaluation at 109 AmeriFlux sites with five land cover types when former works mainly focused on a specific research area. The availability of the 30 m Landsat albedo data set makes it possible for the analysis to be carried out at sites with different land cover types. There are many high-level remote sensing products, in this work, we only focus on evaluating the spatial representativeness of MODIS daily albedo product. Similar work is also worth for other products.

Author Contributions: H.Z. and S.L. conceived and designed the experiments. H.Z. performed the experiments and analyzed the data. J.W. and Y.B. helped the experiment and paper writing. T.H. and D.W. helped technology implementation of the albedo data processing. H.Z. wrote the paper. All the authors reviewed and provided valuable comments for the manuscript.

Funding: This research was supported by the National Natural Science Foundation of China under grants 41801242 and 41771379, the Key research and development program of China under grants 2016YFB0501404, 2016YFB0501502, the Chinese 973 Program under grant 2013CB733403.

Conflicts of Interest: The authors declare no conflict of interest.

References

1. Dickinson, R.E. Land surface processes and climate-surface albedos and energy balance. *Adv. Geophys.* **1983**, *25*, 305–353.
2. Bastiaanssen, W.G.M.; Menenti, M.; Feddes, R.A.; Holtslag, A.A.M. A remote sensing surface energy balance algorithm for land (SEBAL). 1. Formulation. *J. Hydrol.* **1998**, *212–213*, 198–212. [CrossRef]
3. Tang, R.; Li, Z.-L.; Tang, B. An application of the Ts–VI triangle method with enhanced edges determination for evapotranspiration estimation from MODIS data in arid and semi-arid regions: Implementation and validation. *Remote Sens. Environ.* **2010**, *114*, 540–551. [CrossRef]
4. Merlin, O. An original interpretation of the wet edge of the surface temperature–albedo space to estimate crop evapotranspiration (SEB-1S), and its validation over an irrigated area in northwestern Mexico. *Hydrol. Earth Syst. Sci.* **2013**, *17*, 3623–3637. [CrossRef]
5. Oaida, C.M.; Xue, Y.; Flanner, M.G.; Skiles, S.M.; De Sales, F.; Painter, T.H. Improving snow albedo processes in WRF/SSiB regional climate model to assess impact of dust and black carbon in snow on surface energy balance and hydrology over western US. *J. Geophys. Res. Atmos.* **2015**, *120*, 3228–3248. [CrossRef]
6. Hawcroft, M.; Collins, M.; Haywood, J.; Jones, A.; Jones, A.; Stephens, G. Southern Ocean albedo, cross-equatorial energy transport and the ITCZ: Global impacts of biases in a coupled model. *Clim. Dyn.* **2016**, *48*, 2279–2295. [CrossRef]

7. Henderson-Sellers, A.; Wilson, M. Surface albedo data for climatic modeling. *Rev. Geophys.* **1983**, *21*, 1743–1778.
8. Sellers, P.; Meeson, B.; Hall, F.; Asrar, G.; Murphy, R.; Schiffer, R.; Bretherton, F.; Dickinson, R.; Ellingson, R.; Field, C. Remote sensing of the land surface for studies of global change: Models—Algorithms—Experiments. *Remote Sens. Environ.* **1995**, *51*, 3–26. [CrossRef]
9. Maignan, F.; Bréon, F.M.; Lacaze, R. Bidirectional reflectance of Earth targets: Evaluation of analytical models using a large set of spaceborne measurements with emphasis on the Hot Spot. *Remote Sens. Environ.* **2004**, *90*, 210–220. [CrossRef]
10. Mira, M.; Weiss, M.; Baret, F.; Courault, D.; Hagolle, O.; Gallego-Elvira, B.; Olioso, A. The MODIS (collection V006) BRDF/albedo product MCD43D: Temporal course evaluated over agricultural landscape. *Remote Sens. Environ.* **2015**, *170*, 216–228. [CrossRef]
11. Rutan, D.; Rose, F.; Roman, M.; Manalo-Smith, N.; Schaaf, C.; Charlock, T. Development and assessment of broadband surface albedo from Clouds and the Earth's Radiant Energy System Clouds and Radiation Swath data product. *J. Geophys. Res. Atmos.* **2009**, *114*. [CrossRef]
12. Rutan, D.; Charlock, T.; Rose, F.; Kato, S.; Zentz, S.; Coleman, L. Global surface albedo from CERES/TERRA surface and atmospheric radiation budget (SARB) data product. In Proceedings of the 12th conference on atmospheric radiation (AMS), Madison, WI, USA, 9–14 July 2006; pp. 11–12.
13. Muller, J.-P. Algorithm theoretical basis document ATBD 1:4: BRDF/ALBEDO RETRIEVAL. *MERIS AlbedoMaps* **2006**, *1*, 1–19.
14. Wang, D.; Liang, S.; He, T.; Yu, Y. Direct estimation of land surface albedo from VIIRS data: Algorithm improvement and preliminary validation. *J. Geophys. Res. Atmos.* **2013**, *118*, 12577–12586. [CrossRef]
15. Zhou, Y.; Wang, D.; Liang, S.; Yu, Y.; He, T. Assessment of the suomi NPP VIIRS land surface albedo data using station measurements and high-resolution albedo maps. *Remote Sens.* **2016**, *8*, 137. [CrossRef]
16. He, T.; Liang, S.; Wang, D.; Shi, Q.; Tao, X. Estimation of High-Resolution Land Surface Shortwave Albedo From AVIRIS Data. *IEEE J. Sel. Top. Appl. Earth Obs. Remote Sens.* **2014**, *7*, 4919–4928. [CrossRef]
17. Schaaf, C.B.; Gao, F.; Strahler, A.H.; Lucht, W.; Li, X.; Tsang, T.; Strugnell, N.C.; Zhang, X.; Jin, Y.; Muller, J.-P. First operational BRDF, albedo nadir reflectance products from MODIS. *Remote Sens. Environ.* **2002**, *83*, 135–148. [CrossRef]
18. Qu, Y.; Liu, Q.; Liang, S.; Wang, L.; Liu, N.; Liu, S. Direct-Estimation Algorithm for Mapping Daily Land-Surface Broadband Albedo From MODIS Data. *IEEE Trans. Geosci. Remote Sens.* **2014**, *52*, 907–919. [CrossRef]
19. Liang, S.; Zhao, X.; Liu, S.; Yuan, W.; Cheng, X.; Xiao, Z.; Zhang, X.; Liu, Q.; Cheng, J.; Tang, H. A long-term Global LAnd Surface Satellite (GLASS) data-set for environmental studies. *Int. J. Digit. Earth* **2013**, *6*, 5–33. [CrossRef]
20. Masson, V.; Champeaux, J.-L.; Chauvin, F.; Meriguet, C.; Lacaze, R. A global database of land surface parameters at 1-km resolution in meteorological and climate models. *J. Clim.* **2003**, *16*, 1261–1282. [CrossRef]
21. Lawrence, P.J.; Chase, T.N. Representing a new MODIS consistent land surface in the Community Land Model (CLM 3.0). *J. Geophys. Res. Biogeosci.* **2007**, *112*. [CrossRef]
22. Tian, Y.; Dickinson, R.; Zhou, L.; Myneni, R.; Friedl, M.; Schaaf, C.; Carroll, M.; Gao, F. Land boundary conditions from MODIS data and consequences for the albedo of a climate model. *Geophys. Res. Lett.* **2004**, *31*, L05504. [CrossRef]
23. Rechid, D.; Raddatz, T.J.; Jacob, D. Parameterization of snow-free land surface albedo as a function of vegetation phenology based on MODIS data and applied in climate modelling. *Theor. Appl. Climatol.* **2009**, *95*, 245–255. [CrossRef]
24. Ran, L.; Pleim, J.; Gilliam, R.; Binkowski, F.S.; Hogrefe, C.; Band, L. Improved meteorology from an updated WRF/CMAQ modeling system with MODIS vegetation and albedo. *J. Geophys. Res. Atmos.* **2016**, *121*, 2393–2415. [CrossRef]
25. Zhou, X.; Matthes, H.; Rinke, A.; Klehmet, K.; Heim, B.; Dorn, W.; Klaus, D.; Dethloff, K.; Rockel, B. Evaluation of arctic land snow cover characteristics, surface albedo, and temperature during the transition seasons from regional climate model simulations and satellite data. *Adv. Meteorol.* **2014**. [CrossRef]
26. Yin, J.; Zhan, X.; Zheng, Y.; Hain, C.R.; Ek, M.; Wen, J.; Fang, L.; Liu, J. Improving Noah land surface model performance using near real time surface albedo and green vegetation fraction. *Agric. For. Meteorol.* **2016**, *218*, 171–183.

27. Liang, S.; Fang, H.; Chen, M.; Shuey, C.J.; Walthall, C.; Daughtry, C.; Morisette, J.; Schaaf, C.; Strahler, A. Validating MODIS land surface reflectance and albedo products: Methods and preliminary results. *Remote Sens. Environ.* **2002**, *83*, 149–162. [CrossRef]
28. Chen, Y.M.; Liang, S.; Wang, J.; Kim, H.Y.; Martonchik, J.V. Validation of MISR land surface broadband albedo. *Int. J. Remote Sens.* **2008**, *29*, 6971–6983. [CrossRef]
29. Liu, J.; Schaaf, C.; Strahler, A.; Jiao, Z.; Shuai, Y.; Zhang, Q.; Roman, M.; Augustine, J.A.; Dutton, E.G. Validation of Moderate Resolution Imaging Spectroradiometer (MODIS) albedo retrieval algorithm: Dependence of albedo on solar zenith angle. *J. Geophys. Res. Atmos.* **2009**, *114*. [CrossRef]
30. Salomon, J.G.; Schaaf, C.B.; Strahler, A.H.; Gao, F.; Jin, Y. Validation of the MODIS bidirectional reflectance distribution function and albedo retrievals using combined observations from the aqua and terra platforms. *IEEE Trans. Geosci. Remote Sens.* **2006**, *44*, 1555–1565. [CrossRef]
31. Wang, K.; Liu, J.; Zhou, X.; Sparrow, M.; Ma, M.; Sun, Z.; Jiang, W. Validation of the MODIS global land surface albedo product using ground measurements in a semidesert region on the Tibetan Plateau. *J. Geophys. Res. Atmos.* **2004**, *109*, D05107. [CrossRef]
32. Jin, Y.; Schaaf, C.B.; Woodcock, C.E.; Gao, F.; Li, X.; Strahler, A.H.; Lucht, W.; Liang, S. Consistency of MODIS surface bidirectional reflectance distribution function and albedo retrievals: 2. Validation. *J. Geophys. Res. Atmos.* **2003**, *108*. [CrossRef]
33. Klein, A.G.; Stroeve, J. Development and validation of a snow albedo algorithm for the MODIS instrument. *Ann. Glaciol.* **2002**, *34*, 45–52. [CrossRef]
34. Cescatti, A.; Marcolla, B.; Santhana Vannan, S.K.; Pan, J.Y.; Román, M.O.; Yang, X.; Ciais, P.; Cook, R.B.; Law, B.E.; Matteucci, G.; et al. Intercomparison of MODIS albedo retrievals and in situ measurements across the global FLUXNET network. *Remote Sens. Environ.* **2012**, *121*, 323–334. [CrossRef]
35. Fang, H.; Liang, S.; Chen, M.; Walthall, C.; Daughtry, C. Statistical comparison of MISR, ETM+ and MODIS land surface reflectance and albedo products of the BARC land validation core site, USA. *Int. J. Remote Sens.* **2004**, *25*, 409–422. [CrossRef]
36. Susaki, J.; Yasuoka, Y.; Kajiwara, K.; Honda, Y.; Hara, K. Validation of MODIS albedo products of paddy fields in Japan. *IEEE Trans. Geosci. Remote Sens.* **2007**, *45*, 206–217. [CrossRef]
37. Burakowski, E.A.; Ollinger, S.V.; Lepine, L.; Schaaf, C.B.; Wang, Z.; Dibb, J.E.; Hollinger, D.Y.; Kim, J.; Erb, A.; Martin, M. Spatial scaling of reflectance and surface albedo over a mixed-use, temperate forest landscape during snow-covered periods. *Remote Sens. Environ.* **2015**, *158*, 465–477. [CrossRef]
38. Tan, B.; Woodcock, C.E.; Hu, J.; Zhang, P.; Ozdogan, M.; Huang, D.; Yang, W.; Knyazikhin, Y.; Myneni, R.B. The impact of gridding artifacts on the local spatial properties of MODIS data: Implications for validation, compositing, and band-to-band registration across resolutions. *Remote Sens. Environ.* **2006**, *105*, 98–114. [CrossRef]
39. Wang, Z.; Schaaf, C.B.; Strahler, A.H.; Chopping, M.J.; Román, M.O.; Shuai, Y.; Woodcock, C.E.; Hollinger, D.Y.; Fitzjarrald, D.R. Evaluation of MODIS albedo product (MCD43A) over grassland, agriculture and forest surface types during dormant and snow-covered periods. *Remote Sens. Environ.* **2014**, *140*, 60–77. [CrossRef]
40. Campagnolo, M.L.; Montano, E.L. Estimation of effective resolution for daily MODIS gridded surface reflectance products. *IEEE Trans. Geosci. Remote Sens.* **2014**, *52*, 5622–5632. [CrossRef]
41. Campagnolo, M.L.; Sun, Q.; Liu, Y.; Schaaf, C.; Wang, Z.; Román, M.O. Estimating the effective spatial resolution of the operational BRDF, albedo, and nadir reflectance products from MODIS and VIIRS. *Remote Sens. Environ.* **2016**, *175*, 52–64. [CrossRef]
42. Shuai, Y. Tracking Daily Land Surface Albedo and Reflectance Anisotropy with MODerate-Resolution Imaging Spectroradiometer (MODIS). Ph.D. Thesis, Boston University, Boston, MA, USA, 2010.
43. Shuai, Y.; Masek, J.G.; Gao, F.; Schaaf, C.B. An algorithm for the retrieval of 30-m snow-free albedo from Landsat surface reflectance and MODIS BRDF. *Remote Sens. Environ.* **2011**, *115*, 2204–2216. [CrossRef]
44. He, T.; Liang, S.; Wang, D.; Chen, X.; Song, D.X.; Jiang, B. Land surface albedo estimation from Chinese HJ satellite data based on the direct estimation approach. *Remote Sens.* **2015**, *7*, 5495–5510. [CrossRef]
45. He, T.; Liang, S.; Wang, D.; Cao, Y.; Gao, F.; Yu, Y.; Feng, M. Evaluating land surface albedo estimation from Landsat MSS, TM, ETM+, and OLI data based on the unified direct estimation approach. *Remote Sens. Environ.* **2018**, *204*, 181–196. [CrossRef]

46. Baldocchi, D.D. Assessing the eddy covariance technique for evaluating carbon dioxide exchange rates of ecosystems: Past, present and future. *Glob. Chang. Biol.* **2003**, *9*, 479–492. [CrossRef]
47. Román, M.O.; Schaaf, C.B.; Woodcock, C.E.; Strahler, A.H.; Yang, X.; Braswell, R.H.; Curtis, P.S.; Davis, K.J.; Dragoni, D.; Goulden, M.L.; et al. The MODIS (Collection V005) BRDF/albedo product: Assessment of spatial representativeness over forested landscapes. *Remote Sens. Environ.* **2009**, *113*, 2476–2498. [CrossRef]
48. Yuan, F.; Sawaya, K.E.; Loeffelholz, B.C.; Bauer, M.E. Land cover classification and change analysis of the Twin Cities (Minnesota) Metropolitan Area by multitemporal Landsat remote sensing. *Remote Sens. Environ.* **2005**, *98*, 317–328. [CrossRef]
49. Song, C.; Woodcock, C.E.; Seto, K.C.; Lenney, M.P.; Macomber, S.A. Classification and change detection using Landsat TM data: When and how to correct atmospheric effects? *Remote Sens. Environ.* **2001**, *75*, 230–244. [CrossRef]
50. Collins, J.B.; Woodcock, C.E. An assessment of several linear change detection techniques for mapping forest mortality using multitemporal Landsat TM data. *Remote Sens. Environ.* **1996**, *56*, 66–77. [CrossRef]
51. Hansen, M.C.; Roy, D.P.; Lindquist, E.; Adusei, B.; Justice, C.O.; Altstatt, A. A method for integrating MODIS and Landsat data for systematic monitoring of forest cover and change in the Congo Basin. *Remote Sens. Environ.* **2008**, *112*, 2495–2513.
52. Zhu, Z.; Fu, Y.; Woodcock, C.E.; Olofsson, P.; Vogelmann, J.E.; Holden, C.; Wang, M.; Dai, S.; Yu, Y. Including land cover change in analysis of greenness trends using all available Landsat 5, 7, and 8 images: A case study from Guangzhou, China (2000–2014). *Remote Sens. Environ.* **2016**, *185*, 243–257.
53. Wang, Z.; Erb, A.M.; Schaaf, C.B.; Sun, Q.; Liu, Y.; Yang, Y.; Shuai, Y.; Casey, K.A.; Román, M.O. Early spring post-fire snow albedo dynamics in high latitude boreal forests using Landsat-8 OLI data. *Remote Sens. Environ.* **2016**, *185*, 71–83. [PubMed]
54. Wang, Z.; Schaaf, C.B.; Chopping, M.J.; Strahler, A.H.; Wang, J.; Román, M.O.; Rocha, A.V.; Woodcock, C.E.; Shuai, Y. Evaluation of Moderate-resolution Imaging Spectroradiometer (MODIS) snow albedo product (MCD43A) over tundra. *Remote Sens. Environ.* **2012**, *117*, 264–280. [CrossRef]
55. Matheron, G. Principles of geostatistics. *Econ. Geol.* **1963**, *58*, 1246–1266.
56. Noreus, J.; Nyborg, M.; Hayling, K. The gravity anomaly field in the Gulf of Bothnia spatially characterized from satellite altimetry and in situ measurements. *J. Appl. Geophys.* **1997**, *37*, 67–84.
57. Román, M.O.; Schaaf, C.B.; Lewis, P.; Gao, F.; Anderson, G.P.; Privette, J.L.; Strahler, A.H.; Woodcock, C.E.; Barnsley, M. Assessing the coupling between surface albedo derived from MODIS and the fraction of diffuse skylight over spatially-characterized landscapes. *Remote Sens. Environ.* **2010**, *114*, 738–760. [CrossRef]
58. Peng, J.; Liu, Q.; Wen, J.; Liu, Q.; Tang, Y.; Wang, L.; Dou, B.; You, D.; Sun, C.; Zhao, X.; et al. Multi-scale validation strategy for satellite albedo products and its uncertainty analysis. *Sci. China Earth Sci.* **2015**, *58*, 573–588. [CrossRef]

Article

Improving the AVHRR Long Term Data Record BRDF Correction

Jose Luis Villaescusa-Nadal [1,2,*], Belen Franch [1,2], Eric F. Vermote [2] and Jean-Claude Roger [1,2]

[1] NASA Goddard Space Flight Center 8800 Greenbelt Rd, Greenbelt, MD 20771, USA; belen.franchgras@nasa.gov (B.F.); jean-claude.roger@nasa.gov (J.-C.R.)

[2] Department of Geographical Sciences, University of Maryland College Park, 2181 LeFrak Hall, College Park, MD 20740, USA; eric.f.vermote@nasa.gov

* Correspondence: joseluis.villaescusanadal@nasa.gov

Received: 24 January 2019; Accepted: 25 February 2019; Published: 1 March 2019

Abstract: The Long Term Data Record (LTDR) project has the goal of developing a quality and consistent surface reflectance product from coarse resolution optical sensors. This paper focuses on the Advanced Very High Resolution Radiometer (AVHRR) part of the record, using the Moderate Resolution Imaging Spectrometer (MODIS) instrument as a reference. When a surface reflectance time series is acquired from satellites with variable observation geometry, the directional variation generates an apparent noise which can be corrected by modeling the bidirectional reflectance distribution function (BRDF). The VJB (Vermote, Justice and Bréon, 2009) method estimates a target's BRDF shape using 5 years of observation and corrects for directional effects maintaining the high temporal resolution of the measurement using the instantaneous Normalized Difference Vegetation Index (NDVI). The method was originally established on MODIS data but its viability and optimization for AVHRR data have not been fully explored. In this study we analyze different approaches to find the most robust way of applying the VJB correction to AVHRR data, considering that high noise in the red band (B1) caused by atmospheric effect makes the VJB method unstable. Firstly, our results show that for coarse spatial resolution, where the vegetation dynamics of the target don't change significantly, deriving BRDF parameters from 15+ years of observations reduces the average noise by up to 7% in the Near Infrared (NIR) band and 6% in the NDVI, in comparison to using 3-year windows. Secondly, we find that the VJB method can be modified for AVHRR data to improve the robustness of the correction parameters and decrease the noise by an extra 8% and 9% in the red and NIR bands with respect to using the classical VJB inversion. We do this by using the Stable method, which obtains the volumetric BRDF parameter (V) based on its NDVI dependency, and then obtains the geometric BRDF parameter (R) through the inversion of just one parameter.

Keywords: AVHRR; BRDF; MODIS; VJB; LTDR; directional correction

---

1. Introduction

The Advanced Very High-Resolution Radiometer (AVHRR) sensor provides a unique global remote sensing dataset that ranges from the 1980s to the present. Among the different products delivered from this sensor, NASA is currently funding the Long Term Data Record (LTDR) project [1] to develop a quality and consistent Climate Data Record (CDR) of AVHRR data with the use of the Moderate Resolution Imaging Spectrometer (MODIS) instrument as a reference. This data record creates daily global surface reflectance products with a geographic projection at coarse spatial resolution (0.05°). The utility of this long time series has been demonstrated in the literature for a large number of applications such as agriculture [2], burned area mapping [3], Leaf Area Index (LAI) and Fraction of Absorbed Photosynthetically Active Radiation (FAPAR) retrieval [4,5], snow cover estimation [6], global vegetation monitoring [7], and surface albedo estimation products [8–11].

Long-term consistent data records are becoming crucial to provide improved detection, attribution and prediction of global climate and environmental changes, as well as helping decision makers respond and adapt to climate change and other variability in advance [12–14]. The knowledge of surface albedo is of critical importance to monitor land surface processes and plays an important role in energy-budget considerations within climate and weather-prediction models. For this reason, it has been listed as an Essential Climate Variable by the Global Climate Observing System (GCOS) [13,15]. Surface albedo can be derived from satellite data. The most common procedure consists on integrating a BRDF angular model, which can explain the reflectance's anisotropic behavior on different types of surfaces to obtain the black-sky (direct beam) and white-sky (completely diffuse) narrowband albedo. One can then perform narrow-to-broadband conversions to obtain the respective broadband albedos [16] and obtain the actual (blue-sky) albedo by doing a weighted average, using the fraction of diffuse skylight [17].

For this product to be of highest quality, it is critical that the surface reflectance data record has minimal uncertainties in the calibration, geo-location, spectral correction, cloud masking, atmospheric correction and directional effects correction. Issues regarding calibration, cloud masking and atmospheric correction in the AVHRR data have been accurately corrected, after the year 2000, when MODIS data were used as a reference [2]. However, some issues persist for data before this year (1982–2000), such as aerosol and water vapor correction and calibration [1]. These errors propagate resulting in surface reflectance time series with high noise. This is especially true in the red band, where the atmospheric errors are higher, compared to the Near Infrared (NIR) band. Therefore, the BRDF parameters derived from these time series have high uncertainties and might not provide an accurate correction of the directional effects.

To address this issue, the LTDR product uses MODIS retrieved parameters using the Vermote, Justice and Bréon (VJB) method [18,19] and then applies them to AVHRR data. The VJB method uses a pixel's time series (typically 5 years) to compute BRDF parameters (V, R) in a daily manner. This is done with the use of the instantaneous Normalized Difference Vegetation Index (NDVI), which quantifies the vegetation by measuring the difference between NIR and red sunlight. These parameters can later be used to correct the database. This method is based on the assumptions that 1) the target reflectance changes during the year but BRDF shape variations are limited and 2) the difference in surface reflectance between successive acquisitions is mostly explained by directional effects.

Assumption 1 holds true while retrieving the correction parameters with short enough periods, or in other words, if the surface doesn't change significantly over this period. When dealing with AVHRR data, due to the lower number of high-quality observations, a larger number of years is required to make the computation of the BRDF parameters reliable, in which the surface is subject to change. One could argue, however, that because the product is at 0.05° spatial resolution, the stability of the pixel through the years is more likely to be maintained. Assumption 2 holds true for MODIS data because the evolution of the view zenith angle during the year is not gradual, so the difference between successive observations is high, and atmospheric errors are low. This means that the time series have high directional effects, which are higher than the atmospheric correction errors. This is not the case for AVHRR data; the difference of view zenith angle between successive observations is low and atmospheric errors are high, especially for the red band. These facts could justify the use of MODIS parameters to correct AVHRR data, however, these also experience propagated uncertainties caused by the different spectral response of the two sensors or calibration issues of AVHRR.

In this paper we explore different approaches to the AVHRR directional effects correction using the VJB method, to optimize it to AVHRR data. Firstly, we find relationships between BRDF correction parameters using different bands and band combinations derived from MODIS data. With this, we aim to minimize the propagation of atmospheric errors to the correction of directional effects. Secondly, we calculate said parameters using 3 years and the whole time series (15+ years). To test the best method, we compare the noise of the normalized surface reflectance to check which one shows the lowest noise in the time series. Section 2 describes the materials used for the study and the methodology employed.

Section 3 presents the results. Section 4 presents a discussion of the results and Section 5 presents the conclusion.

2. Materials and Methods

2.1. Materials

For this study, we downloaded MODIS data from the MOD09 product [20,21] from 2000–2015 and AVHRR from the LTDR product [22]. We divided the data into AVHRR-pre MODIS (1982-2000) and AVHRR (2000–2015). These data contain surface reflectance at Climate Modelling Grid (CMG) spatial resolution (0.05°). We use the red and NIR surface reflectance bands, as well as the view zenith (θv), solar zenith (θs), and relative azimuth (ϕ) angles, selecting pixels of the highest quality. We extracted the data for individual pixels in the 445 Benchmark Land Multisite Analysis and Intercomparison of Products (BELMANIP2) sites. BELMANIP2 sites are an update of BELMANIP1 [23] and were selected due to their representativeness and variability of vegetation types and climatological conditions around the world. Moreover, they were selected so that the sites are homogeneous over a 10×10 km^2 area, so these sites are homogeneous at CMG resolution (0.05° or 5.6 km). We obtained the NDVI from the red and NIR bands from AVHRR-pre, AVHRR and MODIS data. The product includes no spectral adjustment method, which is necessary considering that the time series is composed of data from several different National Oceanic and Atmospheric Administration (NOAA) satellites. Moreover, an accurate intercomparison with MODIS requires that all data be on the same radiometric scale. For this reason, we perform spectral adjustment using methods from Reference [24], selecting NOAA14 as a reference.

2.2. Methods

An outline of the methodology is shown in Figure 1 We first use MODIS data to obtain relationships between the BRDF parameters, so we can use them to build different models with AVHRR data. They are then applied to MODIS and AVHRR data, using parameters derived every 3 years, and for the full time series. With these, we derive BRDF correction parameters, which are used to calculate the normalized time series using Equation (9). The normalized NDVI is obtained from the normalized red and NIR reflectances. Finally, given the lack of BRDF field measurements that can be used as a reference to evaluate the best method, we derive the noise of the time series before correction and after the model inversions. This allows us to compare the noise improvement depending on the different BRDF model inversion used, and the number of years employed. The noise is calculated using the statistical difference between the center measurement of three successive triplets and the linear interpolation between the two extremes [18]:

$$Noise(y) = \sqrt{\frac{\sum_{i=1}^{n-2}\left(y_{i+1} - \frac{y_{i+2}-y_i}{day_{i+2}-day_i}(day_{i+1} - day_i) - y_i\right)^2}{N-2}}. \quad (1)$$

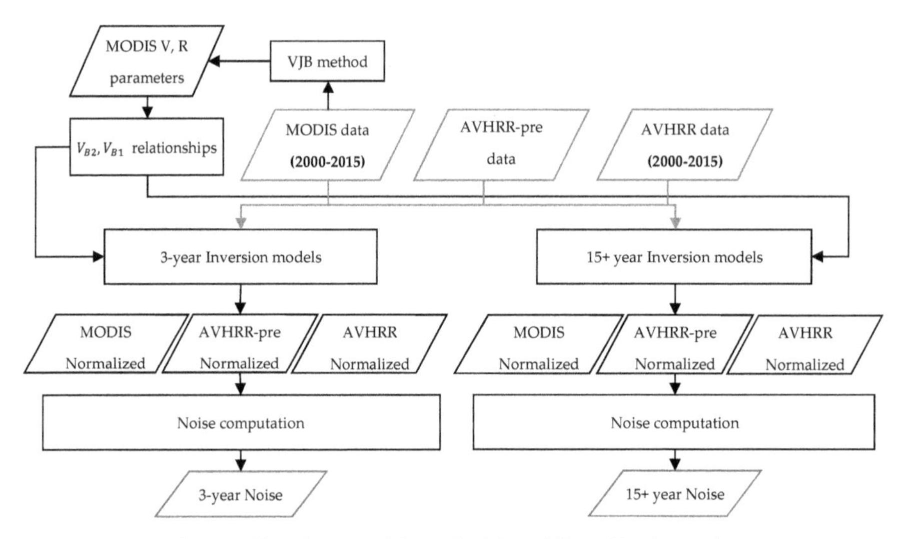

Figure 1. Flow diagram of the methodology followed by this study.

2.2.1. VJB Method

The directional effects correction is done using the VJB method, which is based on the Ross Thick Li-Sparse Reciprocal (RTLSR) BRDF model [25,26]. This model expresses the surface reflectance ρ as:

$$\rho(\theta_v, \theta_s, \phi, t) = k_{iso}(t)\left[1 + \boldsymbol{R}F_{geo}(\theta_v, \theta_s, \phi) + \boldsymbol{V}F_{vol}(\theta_v, \theta_s, \phi)\right] \tag{2}$$

where, (F_{geo}, F_{vol}) are the geometric and volumetric scattering components that provide the basic BRDF shapes for characterizing the heterogeneous scattering of the soil-vegetation system; $(k_{iso}, k_{geo}, k_{vol})$ are the weight coefficients of the isotropic, geometric and volumetric kernel functions; $\mathbf{V} = k_{vol}/k_{iso}$, $\mathbf{R} = k_{geo}/k_{iso}$.

It assumes that the difference between consecutive observations is due principally to directional effects, with small variations of the isotropic weight coefficient k_{iso}.

$$\frac{\rho(t_i)}{1 + \boldsymbol{V}F^{i}_{vol} + \boldsymbol{R}F^{i}_{geo}} \approx \frac{\rho(t_{i+1})}{1 + \boldsymbol{V}F^{i+1}_{vol} + \boldsymbol{R}F^{i+1}_{geo}}. \tag{3}$$

The values of V and R are the unknowns and can be derived using N observations from a certain pixel by minimizing the merit function:

$$M = \sum_{i=1}^{N-1} \frac{\left(\rho_{i+1}\left[1 + VF^{i}_{1} + RF^{i}_{2}\right] - \rho_i\left[1 + VF^{i+1}_{1} + RF^{i+1}_{2}\right]\right)^2}{day^{i+1} - day^{i} + 1}. \tag{4}$$

The minimization is done by the classic derivation:

$$\frac{dM}{dV} = \frac{dM}{dR} = 0. \tag{5}$$

This leads to the system of equations:

$$\begin{pmatrix} \sum\limits_{i=1}^{N-1} \Delta^i\rho F_1 \Delta^i\rho F_1 & \sum\limits_{i=1}^{N-1} \Delta^i\rho F_1 \Delta^i\rho F_2 \\ \sum\limits_{i=1}^{N-1} \Delta^i\rho F_1 \Delta^i\rho F_2 & \sum\limits_{i=1}^{N-1} \Delta^i\rho F_2 \Delta^i\rho F_2 \end{pmatrix} \otimes \begin{pmatrix} V \\ R \end{pmatrix} = \begin{pmatrix} -\sum\limits_{i=1}^{N-1} \Delta^i\rho \Delta^i\rho F_1 \\ -\sum\limits_{i=1}^{N-1} \Delta^i\rho \Delta^i\rho F_2 \end{pmatrix} \tag{6}$$

where:

$$
\begin{aligned}
\Delta^i d &= day_{i+1} - day_i + 1 \\
\Delta^i \rho &= (\rho_{i+1} - \rho_i) / \sqrt{\Delta^i d} \\
\Delta^i \rho F_{1,2} &= \left(\rho_{i+1} F^i_{1,2} - \rho_{i+1} F^{i+1}_{1,2}\right) / \sqrt{\Delta^i d}\,.
\end{aligned} \tag{7}
$$

Through the inversion of Equation (6), we can now obtain a V and R parameter for every pixel. To perform the instantaneous directional correction for every observation, V and R parameters are needed for every date, so the inversion of the V and R parameters is done for five different NDVI populations. A linear regression is performed for V and R as a function of the NDVI (Equation (8)). This retrieves a slope and intercept for every pixel which allow the computation of V and R parameters for a certain date given the surface's NDVI value.

$$
\begin{aligned}
V &= V_0 + V_1 * NDVI \\
R &= R_0 + R_1 * NDVI.
\end{aligned} \tag{8}
$$

These V and R can now be used to calculate the normalized surface reflectance (ρ^N) at θs = 45°, and nadir observation using Equation (9) [18]:

$$
\rho^N(45,0,0) = \rho(\theta_s, \theta_v, \phi) * \frac{1 + VF_1(45,0,0) + RF_2(45,0,0)}{1 + VF_1(\theta_s, \theta_v, \phi) + RF_2(\theta_s, \theta_v, \phi)} \tag{9}
$$

The NDVI can now be obtained using the normalized reflectances:

$$
NDVI^N = \frac{\rho^N_{NIR} - \rho^N_{red}}{\rho^N_{NIR} + \rho^N_{red}}\,. \tag{10}
$$

2.2.2. BRDF Parameters Relationship

To minimize the propagation of the atmospheric errors into the BRDF correction, we analyze the V and R parameters for bands 1 (red) and 2 (NIR). The goal is to find a physical relationship between the BRDF parameters of both bands to avoid using the noisy red AVHRR band. Therefore, we can derive the BRDF parameters of the NIR band and then estimate the red band based on these parameters. To build this physical relationship we use MODIS data since we need data with the least error possible.

To do this, we extract the band 1, band 2 and NDVI time series of one pixel. We then sort them in ascending values of NDVI. We divide these values into five groups, using the 20th, 40th, 60th and 80th percentiles as group edge values. Each of these groups is defined as an NDVI population with its own average. We then extract the V and R values of each one using their band 1, band 2 and NDVI. This means that for every pixel and every band, we obtain 5 different V and R values. When using all the BELMANIP2 sites we obtain a total of 445 × 5 = 2225 points for each band. Finally, we do simple regressions between the obtained parameters for the different bands and with the NDVI itself.

2.2.3. Inversion Period

The idea of this section is to analyze the effect of using a short versus a long period in the computation of the BRDF parameters. For the short period, we use 3 years to make sure the AVHRR time series has enough observations to allow a reliable estimate of parameters. We divided the data from (1982–1999) into three year intervals 1982–1984, 1985–1987, 1988–1990, 1991–1993, 1994–1996, and 1997–1999, and the data from (2000–2015) in 2000–2002, 2003–2005, 2006–2008, 2009–2011, and 2012–2015. For the long period, we use 18 years for AVHRR-pre and 15 years for MODIS and AVHRR. We compute the BRDF parameters using the VJB inversion and the original sensor's data. In other words, we use MODIS data to calculate MODIS BRDF parameters and AVHRR data to calculate AVHRR BRDF parameters.

2.2.4. Inversion Models

In this section, we compare the different inversion models whose choice is based on different possible hypotheses. The aim is to see which of these are valid for the different data employed. The inversion models used in this paper are the following:

MODIS

We calculate the V and R parameters using the VJB method directly from MODIS data. We hypothesize that calculating BRDF parameters using a time series with small atmospheric and calibration errors provide the best correction. These errors are normally propagated to the correction of directional effects, decreasing the uncertainty of the correction parameters.

AVHRR

Analogous to the MODIS approach, but using AVHRR data. Here we hypothesize that calculating V and R from the sensor we are attempting to correct is more representative than using a different sensor, even if said sensor has less noise in the time series. In other words, we theorize that propagated uncertainties from the atmospheric correction are smaller than data harmonization uncertainties and problems in the original model's assumption.

Average

Here we make the broad assumption that given the high noise in the AVHRR time series, average V and R parameters which directly depend on the NDVI can be applied. The correction parameters used are derived from Bréon and Vermote [27] and are shown in Table 1.

Table 1. V(NDVI) and R(NDVI) from [27].

Band 1 (red)	V = NDVI + 0.50	R = 0.20*NDVI + 0.10
Band 2 (NIR)	V = 2.00*NDVI + 0.50	R = −0.05*NDVI + 0.15

B1(B2)

Given that the red AVHRR band (B1) is noisy due to the higher errors in the atmospheric correction, we attempt to correct the red band using the parameters from the NIR band. The relationships between them are estimated using MODIS data, which has a robust atmospheric correction and is now a well-established product [28]. We derive the NIR parameters using normal VJB inversion, so one value of V and R for each of the five NDVI populations in each pixel. Then, we apply Equation (10) and continue with the regular VJB method procedure (dependence with NDVI). These relationships are derived in the results section, but are:

$$V_{B1} = 0.82 * V_{B2} + 0.11$$

$$R_{B1} = 0.91 * R_{B2} + 0.04. \quad (11)$$

Stable

In this inversion, we hypothesize that the instability of AVHRR BRDF parameters is due to performing a matrix inversion using two parameters. This provides occasionally very unstable correction parameters when using noisy data. For this reason, we calculate the V_{B1} parameter using the NDVI, and the V_{B2} parameter from the V_{B1}. We finally solve R for each band from the second equation of Equation (6). Again, this is done for each of the 5 NDVI populations within every pixel, after which a linear regression with the NDVI is performed.

$$V_{B1} = 1.42 * NDVI + 0.63$$

$$V_{B2} = 0.83 * V_{B1} + 0.30. \quad (12)$$

3. Results

3.1. BRDF Parameters Relationship

Figure 2 shows the relationship obtained between V_{B1} and V_{B2} with the NDVI and between V and R of bands 1 and 2. The r^2, Root Mean Square Error (RMSE) and linear fit (red line) regression values are shown in the top left of each subplot. The results show that there is a general dependence of the V parameter with the NDVI. This result was expected considering that the V parameter models the volumetric component of the vegetation. A high V value means a denser vegetation, a higher biomass, and effectively a higher NDVI value. However, the high RMSE values both for band 1 and band 2 (0.42,0.45), suggest that this is not a very precise approximation. In the case of the inter-band relationships, the results show that there is a high correlation ($r^2 > 0.8$) between the parameters derived from bands 1 and 2. The small RMSE values (0.24,0.04) indicate that this approximation is reliable and could provide a smaller error than that derived by the atmospheric effects propagation from AVHRR or the spectral adjustment and calibration errors from MODIS. The cluster of points that show a 1:1 relation belongs to points with a small NDVI (NDVI < 0.2). Figure 3 shows the relationship between the Band 1 and Band 2 parameters for low NDVI values. It has been shown in previous studies that for low vegetation amount, the R_{B1} and R_{B2} values are almost identical [25,29]. We also noticed the same behavior for the V parameters, when the NDVI < 0.1. In this study, we also computed the relationship of R with the NDVI, but there was little or no correlation. This is expected considering that the R parameter is associated with the geometric component, and higher NDVI values such as for forests show a similar value to bare ground.

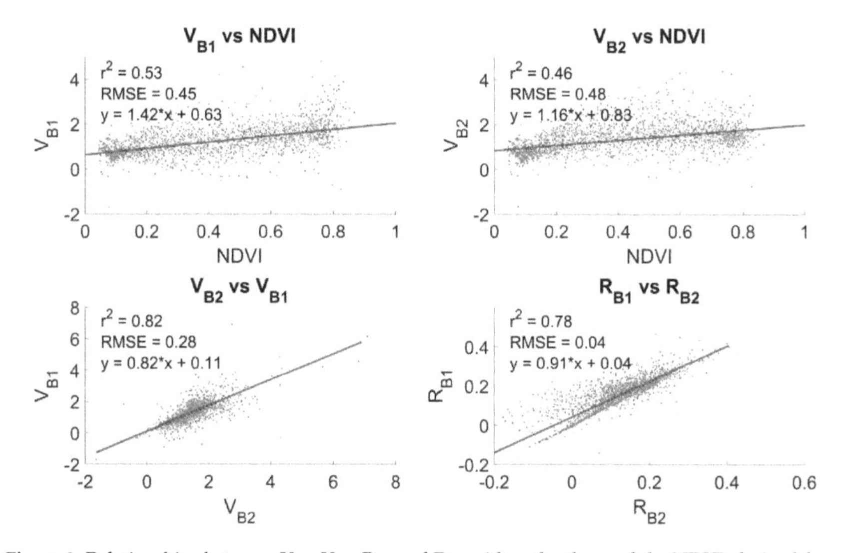

Figure 2. Relationships between V_{B1}, V_{B2}, R_{B1} and R_{B2} with each other and the NDVI, derived from the Moderate Resolution Imaging Spectrometer (MODIS).

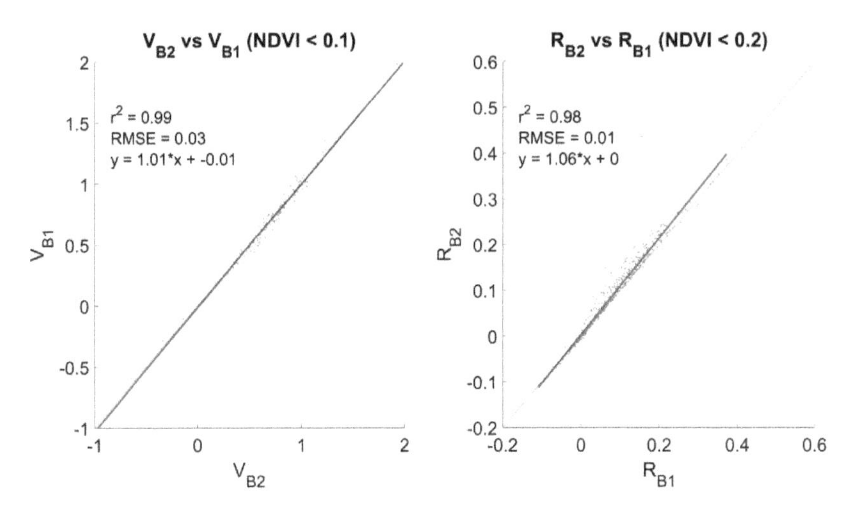

Figure 3. Regression between V_{B1}, V_{B2}, R_{B1} and R_{B2} for low NDVI values. The red line represents the linear regression fit. The blue line shows the 1:1 line.

3.2. Inversion Period

Table 2 shows the average absolute noise ($\times 10^4$) of the BELMANIP sites' time series obtained when the full dataset and 3-year intervals are used to derive the correction parameters for the red, NIR and NDVI. Inversion using full and 3-year parameters is shown in green and brown font respectively. The percentage under every noise value indicates the improvement with respect to the raw data. For almost every band and dataset, the average noise of the time series using full parameters is lower than using 3-year parameters. In MODIS data, the difference between the full and 3-year parameters is of ~2.5%, 8% and 5% in the red, NIR and NDVI time series noise. This difference is lower for AVHRR-pre (~0%, 7% and 6%) and AVHRR (~2%, 5% and 1%) data.

Table 2. Average noise ($\times 10^4$) of the Benchmark Land Multisite Analysis and Intercomparison Products (BELMANIP2) sites' time series obtained before (raw) and after directional effects for the red and Near Infrared bands and the NDVI, using MODIS, the Advanced Very High Resolution Radiometer (AVHRR) pre-MODIS data and AVHRR data from top to bottom. Full and 3-year columns describe the noise after computing the Bidirectional Reflectance Distribution Function (BRDF) parameters using the whole time series, or 3-year intervals, respectively. The percentage under every noise value indicates the improvement with respect to the raw data.

		Red		NIR		NDVI	
		Full	3-Year	Full	3-Year	Full	3-Year
MODIS	Raw	**201.42**	**201.42**	**430.45**	**430.45**	**162.21**	**162.21**
	Corrected	**48.38** (75.98%)	**53.44** (73.47%)	**76.23** (82.29%)	**107.80** (74.96%)	**113.20** (30.22%)	**121.19** (25.29%)
AVHRR pre	Raw	**212.97**	**212.97**	**456.51**	**456.51**	**379.83**	**379.83**
	Corrected	**130.46** (38.74%)	**130.16** (38.88%)	**198.10** (56.61%)	**230.63** (49.48%)	**363.58** (4.28%)	**389.08** (−2.43%)
AVHRR	Raw	**168.22**	**168.22**	**292.62**	**292.62**	**317.32**	**317.32**
	Corrected	**103.73** (38.34%)	**106.58** (36.64%)	**144.02** (50.78%)	**158.68** (45.77%)	**315.33** (0.63%)	**319.48** (−0.68%)

Effectively, these results show that the effect of the gradual surface change in coarse spatial resolution pixels has little impact on the noise of the time series, as compared with the effect of the number of observations used to compute the parameters. Using a larger number of years retrieves information about the surface which is used in the model to retrieve more accurate parameters and reduce the noise of the normalized time series by a significant amount, especially for the NIR and NDVI.

3.3. Inversion Models

Figure 4 shows the time series of two BELMANIP pixels for different inversion models and their noise value using AVHRR-pre-data. In brackets is shown the relative noise of the time series. For visual purposes, the data is shifted in the y-axis by $0.3 \times n$ in the red and $0.6 \times n$ in the NIR and NDVI for the nth model. The first one is a savanna pixel located in Brazil (−14.72, −41.75). The red and NIR band show a significant improvement in the noise after the directional effects' correction of ~60% and ~80% respectively, even when using the average method, which is based on the broadest approximation. The result can be appreciated visually, where the seasonal variation of the data can be distinguished after the correction. In the case of the NDVI, however, there is little or no improvement in the noise. When using MODIS parameters, for example, the noise increases. This is due to the intrinsic directional effects' correction of the NDVI computation [18]. The second pixel is a bare ground pixel located in the Algerian Saharan Desert (28.28, −5.03). In this case, there is also a significant decrease in the red and NIR noise when using the different inversion models, but not in the NDVI. The average method, however, significantly increases the noise, evidencing that the approximation it uses might not be viable for non-vegetated sites. These results also show that the assumption that for low NDVI values, the parameters are the same for both bands is reasonable.

To analyze the performance of every individual model in detail, we plotted the distribution of the noise corrections for every pixel considered in this study. Figure 5 shows these distributions in the form of a boxplot, using the different inversion models and for MODIS, AVHRR-pre and AVHRR data. The top and bottom blue edges of the box represent the 25th and 75th percentiles, respectively, while the middle red line shows the median. Points outside the black bars are considered outliers. The green diamond represents the average of the distribution. These values are shown in Table 3.

Table 3. Average noise ($\times 10^4$) of the BELMANIP sites' time series obtained before (raw) and after directional effects using the models described for the red and NIR bands and the NDVI, and using MODIS, AVHRR-pre and AVHRR data from top to bottom. The percentage next to every noise value indicates the improvement with respect to the raw data.

		Red	NIR	NDVI
MODIS	Raw	201.42	430.45	162.21
	MODIS	48.94 (75.70%)	76.23 (82.29%)	113.20 (30.22%)
	Average	90.19 (55.22%)	163.47 (62.02%)	175.16 (−7.98%)
	B1(B2)	53.28 (73.55%)	76.28 (82.28%)	146.00 (10.00%)
	Stable	53.33 (73.52%)	89.74 (79.15%)	136.81 (15.66%)
AVHRR pre	Raw	212.97	456.51	379.83
	MODIS	115.96 (45.55%)	184.07 (59.68%)	396.65 (−4.43%)
	AVHRR-pre	126.98 (40.37%)	198.10 (56.61%)	363.58 (4.28%)
	Average	132.78 (37.65%)	210.16 (53.96%)	394.81 (−3.94%)
	B1(B2)	134.41 (36.89%)	199.32 (56.34%)	397.18 (−4.57%)
	Stable	108.75 (48.93%)	163.30 (64.23%)	368.93 (2.87%)
AVHRR	Raw	168.22	292.62	317.32
	MODIS	103.42 (38.52%)	145.35 (50.33%)	314.67 (0.84%)
	AVHRR	104.08 (38.13%)	144.02 (50.78%)	315.33 (0.63%)
	Average	116.37 (30.82%)	166.03 (43.26%)	329.07 (−3.70%)
	B1(B2)	105.14 (37.50%)	144.74 (50.53%)	315.00 (0.73%)
	Stable	99.28 (40.98%)	140.84 (51.87%)	314.92 (0.76%)

Figure 4. Time series of two BELMANIP pixels (savanna and barren) for different inversion models and their noise value using AVHRR-pre-data. In brackets is shown the relative noise of the time series. For visual purposes, the data is shifted in the y-axis by 0.3n in the red band, and 0.6n in the NIR and NDVI for the nth model.

Firstly, we can see that the average raw noise of the red and NIR time series is similar for MODIS data (0.020, 0.043) and AVHRR-pre data (0.021, 0.045), but higher than the AVHRR data (0.017, 0.029). This is caused by the large directional errors in MODIS data and the high atmospheric errors in AVHRR-pre-data. Evidence of this can be seen on the NDVI noise. In MODIS, its value is very low (0.016), meaning that the intrinsic directional correction of this index has corrected most of the directional effects. In AVHRR-pre, because the directional effects are not as high, the NDVI correction errors are mostly because of atmospheric uncertainty propagation (0.038).

Secondly, looking at MODIS data (Figure 5, column 1) gives an indication of the quality of the approximations considered by the different models. As is expected, the MODIS inversion provides the best correction (75.7%, 82.3% and 30.2% for the red, NIR and NDVI respectively). Using the Average model, not only is the red and NIR noise improvement lower by ~20%, as compared to the MODIS inversion, but it also has a higher spread of the noise distribution. This is expected considering the broad generalizations of the model. The B1(B2) model shows the second-best performance, only 2% worse than the MODIS inversion, indicating that this a valid approximation that could reduce the computational time while achieving high-quality directional correction. For the NDVI, however, this method shows a significantly smaller improvement (10.0%) than the MODIS model (30.2%). Finally, the Stable method shows to be a valid alternative for the red band, but not for the NIR band. For the NDVI value, it can correct ~5% better than the B1(B2) method.

Analyzing the effects of these models on AVHRR-pre-data (Figure 5, column 2) can now show whether the error provided by their approximations is smaller or higher than the propagated atmospheric error. The MODIS model on AVHRR-pre provides a good correction of the directional effects with ~45.6%, 59.7% in the red and NIR bands, 5% and 3% better than the AVHRR-pre model. This shows that MODIS parameters are preferable to AVHRR-pre derived parameters. The opposite is true for the NDVI. The Average and the B1(B2) model show the worst performances among all of them, with a significant difference with the MODIS model in the red (~8% and 9%) and the NIR (~6% and 3%). This is expected for the Average method but is surprising for the B1(B2) model considering its good performance with MODIS data. It seems like the inversion with two parameters still isn't good enough, despite having reliable assumptions. In the case of the NDVI, the three of them provide a negative improvement (~4%) given that the noise on the raw data is already low. Finally, the Stable method provides the best improvement in the red and NIR bands, with differences of ~9% and 8% respectively, compared to the AVHRR-pre model. This shows that performing the inversion with one parameter provides higher stability to the parameters and therefore a smaller distribution in the corrected noise values, as can be appreciated by the width of the boxplots in Figure 5. In the NDVI, it provides a positive improvement of 2.87%, but it's still lower than using the AVHRR-pre-data.

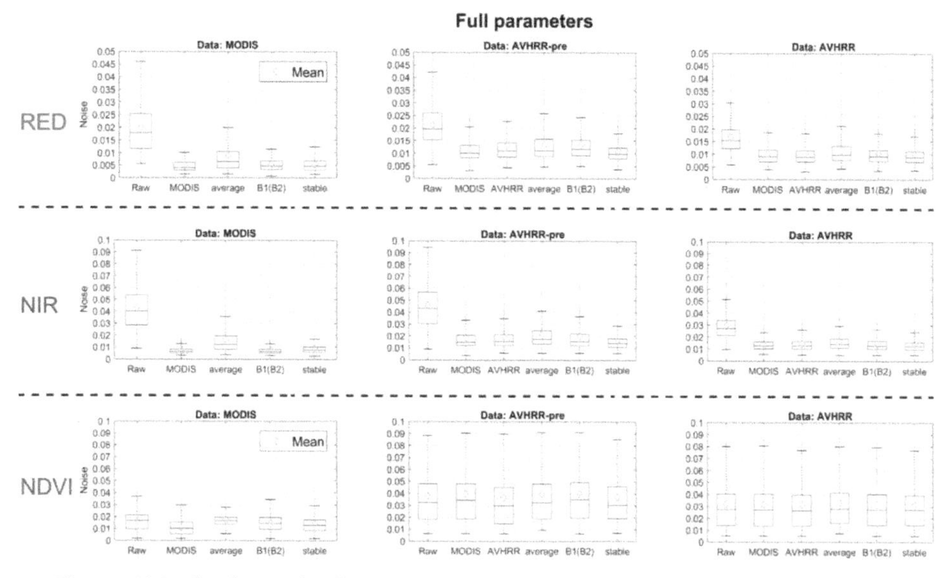

Figure 5. Noise distributions for all the bands considered (rows), using the different inversion models and for MODIS, AVHRR-pre and AVHRR data, (columns 1, 2, and 3, respectively). The green diamonds represent the average of the distribution. The top and bottom blue edges of the box represent the 25th and 75th percentiles, respectively, while the middle red line shows the median. Points outside the black bars are considered outliers.

These results are analogous to the AVHRR data (Figure 5, column 3), but with a significantly smaller difference between the methods. The Stable method, for example, only provides a ~3% and 2% improvement difference with the AVHRR model in the red and NIR bands respectively. The average method of AVHRR data improves significantly less than with AVHRR-pre-data. This is because atmospheric errors are not as high in this time series and, therefore, the broad assumptions made by the model provide more uncertainty than the propagated atmospheric perturbations.

4. Discussion

4.1. BRDF Parameters Relationships

The V parameter shows good correspondence with the NDVI, which is expected considering that it models the volumetric component of vegetation. Therefore, it is not surprising that we find a relationship between them. The R parameter can be physically interpreted as the aerodynamic roughness [29]. The relationship between the aerodynamic roughness and the NDVI is not that evident, in fact, we didn't find any meaningful relationship between the parameters. Given a certain pixel and its geometry, it's reasonable to think that for higher amounts of vegetation, the aerodynamic roughness is bound to change. In fact, Franch et al., studied the possibility of a quadratic evolution of R with the NDVI within a pixel [30].

Considering that the R parameter is associated with a geometrical component, one would expect it to be independent of the spectral band. Experimental results have shown that this isn't true and that there is a difference between both values for medium to high NDVI values [25,31,32]. We did find, nevertheless, a good relationship between the R parameter for both bands, which have a slope relatively close to 1 and an intercept close to 0. For low NDVI values, the R parameters are independent of the spectral band. This was also the case for V parameters with NDVI < 0.1, which as far as we are concerned, has never been observed in previous literature.

4.2. Inversion Period

Originally, the VJB method was performed using 5-years of data. The results showed an improvement of ~30% in the NDVI using MODIS data [18], which agree with our results. When including a higher number of years in the inversion, we are increasing the number of observations that are used in the model. This means increasing the range of observation geometries and of possible NDVI values that the pixel can have during the years. However, the accuracy gained by including these observations in the models might be counterbalanced by the change in vegetation characteristics of the target over the years. The larger the amount of years, the more likely it is that this change occurs, and the less likely it is for the assumptions behind the VJB method to hold. Nonetheless, these land cover changes are contemplated to some extent through the NDVI regression implicit in the VJB method. The analysis from this section on the different bands and sensors has shown that this change in vegetation is insignificant compared to the information gained by the increased number of observations. This might only be true, however, so long as the vegetation dynamics or human factors such as deforestation or agricultural practices don't change the surface value significantly. In these cases, the noise of the time series after BRDF normalization is likely to be high, and a shorter inversion period is recommended that correctly quantifies these land cover changes. When using high spatial resolution, these land cover changes are more discernible than at moderate or high spatial resolution. In this case, a further analysis is required to determine what the adequate inversion period to retrieve the V and R parameters would be.

4.3. Inversion Models

The results of this section show that the conclusions are different for the NDVI correction than for the individual bands. For the red and NIR bands, the Stable model proved to be the best method to use when the data's time series has too much noise to derive the correction parameters from it. This is true especially for AVHRR-pre, for which the Stable method provided a significant improvement in the noise correction. This is, therefore, the recommended model for the derivation of surface albedo using AVHRR-pre and AVHRR data, provided that the NDVI computation is not required.

For MODIS data, the use of the MODIS inversion is still preferable, as assumptions made by the models devised introduce uncertainty. In the case of the NDVI, however, the use of the regular VJB model using parameters derived from the same data is preferred. For users with a large computational burden that look for a compromise between computational time and accuracy, the use of the Average method provides very fast correction parameters at the expense of ~10% noise improvement. If the BRDF correction parameters are already available from MODIS data or another sensor with little atmospheric perturbation at the same spatial resolution, the use of these parameters provides the second best correction after the Stable method with virtually no significant computational time required in comparison.

5. Conclusions

In this paper we explore, different approaches to find the most robust way of applying the VJB correction to AVHRR data. To do this we use AVHRR data from 1982-2015 divided in AVHRR-pre (1982–2000) and AVHRR (2000–2015) and MODIS data (2000–2015) at CMG spatial resolution (0.05°) in 445 BELMANIP2 sites. In the first place, we compare the effect of using 3-years or 15+ years to derive the BRDF parameters. We found that for coarse spatial resolution, where the vegetation characteristics of the target don't appear to change significantly, it's preferable to use 15+ years of observation. This result was true both for AVHRR and for MODIS data. The differences were higher for the NIR band, where the average noise was reduced by 7% when using the whole time series from AVHRR-pre-data instead of 3-year parameters.

Secondly, we used MODIS data to retrieve relationships between the BRDF parameters. With the information derived from this sensor, we built different models based on the VJB method, which aim to minimize the propagation of the atmospheric errors present in the AVHRR data to the correction of directional effects. We found that for the red and NIR bands, the Stable method provides a robust correction in terms of reducing both the average noise of the pixels considered and the width of the noise distribution. This is the recommended model for surface albedo retrieval for the VJB method using AVHRR data. For the NDVI, however, we found that the lowest average noise is obtained by correcting MODIS data with MODIS derived parameters and AVHRR data with AVHRR derived parameters. These results are true both for AVHRR-pre which has no aerosol or water vapor correction and for AVHRR, which uses MODIS information for the atmospheric correction.

Further studies should focus on the effect of the spatial resolution on the assumptions made by the VJB method, such as the invariability of a certain site during the composite period of the model. This information is especially useful for the derivation of surface albedo. A surface albedo using AVHRR data from 1982–2018 is of great interest to the scientific community, especially one which relies heavily on observations and on semi-empirical physical models.

Author Contributions: Conceptualization, J.L.V.-N., B.F. and E.F.V.; Formal analysis, J.L.V.-N. and E.F.V.; Investigation, J.L.V.-N.; Methodology, J.L.V.-N., B.F. and E.F.V.; Supervision, B.F., E.F.V. and J.-C.R.; Writing—original draft, J.L.V.-N.; Writing—review & editing, J.L.V.-N., B.F., E.F.V. and J.-C.R.

Funding: This research received no external funding.

Conflicts of Interest: The authors declare no conflict of interest.

References

1. Vermote, E.; Claverie, M. *Climate Algorithm Theoretical Basis Document (C-ATBD) AVHRR Land Bundle—Surface Reflectance and Normalized Difference Vegetation Index*; University of Wisconsin-Madison: Madison, WI, USA, 2013.
2. Franch, B.; Vermote, E.; Roger, J.-C.; Becker-Reshef, I.; Justice, C.O. A 30+ year AVHRR Land Surface Reflectance Climate 2 Data Record and its application to wheat yield 3 monitoring. *Remote Sens.* **2016**, *9*, 296. [CrossRef]

3. Moreno Ruiz, J.A.; Riaño, D.; Arbelo, M.; French, N.H.F.; Ustin, S.L.; Whiting, M.L. Burned area mapping time series in Canada (1984–1999) from NOAA-AVHRR LTDR: A comparison with other remote sensing products and fire perimeters. *Remote Sens. Environ.* **2012**, *117*, 407–414. [CrossRef]
4. Claverie, M.; Matthews, J.L.; Vermote, E.F.; Justice, C.O. A 30+ Year AVHRR LAI and FAPAR Climate Data Record: Algorithm Description and Validation. *Remote Sens.* **2016**, *8*, 263. [CrossRef]
5. Verger, A.; Baret, F.; Weiss, M.; Lacaze, R.; Makhmara, H.; Vermote, E. Long term consistent global GEOV1 AVHRR biophysical products. In Proceedings of the 1st EARSeL Workshop on Temporal Analysis of Satellite Images, Mykonos, Greece, 23–25 May 2012; Volume 2325, p. 2833.
6. Wang, S.; Yin, H.; Yang, Q.; Yin, H.; Wang, X.; Peng, Y.; Shen, M. Spatiotemporal patterns of snow cover retrieved from NOAA-AVHRR LTDR: a case study in the Tibetan Plateau, China. *Int. J. Dig. Earth* **2017**, *10*, 504–521. [CrossRef]
7. Julien, Y.; Sobrino, J.A. Monitoring global vegetation with the Yearly Land Cover Dynamics (YLCD) method. In Proceedings of the 2011 6th International Workshop on the Analysis of Multi-Temporal Remote Sensing Images (Multi-Temp), Trento, Italy, 12–14 July 2011; pp. 121–124.
8. Hu, B.; Lucht, W.; Strahler, A.H.; Barker Schaaf, C.; Smith, M. Surface Albedos and Angle-Corrected NDVI from AVHRR Observations of South America. *Remote Sens. Environ.* **2000**, *71*, 119–132. [CrossRef]
9. Saunders, R.W. The determination of broad band surface albedo from AVHRR visible and near-infrared radiances. *Int. J. Remote Sens.* **1990**, *11*, 49–67. [CrossRef]
10. Strugnell, N.C.; Lucht, W.; Schaaf, C. A global albedo data set derived from AVHRR data for use in climate simulations. *Geophys. Res. Lett.* **2001**, *28*, 191–194. [CrossRef]
11. Trishchenko, A.P.; Luo, Y.; Khlopenkov, K.V.; Wang, S. A Method to Derive the Multispectral Surface Albedo Consistent with MODIS from Historical AVHRR and VGT Satellite Data. *J. Appl. Meteor. Climatol.* **2008**, *47*, 1199–1221. [CrossRef]
12. Bates, J.J.; Privette, J.L.; Kearns, E.J.; Glance, W.; Zhao, X. Sustained Production of Multidecadal Climate Records: Lessons from the NOAA Climate Data Record Program. *Bull. Am. Meteorol. Soc.* **2015**, *97*, 1573–1581. [CrossRef]
13. Hollmann, R.; Merchant, C.J.; Saunders, R.; Downy, C.; Buchwitz, M.; Cazenave, A.; Chuvieco, E.; Defourny, P.; de Leeuw, G.; Forsberg, R.; et al. The ESA Climate Change Initiative: Satellite Data Records for Essential Climate Variables. *Bull. Am. Meteorol. Soc.* **2013**, *94*, 1541–1552. [CrossRef]
14. Schulz, J.; Albert, P.; Behr, H.-D.; Caprion, D.; Deneke, H.; Dewitte, S.; Dürr, B.; Fuchs, P.; Gratzki, A.; Hechler, P.; et al. Operational climate monitoring from space: The EUMETSAT satellite application facility on climate monitoring (CM-SAF). *Atmos. Chem. Phys. Discuss.* **2008**, *8*, 8517–8563. [CrossRef]
15. Bojinski, S.; Verstraete, M.; Peterson, T.C.; Richter, C.; Simmons, A.; Zemp, M. The Concept of Essential Climate Variables in Support of Climate Research, Applications, and Policy. *Bull. Am. Meteorol. Soc.* **2014**, *95*, 1431–1443. [CrossRef]
16. Qu, Y.; Liang, S.; Liu, Q.; He, T.; Liu, S.; Li, X. Mapping Surface Broadband Albedo from Satellite Observations: A Review of Literatures on Algorithms and Products. *Remote Sens.* **2015**, *7*, 990–1020. [CrossRef]
17. Schaaf, C.B.; Gao, F.; Strahler, A.H.; Lucht, W.; Li, X.; Tsang, T.; Strugnell, N.C.; Zhang, X.; Jin, Y.; Muller, J.-P.; et al. First operational BRDF, albedo nadir reflectance products from MODIS. *Remote Sens. Environ.* **2002**, *83*, 135–148. [CrossRef]
18. Vermote, E.; Justice, C.O.; Breon, F.M. Towards a Generalized Approach for Correction of the BRDF Effect in MODIS Directional Reflectances. *IEEE Trans. Geosci. Remote Sens.* **2009**, *47*, 898–908. [CrossRef]
19. Franch, B.; Vermote, E.; Skakun, S.; Roger, J.-C.; Santamaria-Artigas, A.; Villaescusa-Nadal, J.L.; Masek, J. Toward Landsat and Sentinel-2 BRDF Normalization and Albedo Estimation: A Case Study in the Peruvian Amazon Forest. *Front. Earth Sci.* **2018**, *6*, 185. [CrossRef]
20. Vermote, E.F. *MOD09A1 MODIS Surface Reflectance 8-Day L3 Global 500m SIN Grid V006*; NASA: Washington, DC, USA, 2015.
21. Vermote, E.; Vermeulen, A. *Atmospheric Correction Algorithm: Spectral Reflectances (MOD09)*; ATBD: New York, NY, USA, 1999.
22. LTDR (Land Long Term Data Record) Home. Available online: https://ltdr.modaps.eosdis.nasa.gov/cgi-bin/ltdr/ltdrPage.cgi?fileName=products (accessed on 11 January 2019).
23. Baret, F.; Morissette, J.T.; Fernandes, R.A.; Champeaux, J.L.; Myneni, R.B.; Chen, J.; Plummer, S.; Weiss, M.; Bacour, C.; Garrigues, S.; et al. Evaluation of the representativeness of networks of sites for the global

validation and intercomparison of land biophysical products: proposition of the CEOS-BELMANIP. *IEEE Trans. Geosci. Remote Sens.* **2006**, *44*, 1794–1803. [CrossRef]
24. Villaescusa-Nadal, J.L.; Franch, B.; Roger, J.; Vermote, E.F.; Skakun, S.; Justice, C. Spectral Adjustment Model's Analysis and Application to Remote Sensing Data. *IEEE J. Sel. Top. Appl. Earth Obs. Remote Sens.* **2019**, 1–12. [CrossRef]
25. Roujean, J.-L.; Leroy, M.; Deschamps, P.-Y. A bidirectional reflectance model of the Earth's surface for the correction of remote sensing data. *J. Geophys. Res.* **1992**, *97*, 20455–20468. [CrossRef]
26. Maignan, F.; Bréon, F.-M.; Lacaze, R. Bidirectional reflectance of Earth targets: evaluation of analytical models using a large set of spaceborne measurements with emphasis on the Hot Spot. *Remote Sens. Environ.* **2004**, *90*, 210–220. [CrossRef]
27. Bréon, F.-M.; Vermote, E. Correction of MODIS surface reflectance time series for BRDF effects. *Remote Sens. Environ.* **2012**, *125*, 1–9. [CrossRef]
28. Vermote, E.F.; El Saleous, N.Z.; Justice, C.O. Atmospheric correction of MODIS data in the visible to middle infrared: First results. *Remote Sens. Environ.* **2002**, *83*, 97–111. [CrossRef]
29. Marticorena; Chazette, P.; Bergametti, G.; Dulac, F.; Legrand, M. Mapping the aerodynamic roughness length of desert surfaces from the POLDER/ADEOS bi-directional reflectance product. *Int. J. Remote Sens.* **2004**, *25*, 603–626. [CrossRef]
30. Franch, B.; Vermote, E.F.; Sobrino, J.A.; Julien, Y. Retrieval of Surface Albedo on a Daily Basis: Application to MODIS Data. *IEEE Trans. Geosci. Remote Sens.* **2014**, *52*, 7549–7558. [CrossRef]
31. Marticorena, B.; Kardous, M.; Bergametti, G.; Callot, Y.; Chazette, P.; Khatteli, H.; Le Hégarat-Mascle, S.; Maillé, M.; Rajot, J.-L.; Vidal-Madjar, D.; et al. Surface and aerodynamic roughness in arid and semiarid areas and their relation to radar backscatter coefficient. *J. Geophys. Res.* **2006**, *111*, F03017. [CrossRef]
32. Franch, B.; Vermote, E.F.; Sobrino, J.A.; Fédèle, E. Analysis of directional effects on atmospheric correction. *Remote Sens. Environ.* **2013**, *128*, 276–288. [CrossRef]

Article

The VIIRS Sea-Ice Albedo Product Generation and Preliminary Validation

Jingjing Peng [1,*], Yunyue Yu [2], Peng Yu [1,2] and Shunlin Liang [3]

[1] Earth System Science Interdisciplinary Center/Cooperative Institute for Climate and Satellites-Maryland, University of Maryland, College Park, MD 20740, USA; peng.yu@noaa.gov

[2] NOAA NESDIS Center for Satellite Applications and Research, College Park, MD 20740, USA; yunyue.yu@noaa.gov

[3] Department of Geographical Sciences, University of Maryland, MD 20740, USA; sliang@umd.edu

* Correspondence: jjpeng@umd.edu; Tel.: +1-240-825-6625

Received: 14 September 2018; Accepted: 15 November 2018; Published: 17 November 2018

Abstract: Ice albedo feedback amplifies climate change signals and thus affects the global climate. Global long-term records on sea-ice albedo are important to characterize the regional or global energy budget. As the successor of MODIS (Moderate Resolution Imaging Spectroradiometer), VIIRS (Visible Infrared Imaging Radiometer Suite) started its observation from October 2011 on S-NPP (Suomi National Polar-orbiting Partnership). It has improved upon the capabilities of the operational Advanced Very High Resolution Radiometer (AVHRR) and provides observation continuity with MODIS. We used a direct estimation algorithm to produce a VIIRS sea-ice albedo (VSIA) product, which will be operational in the National Oceanic and Atmospheric Administration's (NOAA) S-NPP Data Exploration (NDE) version of the VIIRS albedo product. The algorithm is developed from the angular bin regression method to simulate the sea-ice surface bidirectional reflectance distribution function (BRDF) from physical models, which can represent different sea-ice types and vary mixing fractions among snow, ice, and seawater. We compared the VSIA with six years of ground measurements at 30 automatic weather stations from the Programme for Monitoring of the Greenland Ice Sheet (PROMICE) and the Greenland Climate Network (GC-NET) as a proxy for sea-ice albedo. The results show that the VSIA product highly agreed with the station measurements with low bias (about 0.03) and low root mean square error (RMSE) (about 0.07) considering the Joint Polar Satellite System (JPSS) requirement is 0.05 and 0.08 at 4 km scale, respectively. We also evaluated the VSIA using two datasets of field measured sea-ice albedo from previous field campaigns. The comparisons suggest that VSIA generally matches the magnitude of the ground measurements, with a bias of 0.09 between the instantaneous albedos in the central Arctic and a bias of 0.077 between the daily mean albedos near Alaska. The discrepancy is mainly due to the scale difference at both spatial and temporal dimensions and the limited sample size. The VSIA data will serve for weather prediction applications and climate model calibrations. Combined with the historical observations from MODIS, current S-NPP VIIRS, and NOAA-20 VIIRS observations, VSIA will dramatically contribute to providing high-accuracy routine sea-ice albedo products and irreplaceable records for monitoring the long-term sea-ice albedo for climate research.

Keywords: albedo; sea ice; VIIRS; Arctic; PROMICE; GC-NET; validation

---

1. Introduction

Recently, more evidence reveals the shrinking trend of Arctic sea ice [1,2]. The alteration from high-albedo sea ice to a low-albedo ocean would increase the amount of absorbed solar radiation, leading to a warmer effect and further accelerating the ice melting. Therefore, sea-ice albedo variation has attracted more attention when studying Arctic and global climate change. Satellite observations are essential for providing sustained, consistent, and near real-time albedo over large, remote, and sparsely populated areas such as sea ice [3,4].

Despite the unprecedented demand for authoritative information on sea-ice albedo, local data resources are limited since the polar region is one of the most under-sampled domains in the climate system. Satellite observations are essential for providing sustained, consistent, and near real-time albedo estimates over large, remote, and sparsely populated areas. Since the last century, based on the reliable operational global imagery from Advanced Very High Resolution Radiometer (AVHRR) data, several studies have discussed the algorithms for mapping broadband albedo of sea ice [5–9]. The available sea-ice albedo products include the AVHRR Polar Pathfinder (APP) albedo product [10], the APP-extended (APP-x) albedo product [11,12], and the Satellite Application Facility on Climate Monitoring (CM-SAF) surface albedo (SAL) product [13]. However, only two spectral visible/near-infrared bands of AVHRR exist, which has limited its accuracy and sensitivity of broadband albedo [14].

With its higher spectral/spatial characteristics and measurement precision as compared to AVHRR, Moderate Resolution Imaging Spectroradiometer (MODIS) produces an operational 500 m/1 km daily albedo product. However, the operational MODIS albedo product left the sea-ice pixels blank due to the high cloud coverage in the Arctic region [12] and there is increasing uncertainty of the MODIS atmospheric correction algorithm on surfaces without dense vegetation coverage [15,16]. Moreover, the traditional bidirectional reflectance distribution function (BRDF) integration algorithm applied for MODIS albedo production assumes a 16-day stable earth surface BRDF pattern to accumulate enough multi-angle observations for BRDF retrieval. This requirement may be easily violated due to the dynamic coverage of sea ice, especially as the Arctic Ocean is experiencing a shift from perennial to seasonal sea ice [17,18]. Although some studies are trying to combine the observations from different satellites to form the multi-angle dataset [19,20], the rolling window length is still too long to grasp the dynamic sea-ice change.

Alternatively, Liang et al. [21] developed a direct estimation algorithm for land surface albedo estimation over the Greenland ice sheet from MODIS top-of-atmosphere (TOA) reflectance data. Instead of the atmospheric correction and BRDF inversion approaches, the direct retrieval algorithm uses an explicit multiple linear regression method for the albedo estimation. Later, it was further improved by applying surface BRDF models in the regression computation. To grasp the influence of BRDF anisotropy, the incident/reflected hemisphere was divided into regular grids, and linear regression relationships were developed for each grid to represent specific solar-viewing geometries. This prototype of the direct regression algorithm was operationally applied for the land surface albedo productions in the mission of Global LAnd Surface Satellite (GLASS) land surface products from MODIS observations and in the mission of S-NPP (Suomi National Polar-orbiting Partnership) from the VIIRS (Visible Infrared Imaging Radiometer Suite) dataset [15,22]. However, the sea-ice albedo production was still a problem due to the lack of sea-ice surface BRDF observations available for training the dataset built-up in the regression approach. No sea-ice BRDF records were available from satellite observations due to the limited solar zenith angles and large cloud fraction in polar regions. Thus, the sea-ice albedo of VIIRS was not released simultaneously. Currently, no operational sea-ice albedo at a 1 km scale is available to the user community.

Qu et al. [23] simulated sea-ice surface BRDF from physical models, which has been proved to be able to represent different sea-ice types and mixed pixels of ice/snow/seawater with different fractions. In this study, their BRDF simulation method was applied to the VIIRS instrument to produce a VIIRS sea-ice albedo (VSIA) product.

As the successor of MODIS, VIIRS started its observation from October of 2011. VIIRS has improved its design upon the capabilities of the operational AVHRR and provides observation continuity with MODIS. VSIA will service weather prediction applications and climate model calibrations. Moreover, combined with the historical observations from MODIS and from the future Joint Polar Satellite System (JPSS), VSIA will dramatically contribute to providing high-accuracy routine sea-ice albedo products and irreplaceable records for monitoring the long-term sea-ice albedo for climate studies. Recently, VSIA has passed its operational readiness review by the Satellite Products and Services Review Board at National Oceanic and Atmospheric Administration's (NOAA) National Environmental Satellite, Data, and Information Service (NESDIS). It will be produced operationally through the NOAA S-NPP data Exploration (NDE) system.

In this study, we aim to evaluate the VSIA accuracy using ground measurements, regarded as the "true value" of the ice/snow albedo. The accuracy assessment attempts to provide the community and users with information about the VSIA performance and its reliability in applications. The rest of the paper is organized as follows. Section 2 introduces the VIIRS albedo product, VSIA algorithm, and the operational production framework. Ground measurements and the validation results are presented in Section 3. More information about the VSIA algorithm is shown in Section 4, including the illustration of large-scale time-series retrieval instances from the VSIA algorithm, probing of look up table (LUT) profiles at specific angles, and highlights of limitations in this validation attempt. Section 5 summarizes this study.

2. Method and Dataset

2.1. VIIRS Albedo Product

The VIIRS sensor is a component of the Suomi National Polar-orbiting Partnership (S-NPP) satellite and the Joint Polar Satellite System (JPSS). S-NPP was launched on 28 October 2011, and JPSS-1 (NOAA-20) was launched on 18 November 2017. The VIIRS was designed to improve the series of measurements initiated by the Advanced Very High Resolution Radiometer (AVHRR) and the Moderate Resolution Imaging Spectroradiometer (MODIS) [24].

The VIIRS albedo product has been designed as an environmental data record (EDR) generated in a granule-based format and is currently processed in the NOAA near real-time Interface Data Processing Segment (IDPS); archived and distributed by NOAA's Comprehensive Large Array-data Stewardship System (CLASS). The next version of the VIIRS albedo product will be generated by the S-NPP Data Exploration (NDE) system, in which sea-ice albedo retrievals will be included.

The VIIRS albedo deploys a direct estimation algorithm from single-date/angular observations, which is capable of grasping the dynamic variation of surface BRDF change and is critical for the accumulation and melting seasons of sea ice. Due to the high spatial resolution (750 m at nadir) and high temporal resolution (crosses the equator about 14 times daily in the afternoon orbit), the VIIRS can repeatedly observe the polar regions in several paths. The overlapped observations have a similar field of view owing to the design features of VIIRS; by its controlling of pixel growth rate at the edge of the scan to minimize the bow-tie effect. The VIIRS pixel size at the edge of the scan is 2.1 times that of the nadir pixel, while the MODIS pixel size at the edge of the scan is 4.8 times of the nadir pixel [25]. Note that the VIIRS albedo product is defined as the blue-sky albedo, which can be directly compared with the simultaneous ground measurements.

2.2. The BRDF-Based Direct Regression LUT Generation

The broadband instantaneous blue-sky sea-ice albedo is calculated from multispectral top-of-atmosphere (TOA) reflectance using an angular-specific linear regression relationship.

$$\alpha(\theta_s) = c_0(\theta_s, \theta_v, \varphi_s) + \sum_b c_i(\theta_s, \theta_v, \varphi_s)\rho_i(\theta_s, \theta_v, \varphi_s), \quad (1)$$

where $\alpha(\theta_s)$ is the broadband blue-sky albedo; θ_s is the solar zenith angle (SZA); θ_v is the view zenith angle (VZA); and φ_s is the relative azimuth angle (RAA). i (1,2,3,4,5,7,8,10,11) represent the nine VIIRS moderate resolution bands used in sea-ice albedo retrieval. ρ_i is the TOA reflectance from sensor data records (SDRs). c_0 and c_i are the retrieval coefficients; c_0 is the constant term. The coefficients are stored in a pre-defined look up table (LUT) for evenly spaced angular bins in SZA, VZA, and RAA.

The pre-defined LUT trained from the representative dataset renders the algorithm highly efficient and accurate. The algorithm will get the coefficients for an actual $(\theta_s, \theta_v, \varphi_s)$ combination through linear interpolation in the surrounding angular bin, which runs fast in operational practices and avoids the discontinuity in neighboring albedo values. The LUT configuration covers SZA from 0° to 80° with an increment of 2°, VZA from 0° to 64° with an increment of 2°, and RAA from 0° to 180° with an increment of 5°.

Figure 1 shows the data flow used to generate the VSIA LUT. The BRDF database is the basis for deriving TOA reflectance and broadband albedo, respectively. This database can be retrieved from satellite data for land surface pixels but has to be simulated from physical models for sea-ice pixels due to the lack of clear and low-SZA satellite observations in polar regions. A satellite pixel in the sea-ice zone was simplified as a combination of snow, ice, pond, and seawater [23]. The sea-ice BRDF can be regarded as a linear composition of the component BRDFs. The contributions of some distributed surface impurities containing dust and soot are expressed in the snow/ice BRDF model [23,26]. The BRDFs of the snow/ice component are simulated using the asymptotic radiative transfer (ART) model, with the input of inherent optical properties (IOPs) calculated from a variety of snow/ice physical parameters [26]. For ocean water, each BRDF is a linear combination of its three components (sun glint, whitecaps, and water-leaving reflectance from just beneath the air-water interface) [27]. The pond BRDF simulation in VSIA LUT deploys the analytical model proposed by Zege et al. [28] with the optical characteristic values referred to by Morassutti and LeDrew [29].

Sea-ice BRDF is considered as the linear mixing of these components' BRDFs. The fractions of different components in sea-ice pixels vary through time, thus the inherent heterogeneity of sea-ice BRDF should be considered in the simulation process. The Monte Carlo simulation method is used to generate samples of fractions in assembling the sea-ice BRDF for efficiency. The fraction of the first three components is determined by a uniform random number within 0~1. The fractions of the four components used sum to 1 in each BRDF item.

We generated a sea-ice BRDF dataset consisting of 120,000 simulated sea-ice BRDF items. The next key step is to generate the surface albedos. For each sea-ice BRDF item, the surface broadband albedo, including the black-sky albedos (BSAs) and white-sky albedo (WSA), are derived through an angular integration and narrowband-to-broadband conversion. Another method to achieve the sea-ice albedos is by aggregating the BSAs and WSA of each component using the simulated fractions. Its convenience embodies in the direct acquisition of snow/ice/seawater albedo from their BRDF models. The results of the two methods are consistent. The second one was used in our calculation.

The direct estimation algorithm was to directly infer surface albedo from TOA reflectance. Another key step is to simulate the TOA reflectance and diffuse skylight factor in each angular bin from sea-ice BRDF through atmospheric simulation using the 6S (Second Simulation of the Satellite Signal in the Solar Spectrum) tool. To eliminate the uncertainty resulting from atmospheric effects, multiple possible atmospheric conditions have been considered in the training data setup. Sea ice is mainly distributed in the Arctic, Southern Ocean, and Antarctic. The possible atmospheric influence in these regions can be represented in 6S using three predefined atmospheric models including sub-Arctic winter (SAW), mid-latitude winter (MLW), and sub-Arctic summer (SAS), while the typical aerosol types can be described using rural and maritime [30]. In practice, the atmospheric parameters are pre-calculated and stored in an atmospheric LUT. We transferred the surface reflectance spectra under each atmospheric condition type, which is the combination of an aerosol model and an atmospheric model. Then the number of simulated TOA spectra is much expanded than that of the surface BRDF spectra.

For the convenience of users, our retrieval object parameter was set as blue-sky albedo, which is defined as the ratio of up-welling radiation fluxes to down-welling radiation fluxes in a given wavelength range and is comparable with the in situ albedo observations. The blue-sky albedo is estimated using a linear relationship of black-sky albedo (directional-hemispherical surface reflectance) and the white-sky albedo (bi-hemispherical surface reflectance). The diffuse skylight factor has been recorded in the atmospheric simulation and is used as the regulatory factor in combining the BSA and WSA into blue-sky albedo [31] as the regulatory factor (2).

$$\alpha_{blue-sky} = (1-\beta)\alpha_{black-sky} + \beta\alpha_{white-sky}, \tag{2}$$

It is assumed that the BRDF dataset is representative to all possible sea-ice pixels in the polar region. Based on the calculation results above, we built linear relationships between TOA reflectance and surface blue-sky broadband albedo for different solar/viewing angular bins using a least squares method. The assumption is that the regression relationship in each angular bin is applicable to various sea-ice surface types with a minimum overall square error over the BRDF database. The regressed coefficients form the sea-ice albedo LUT.

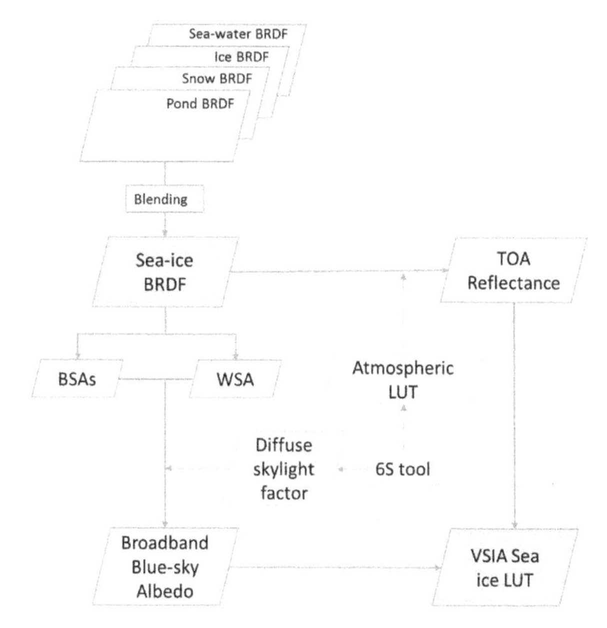

Figure 1. The data flow of VIIRS sea-ice albedo (VSIA) look up table (LUT) development. VIIRS: Visible Infrared Imaging Radiometer Suite; BRDF: bidirectional reflectance distribution function; TOA: top-of-atmosphere; BSAs: black-sky albedos; WSA: white-sky albedo; 6S: Second Simulation of the Satellite Signal in the Solar Spectrum.

2.3. The Operational VSIA Algorithm

The NDE albedo process consists of two components, as shown in Figure 2. The granule albedo, i.e., the published VIIRS albedo product, is computed online from a combination of the directly estimated albedo and a historical temporally filtered gap-free albedo; the historical albedo is computed offline from the granule albedo of previous days.

The instant retrieval of sea-ice albedo is calculated online from VIIRS TOA reflectance stored in the sensor data record (SDR), with latitude/longitude, solar zenith/azimuth angle, and view zenith/azimuth angle that are stored in a geolocation file. VIIRS NDE cloud mask EDR is used to remove the cloud contaminated pixels. The algorithm also needs the VIIRS Ice Concentration intermediate product (IP) [32] to distinguish the ocean pixels covered by ice.

Instantly-retrieved albedo values are only available for clear-sky pixels which results in gaps in the online generated albedo due to cloud contamination. Therefore, an offline process was designed to provide noise-eliminated albedo for online gap-filling. The offline branch deploys a temporal filtering algorithm [33], which deduces the albedo from a time series of instant retrievals (including the previous 8 day retrieval plus the current day) based on maximum likelihood estimation. Albedo climatology acts as a critical static input to provide statistics about inter-annual trends and variation of albedo for weight calculation and backup value estimation.

To solve the geometric registration problem between multi-day historical albedo images, temporal filtering is conducted on a fixed grid of sinusoidal projection. The global grid is evenly divided into 72 (horizontal) by 72 (vertical) tiles. After online instant retrieval, the albedo granules will go through a forward granule-to-tile transformation prior to offline filtering and a backward tile-to-granule transformation subsequently. The online albedo after gap-filling with the filtered albedo can provide continuous and reliable sea-ice albedo for users.

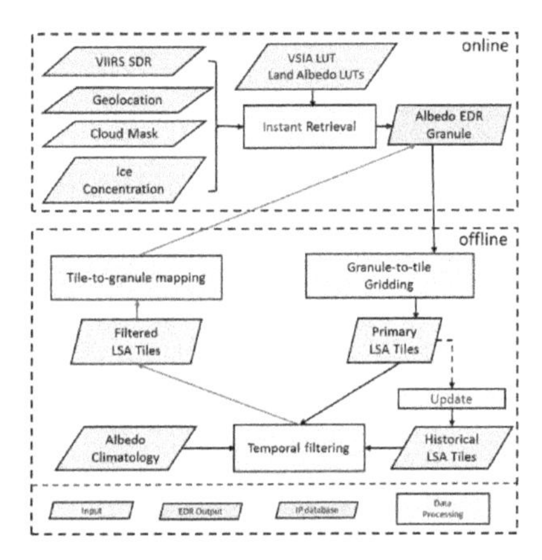

Figure 2. The flowchart of operational S-NPP Data Exploration (NDE) VIIRS sea-ice albedo product (legends shown at the bottom of the figure). S-NPP: Suomi National Polar-orbiting Partnership; EDR: environmental data record; IP: intermediate product; LSA: VIIRS Surface Albedo Product.

3. Validation of VSIA Using Ground Measurements

3.1. Ground Measurement Dataset

In situ albedo observation of sea ice is expected to verify the algorithm. However, this type of measurement is rare and its match-up with VIIRS observations is more so. Therefore, we firstly used Programme for Monitoring of the Greenland Ice Sheet (PROMICE) and Greenland Climate Network (GC-NET) measurements as substitutes. Greenland is the world's second-largest ice mass. Its ice sheet has existed for more than 2.4 million years and covers more than 80% of the island area. Considering the comparability of samples from different seasons, the match-ups around local solar noon (13:30~15:30 UTC time) were used for the comparison. Only the clear-sky observations were retained. The term "clear-sky" is used to distinguish a sky without cloud influence. The cloud effect on the radiation budget, referred to as 'cloud forcing', is one of the largest uncertainties in radiative transfer models. It is acceptable to screen out the cloudy-sky days before the comparison.

We also compared the VSIA with some in situ datasets from existing studies. Although the sample size is limited and the measurement footprint is not enough to catch the spatial heterogeneities within the satellite pixel, the comparison would embody the performance of VSIA in actual sea-ice surfaces.

3.1.1. PROMICE Measurements

PROMICE AWS (automatic weather station) is distributed in the ablation zone of Greenland. The surface types viewed are a mixture of snow patches, ice with time-varying physical parameters, and surface ponds, especially during the summer melt season [31,34–37], which may optically resemble sea-ice surfaces under certain conditions. Therefore, the data was used as a substitute for sea-ice albedo measurements.

The newly-developed PROMICE sites provide high-quality measurements due to their continuous maintenance. The PROMICE automatic weather station (AWS) network currently consists of eight regions each with a minimum of two stations at a variety of elevations on the Greenland ice sheet (the lower one is labeled '_L' and the upper one '_U'). Most stations are located in the ablation area and are thus transitioning from snow-covered to bare ice surfaces through the melt season. Our comparison was applied to 18 PROMICE sites, as shown in Table 1, from January 2012 to June 2017. In each month, the VSIA granules during 11:00~13:00 (local time) on the 15th day were compared with the ground observations closest in time.

Table 1. The information of the Programme for Monitoring of the Greenland Ice Sheet (PROMICE) stations.

Site Name	Lat (N, deg)	Lon (W, deg)	Elevation (m)	Site Name	Lat (N, deg)	Lon (W, deg)	Elevation (m)
KPC_U	79.8347	25.1662	870	NUK_U	64.5108	49.2692	1120
SCO_L	72.223	26.8182	460	NUK_N	64.9452	49.885	920
SCO_U	72.3933	27.2333	970	KAN_L	67.0955	49.9513	670
TAS_L	65.6402	38.8987	250	KAN_M	67.067	48.8355	1270
TAS_U	65.6978	38.8668	570	KAN_U	67.0003	47.0253	1840
QAS_L	61.0308	46.8493	280	UPE_L	72.8932	54.2955	220
QAS_U	61.1753	46.8195	900	UPE_U	72.8878	53.5783	940
QAS_A	61.243	46.7328	1000	THU_L	76.3998	68.2665	570
NUK_L	64.4822	49.5358	530	THU_U	76.4197	68.1463	760

3.1.2. GC-NET Measurements

The GC-NET network has produced a time series of radiation measurements covering the Greenland region since 1995. A detailed description of the GC-NET network and associated instrumentation can be found in Steffen et al. [38]. Although the land cover types of GC-NET sites are not that close to sea ice as PROMICE sites, GC-NET is also a good supplemental data source for VSIA assessment in snow/ice covered regions. We used the available GC-NET data since 2012 for the 13 stations, as shown in Table 2.

Table 2. The information of the Greenland Climate Network (GC-NET) stations used in this study.

Site Name	Lat (N, deg)	Lon (W, deg)	Elevation (m)	Site Name	Lat (N, deg)	Lon (W, deg)	Elevation (m)
CrawfordPt1	69.8783	−46.9967	1958	Saddle	65.9997	−44.5017	2460
GITS	77.1378	−61.04	1925	SouthDome	63.1489	−44.8172	2850
Humboldt	78.5267	−56.8306	1995	NASA-E	75.0006	−29.9972	2631
Summit	72.5794	−38.5053	3150	NASA-SE	66.475	−42.4986	2400
Tunu-N	78.0164	−33.9833	2113	PetermanELA	80.0831	−58.0728	965
DYE-2	66.4806	−46.2831	2053	NEEM	77.5022	−50.8744	2454
JAR1	69.495	−49.7039	962				

3.1.3. Field Measurements

We used the data from previous studies [39,40] to evaluate the VSIA performance under various sea-ice conditions in the Arctic, such as the thin first year sea ice and the mixed scenario of sea ice and open water.

Istomina et al. [39] measured the spectral albedo at six ice stations in the central Arctic using ASD FieldSpec Pro 3 during the POLARSTERN cruise ARK-XXVII/3 (IceArc) in 2012. At each station, they obtained surface albedo measurements every 10 m along 200 m transects at 1 m height. A variety of land cover types were observed, including ice, snow, pond, and their mixtures. The spectral albedo was converted to broadband albedo using the solar spectral radiation distribution observed by

Hudson et al. [41]. The solar irradiance was corrected to the local measurement time according to the solar zenith angle. All samples at each station were averaged to represent the overall albedo at the pixel scale. The temporally averaged VIIRS albedo during the measuring period was calculated to compare with the average broadband albedo.

Dou et al. [40] measured the spectral snow albedo on the frozen gulf nearby Barrow, Alaska, in 2015. Their measurements were carried out from April to June. The covered land surface types included snow, melting snow, and refrozen snow crust on ice. Spectral albedo was sampled every 5 m along a 100 m line using the ASD FieldSpec Pro 3. Four days of daily mean broadband albedo values were reported, which were used to validate the daily averaged satellite albedo. The dates were 26 April, 18 May, 22 May, and 25 May, respectively.

3.2. Comparison between PROMICE and VSIA Match-Ups

The PROMICE-VSIA match-ups span more than five years. The overall scatter plot is illustrated in Figure 3. Comparisons of albedo measurements at each PROMICE station are presented in Figure 4 to show further detail. Additionally, we present the box plot whiskers to show the distribution of the residuals at each station in Figure 5.

Figure 3 demonstrates reasonable agreement between VSIA albedo and the ground reference. Statistical measures that are used to describe these comparisons include the correlation coefficient, the mean difference error (bias), the standard deviation error (precision), and the root mean square error (RMSE) between the two. 1) The correlation coefficient R is 0.947, which indicates an obvious consistency between the two datasets. 2) The bias $\overline{e}$ is 0.028 (about 4.56% relative bias) and the precision $std(e)$ is 0.066. Here, the bias is calculated as the mean difference between the VSIA and the PROMICE albedo; the precision is calculated as the standard deviation of the difference, showing the spread of the error. The small positive bias shows that VSIA provides slightly higher albedo estimates than the PROMICE measurements. 3) The overall RMSE (root mean square error) is 0.072 and the relative RMSE (Se/Sy) is 0.344. Se/Sy, i.e., the ratio of RMSE to the unbiased standard deviation of the in situ albedo. The goodness of fit indicates a high accuracy of the albedo retrievals using the rule of thumb that Se/Sy < 0.5 represents good accuracy.

The albedo values span from 0.2 to 1, which means that the dataset might contain observations of ice/snow, water ponds, bare ground, or a mixture of those, indicating the extensive representativeness of the sample. The goodness of fit is comparable with the previous validation result of MOD10A1, a regional albedo product over Greenland, Iceland and the Canadian Arctic Region, by Ryan et al. [37]. After post-processing including de-noising and calibration of the MOD10A1 albedo (Collection 6), its RMSE with PROMICE ranges from 0.017~0.1 over three KAN sites between 2009 and 2016. In summary, the validation of VSIA shortwave albedo against ground measurements showed promise at the PROMICE sites.

At many sites, an overestimation of VSIA albedo is exhibited in high albedo regions, as shown in Figures 4 and 5, such as SCO-L, TAS-L, QAS-L, QAS-U, KAN-U. These sites contain both the south (TAS, QAS, KAN) and north (SCO) stations, covering a broad latitude range. In combination with the box plot of error between VSIA and in situ albedo, as shown in Figure 5, it is shown that most stations overestimate albedo by about 0.05. An exception of underestimation happens at NUK_L and KAN_L as a large fraction of their observations lie in the lower value regions. These two sites are located in the southern Greenland ice sheet at 500~700 m elevation, having more obvious characteristics of the ablation zone. The land cover type transitions from dry snow to melting snow to glacier ice at melting season.

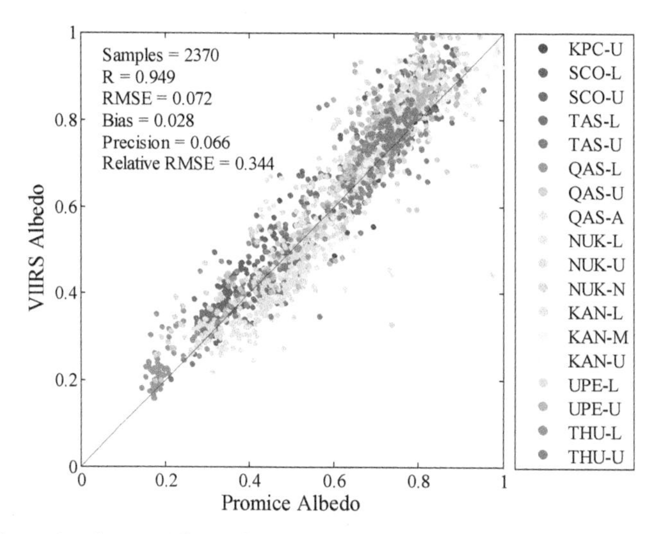

Figure 3. Comparison between VSIA and PROMICE clear-sky in situ albedo over 18 automatic weather stations. The match-up from each site is assigned as one specific color. RMSE: root mean square error.

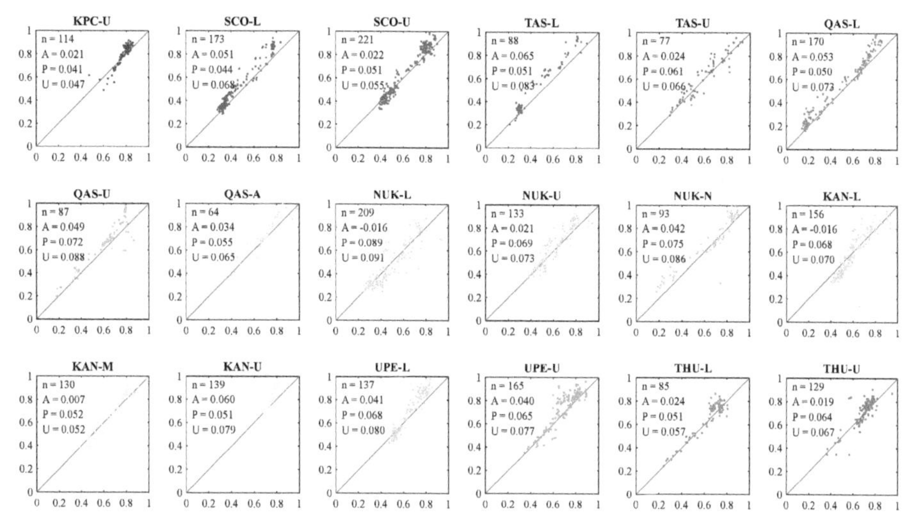

Figure 4. Comparison between VSIA and PROMICE-measured albedo at each station. Each sub-scatterplot corresponds to one station (station name as the title). The horizontal axis represents the PROMICE-measured albedo, and the vertical axis corresponds to the VIIRS-retrieved albedo. Labels: n-sample size, A-accuracy (bias), P-precision (the standard deviation of the difference between retrieved albedo and the corresponding measured albedo), U-uncertainty (the root mean square error).

The boxplot also suggests that 14 of the 19 stations contain the zero bias within $\mu \pm \sigma$ data range, demonstrating that the differences between the VSIA albedo and the PROMICE station data are in general indistinguishable from zero. The sites with a larger spread of albedo error suffer from the smaller sample size, making it difficult to reach solid conclusions on the product's performance at these sites. It is also illustrated that the mean values of most stations are larger than the median value, suggesting the distribution of residuals are skewed and the higher albedo points are dominant.

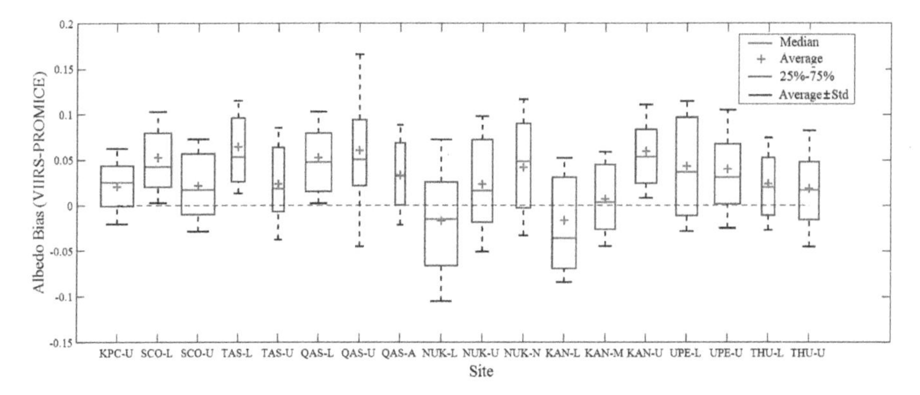

Figure 5. Distribution of residuals between VIIRS-retrieved albedo and PROMICE-measured albedo at each site. The box plots denote the distribution of the difference between VIIRS LSA and PROMICE LSA at each station. It shows lower quartiles, medians, and upper quartiles in the central boxes. Whiskers extend from each central box to show the standard deviation spread of the difference. The width of the box is proportional to the sample size at each site

According to the site-specific validation results, VSIA albedo demonstrates strong sensitivity to in situ surface albedo fluctuation and anomaly. (1) VSIA shows lower albedo value at NUK_L and KAN_L, which are the lower stations on the southwest Greenland coast, while demonstrating higher albedo values at other sites. NUK and KAN are two typical regions reported of albedo anomaly due to darker-than-average ice caused by a stronger warming trend, higher melt, and less winter accumulation [42–44]. Their bias derives from many factors. First, it is mainly attributable to the scale difference. The in situ measurements have a much smaller footprint than the satellite retrievals. The higher impurity concentration around these two sites leads to stronger surface spatial variability, especially in melt seasons [45,46]. Second, the bias might be ascribed to the absence of some dark zone constituents in the VSIA algorithm, including algae, crevassing, supraglacial water, and cryoconite distributed in the Greenland ice sheet [45]. The third attribute is the topography's effect. Some of the PROMICE stations are located on slightly sloping terrain. Sloping terrain alters the incident radiation composition and solar/view zenith angles, which would introduce wavelength dependent uncertainties in satellite albedo retrievals [47]. (2) For most regions, the upper station has lower bias than the lower station, such as SCO_U and SCO_L, TAS_U and TAS_L, UPE_U and UPE_L, and THU_U and THU_L. Lower sites have earlier melt onset and suffer from severe fluctuations. The scale difference between satellite retrieval and ground measurements is amplified. (3) The largest RMSE appears at the north tip site and the smallest RMSE corresponds to the south tip site. RMSE shows a slightly negative relationship with the latitude belt since southern Greenland has a much warmer summer and earlier melt than the northern regions.

3.3. Residuals Analysis

The main purpose of validation is to evaluate and improve the model accuracy. The residual between VSIA and PROMICE albedo measurements were analyzed with the related factors. First, the seasonal change of albedo values and the bias trend were observed. Second, the annual trend of the most important environmental factor for snow/ice albedo (SZA) was analyzed with the albedo time series and residuals. Finally, the influence of another important factor for in situ validation—ground heterogeneity—was also considered.

3.3.1. Temporal Continuity and Variability of Albedo

Figure 6 demonstrates the annual variation curves of the VSIA and PROMICE albedo. It is shown that the VSIA albedo and PROMICE albedo time series agree well with a cross-correlation coefficient of 0.9554, illustrating a significantly strong correlation.

The surface albedo of the ablation area changes with the coverage fractions of snow, bare ice, and dark impurity-rich surfaces. Accordingly, the corresponding difference fluctuates at different seasons along with the ice/snow melting and accumulating cycle. Albedo value fluctuates intensely in winter responding to the snow events. Snowfall will cause a large albedo value and its melting will decrease the surface albedo. When summer approaches, the ice melts first and the pond fraction increases, causing the albedo to drop to its minimum value at the end of July. Then, the freezing of ponds increases the albedo again.

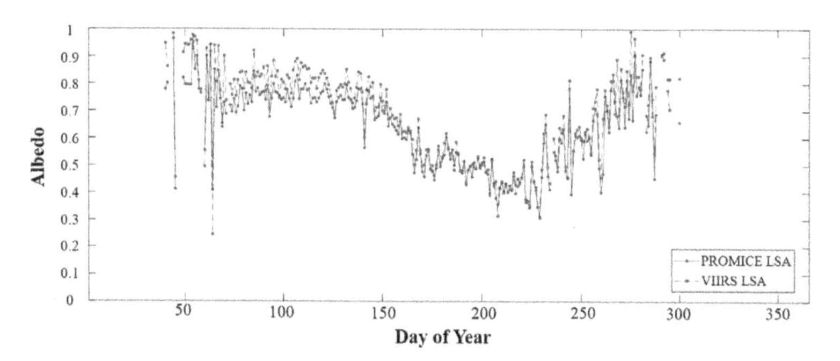

Figure 6. Seasonal cycle of clear-sky noon time albedo averaged over all PROMICE stations and the corresponding VSIA. The line is the 6-year average from 2012 to 2017

3.3.2. Influence of Heterogeneity

Considering the complex land cover types and the topography variation that surrounds many of the PROMICE stations, the standard error of albedo in the 3 × 3 neighboring pixels was analyzed to assess the effect of the local heterogeneity, as shown in Figure 7.

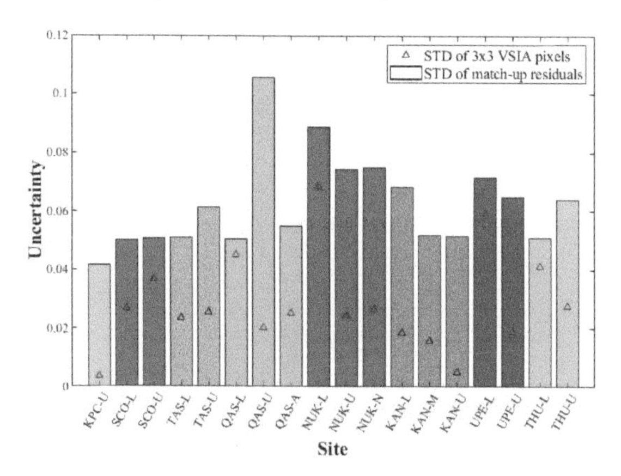

Figure 7. The average standard deviation (STD) of VSIA albedo within 3 × 3 neighboring pixels at each station and the standard deviation of albedo residuals over all match-ups. Each color corresponds to one region.

Here, a range of 3 × 3 VIIRS neighboring pixels is chosen to assess the spatial heterogeneity at satellite level considering the maximum geolocation error of VIIRS M-band SDR data [48]. Figure 7 illustrates the mean standard deviation of albedo within 3 × 3 neighboring pixels and the standard deviation of the residuals between VSIA and PROMICE albedo at the center pixel. It is shown that the albedo heterogeneity is correlated with the spread of the match-up residuals within the same region. This inference is not applicable for cross-region comparison due to the interference from the different albedo magnitudes and distinct land cover compositions. Exceptions include stations in the QAS region and the THU region, which may be due to the limitation of the albedo map used in our analysis. A higher resolution albedo map is preferred to assess the around-site heterogeneity.

3.3.3. Influence of Solar Zenith Angle

Due to the high latitude of the Greenland region, the solar zenith angle corresponding to the observations are distributed from 38° to 82° and peak around 50°~55°, as shown in Figure 8.

The bias between the VIIRS albedo and PROMICE observations exhibit a slight increasing trend with SZA, as shown in Figure 9. SZA determines the fraction of the direct incident radiation and the directional-hemispherical albedo component. It influences the diurnal/seasonal variation of surface albedo. For clear-sky snow/ice albedo, the photons interact with the snow grains over a longer path at larger SZAs, which will result in more interaction with snow surface and higher absorption [49]. At larger SZAs, the retrieval uncertainty of VSIA increases due to the strong anisotropy of the snow/ice surface. Meanwhile, the measurement error of the cosine instrument increases at larger SZAs. They together lead to the larger standard albedo error at larger SZAs.

It is known that SZA at solar noon time is larger in winter than summer. Therefore, a larger bias would be observed in winter.

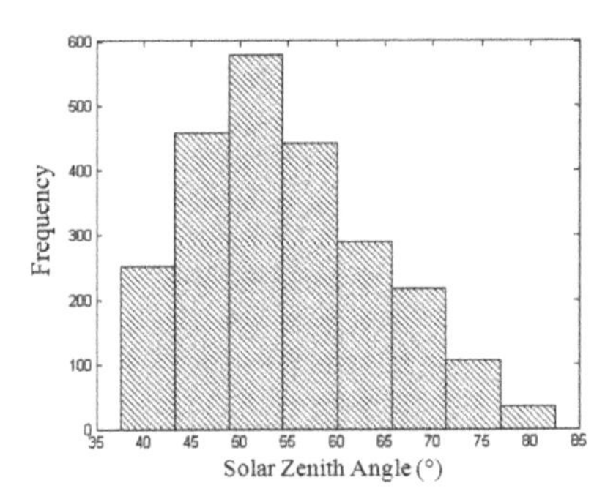

Figure 8. The distribution of SZAs (solar zenith angles) of all match-ups.

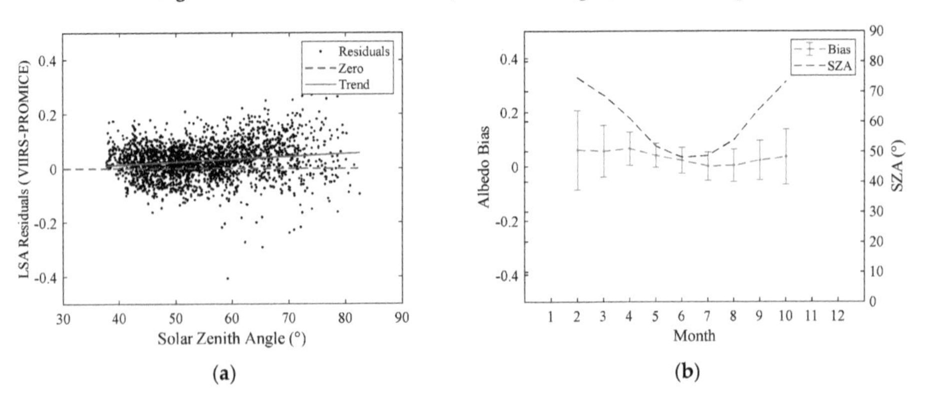

Figure 9. (**a**) Variation of albedo residuals along with SZA and the monthly mean albedo bias and solar zenith angle (SZA); (**b**) Error bars show the standard deviation of the albedo bias.

3.4. Validation Using GC-NET Match-Ups

The validation results over GC-NET sites showed a similar estimation accuracy of VSIA albedo to that over PROMICE sites. The absolute value of overall accuracy is 0.025 with a precision of 0.065, as shown in Figure 10. This is acceptable as the precision of GC-NET observations is around 0.05 [38]. The overall root mean square error is 0.07, while the relative RMSE of 0.661 is showing 66% of the unexplained variance. Similar to the PROMICE validation result, VSIA and the GC-NET measured albedo is generally consistent with better agreement at the ablation zone of lower albedo. Note that some outliers appear at the PetermanELA site, showing larger observations from GC-NET than VIIRS retrievals, that are caused by surface heterogeneity and geometric match uncertainty.

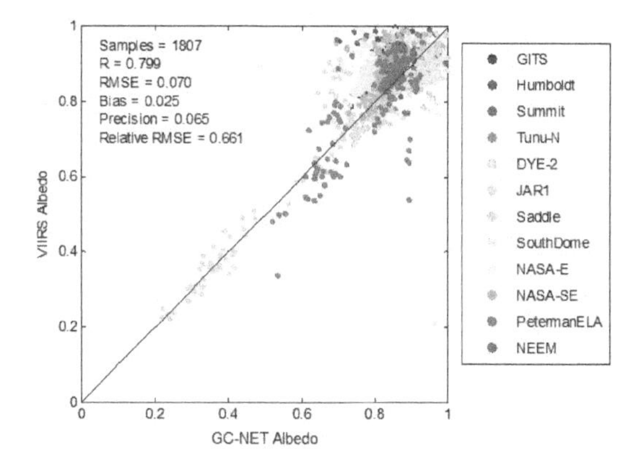

Figure 10. Comparison between VSIA and GC-NET clear-sky in situ albedo at 18 automatic weather stations.

The goodness of fit at different sites displays some spatial patterns, as shown in Figure 11. This implies that the validation results are influenced by the elevation, latitude, and land cover types of different sites. The validation results at red-colored sites show that VSIA has an underestimation of albedo compared with GC-NET observations, with a higher RMSE. These sites distribute in the northernmost and southernmost regions and suffer more from surface heterogeneity. At the green-colored sites, VSIA has more accurate estimations and shows lowest bias and RMSE. At the blue-colored sites, VSIA slightly overestimates ground albedo and shows moderate RMSE.

3.5. Evaluation of VSIA Using In Situ Sea-Ice Albedo

Figure 12 shows the comparison between VSIA and Istomina's measurements [39]. The VSIA values all fell in the range of the ground samples at all the six ice stations. The bias between the two datasets is 0.09. It should be noted that the satellite-derived albedo is instantaneous and averaged spatially; surface measurements are local and averaged temporally. The strong spatial heterogeneity and the albedo variation around the station have introduced large uncertainties to the comparison. Moreover, the in situ measurements were directly averaged without considering the contribution weight of each land cover type due to the lack of auxiliary data. Based on the comparison, we can infer that the VSIA correctly reflects the albedo magnitude of the sea-ice regions covered in the experiment.

The daily mean albedo [40] was collected from one site on different dates. Figure 13 illustrates the time-series plots of VSIA and ground observations. The albedo values generally match with a bias of 0.077. The albedo discrepancy varies along with the sea-ice evolution. (1) The sign of the discrepancy changed at the third match-up because there was a snow event on 20 May so that the spatial distribution of surface albedo changed. (2) The last match-up happens in the sharp snow-melting period. The strong spatial heterogeneity with snow accumulation and melting mainly contributes to the large albedo discrepancy.

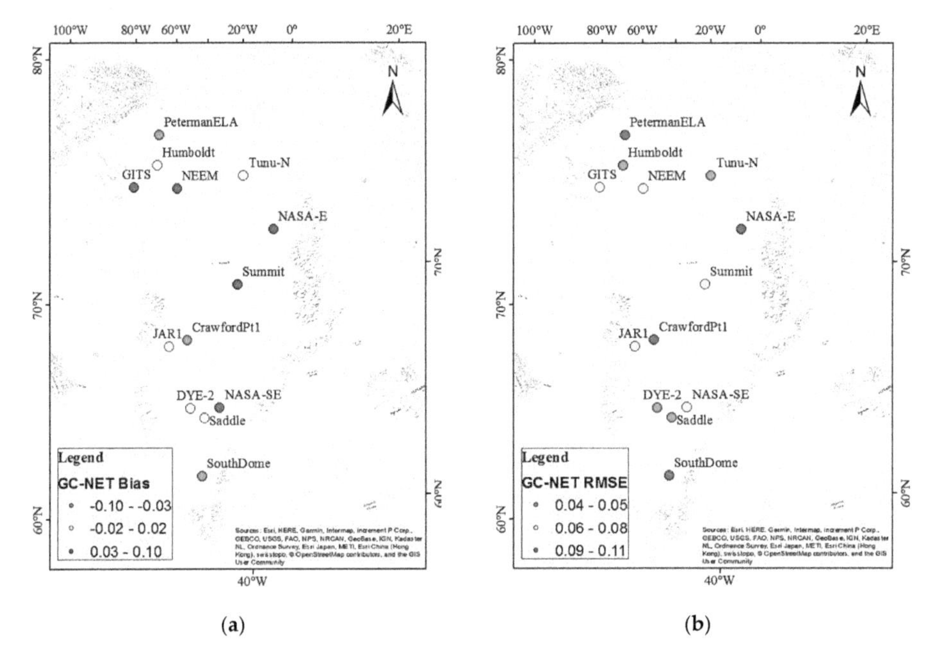

Figure 11. The distribution of the (**a**) ordinal scaled bias and (**b**) RMSE of all GC-NET sites.

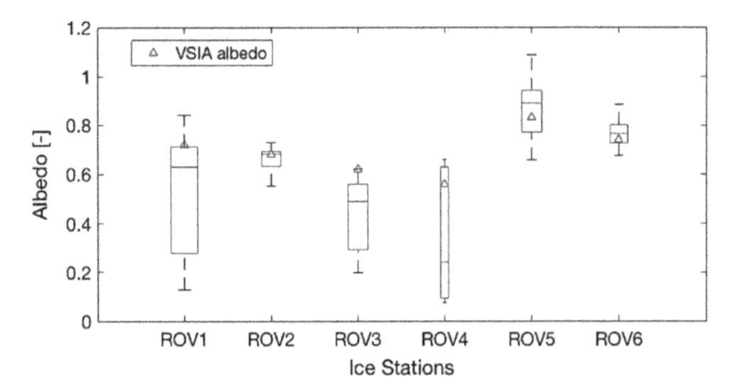

Figure 12. Comparison between the VSIA albedo and the in situ measurement in the central Arctic. ROV: Remotely Operated Vehicle. The box plot illustrates the distribution of the in situ sample at each station. The triangles mark the VSIA albedo.

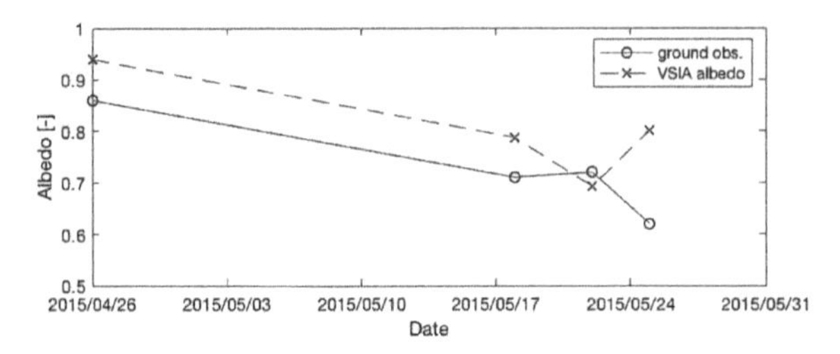

Figure 13. Comparison between the VSIA albedo and the in situ measurement near Alaska. The time-series measurements were collected at one station.

4. Discussion

4.1. Northern Hemisphere Albedo from VSIA

VSIA albedo offers a fine-resolution, large-scale albedo data source. Figure 14 shows monthly albedo maps in 2014 for the Arctic region as a sample of the algorithm performance. The dynamic evolution of albedo over time is mainly caused by the melt timing and intensity. The coverage of the retrieved albedo increases from January to April after the winter solstice. Then albedo stays at a high value because of the covered, cold, optically thick snow. As temperature increases from May to August, the snow begins to melt, thereby decreasing the surface albedo. Due to the formation of larger melt ponds, the albedo value decreases rapidly during this period. Once the new seasonal ice begins to freeze from ponds and open water, albedo increases again. From October, the data in the central Arctic is missing during the polar night period.

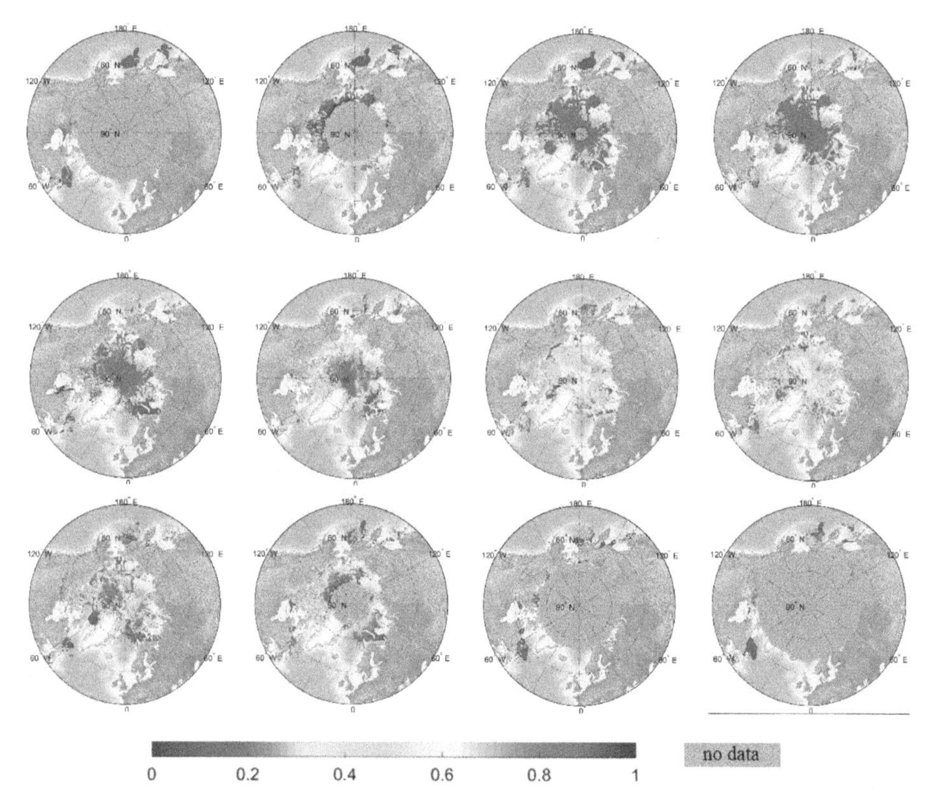

Figure 14. VSIA over 50°~90°N on the middle day of each month in 2014. The gray-colored region, including the polar night, has no data due to SZA cut-off of the LUT. The land and sea water background adopt the Cross Blended Hypso with Shaded Relief and water [50].

4.2. Analysis of VSIA LUT

The VSIA LUT provides mathematical weights used to convert TOA reflectance to surface broadband albedo, reflecting the energy contribution among different bands at each specific angular bin. Due to the numerical regression process, the coefficients' magnitudes have lost their direct link to the BRDF directional reflectance intensity, but still reflect some directional variation patterns. For sea-ice covered regions, the SZA is normally larger than 40°, so we picked 60° as an instance to observe the hemispherical variation trend of the coefficients at each SDR band. Figure 15 demonstrates the polar plots of the band coefficients at various VZAs and RAAs with constant SZA. Generally, the polar plots demonstrate the continuity and rationality of VSIA LUT at angular dimensions.

The most apparent feature of all plots is the bright/dark spot in the forward scattering direction, which is formed due to specular reflection over the snow/ice/seawater surface. The specular component in BRDF is in accordance with geometric optics. Its contribution varies with the sea-ice surface physical characteristics and the solar zenith angle. It is shown that the center points of these spots are all around the symmetry point of solar incident direction. The size of the specular reflection spot varies among different bands due to the angular distribution of the spectral reflected flux. For instance, the center wavelengths of M02 and M03 are close, thus their coefficients show similar hemispherical variation patterns.

The phenomenon that the albedo uncertainty increases at larger SZAs shown in Section 3.2.4 is partly due to the increasing spread of LUT coefficients. Here we calculated the coefficient of variation of LUT coefficients at consecutive SZA intervals, as shown in Figure 16. The coefficient of variation measures the relative variability, which is the ratio of the standard deviation to the mean value. The samples cover the whole VZA range. To eliminate the influence of the specular reflectance spot regions, the RAA was divided into two ranges, 0°~150° and 150°~180°, shown separately in Figure 16. The left figure contains a larger portion of RAA values and represents the majority of observations. It shows that the spread of coefficients in the SZA range of 70°~79° is more significant than other ranges at most visible bands. This causes the larger spread of LUT coefficients at larger SZA values. In the forward scattering cases shown in the right figure, the SZA range of 50°~69° corresponds to larger coefficients of variation. But its influence is limited due to the smaller RAA range. The magnitude of bar plots at different bands is related to the spectral sensitivity of albedo to SDR reflectance. The highest value is shown at M05 with a center wavelength of 0.672 μm (light red).

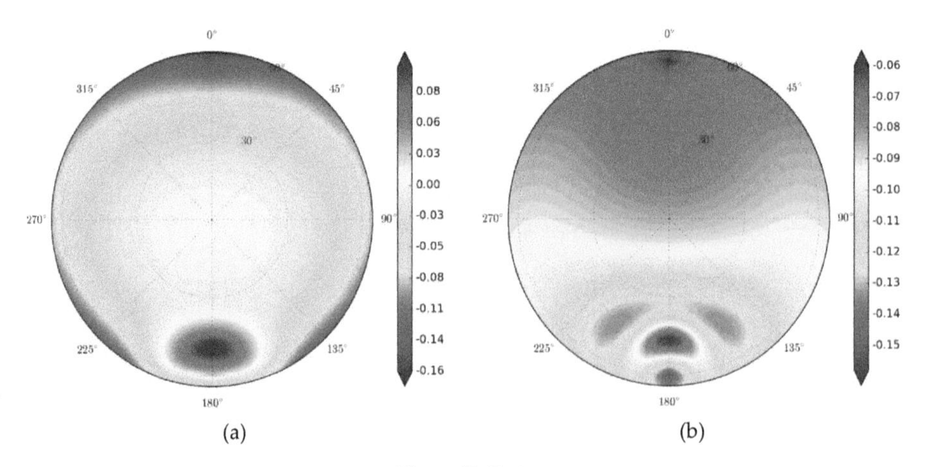

Figure 15. *Cont.*

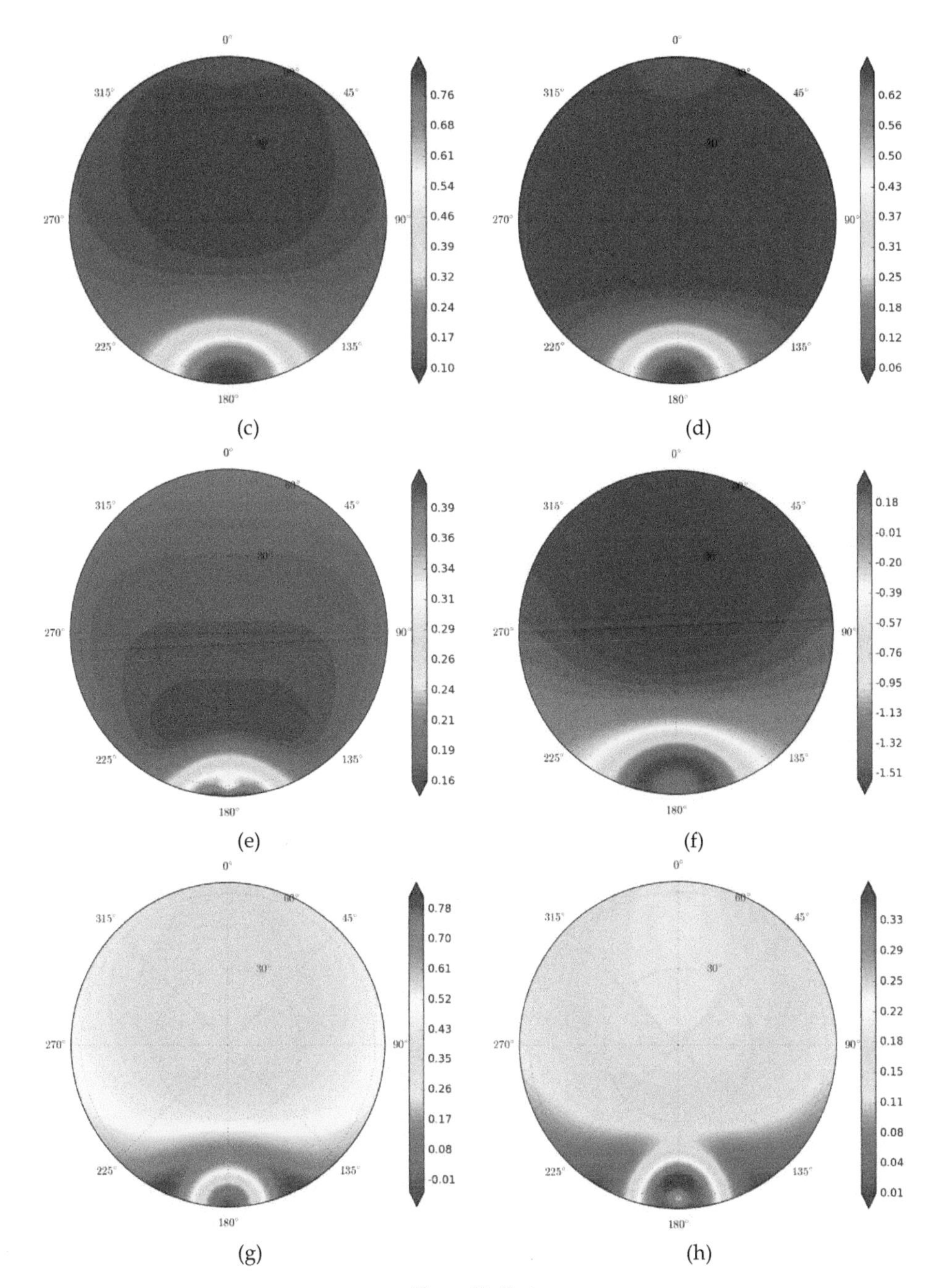

Figure 15. *Cont.*

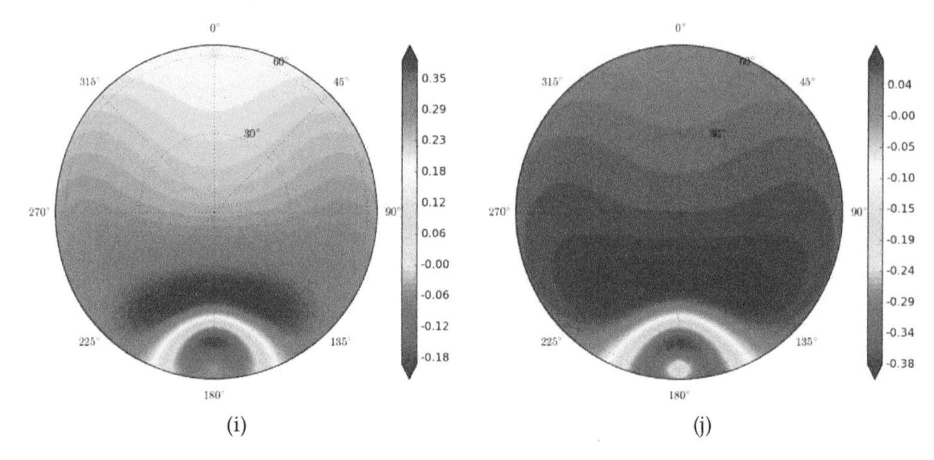

Figure 15. The polar plots (using a polarBRDF tool developed by Singh [51]) shows the coefficients corresponding to the horizon-to-nadir coverage. The bands and central wavelengths are: (**a**) the constant term; (**b**) M01 (0.412 μm); (**c**) M02 (0.445 μm); (**d**) M03 (0.488 μm); (**e**) M04 (0.555 μm); (**f**) M05 (0.672 μm); (**g**) M07 (0.865 μm); (**h**) M08 (1.24 μm); (**i**) M10 (1.61 μm); (**j**) M11 (2.25 μm). Relative azimuth angle represents the angle between lines joining the point and the center of the polar plot. An azimuth of 0° (360°) represents backward scattering while 180° indicates forward scattering. The actual relative azimuth angle (RAA) range of VSIA LUT is [0°~180°]. The 180°~360° hemispherical image mirrored the 0°~180° hemispherical image for illustration integrity.

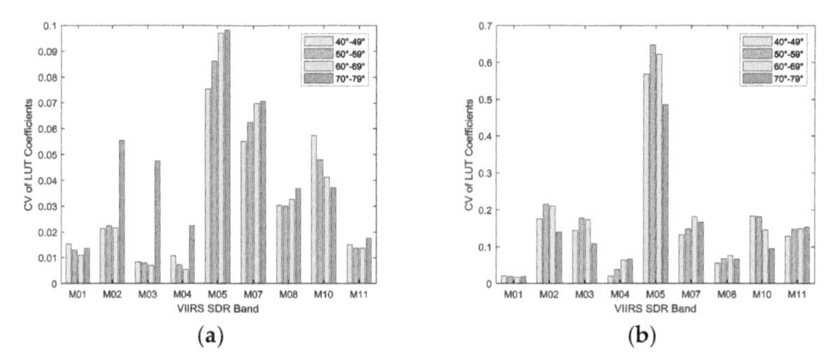

Figure 16. The variation of LUT coefficients at adjacent SZA ranges of each band (corresponding to one specific color). (**a**) The coefficient samples cover all view zenith angle (VZA) ranges and the RAA range of 0°~150°; (**b**) the samples cover all VZA ranges and the RAA range of 150°~180°.

4.3. Limitations in Current Algorithm and Validation

Currently, sea-ice albedo data is going to be produced in the VIIRS albedo environmental data record (EDR). Reliable albedo values were reported through the validation and test in the algorithm readiness review; however, the current VIIRS albedo product over the sea-ice surface still suffers from several issues:

1. Limited validation data on the sea-ice surface

Admittedly, the power of this study is inevitably restricted by the limited sea-ice surface measurements due to rare physical access. The albedo of glacier and sea ice is influenced by the same factors [52], and most of the sea-ice components or their proxies, the ice, snow, and pond, can be found around the Greenland AWS sites, except the sea water surface, which has a very low and relatively constant albedo. Therefore, the long-term monitoring data from the AWS on Greenland has provided substitute reference data for a variety of sea-ice surface conditions, such as the optically thick sea ice with snow cover or melt ponds. It can be seen that the albedo evolution trend over the PROMICE stations, as referred to in the time-series plots in Figure 6, is consistent with the multi-year Arctic sea ice [53].

For seasonal young sea ice, its albedo is typically less than the multiyear sea ice during the melting season and pond evolution [54]. Considering the lack of representativeness of Greenland AWS measured albedo to seasonal sea ice, we also cited some sea-ice albedo measurements as complementary data, which cover the surface conditions such as young sea ice and thin melt ponds. Considering the sample number of the in situ sea ice measurements is limited, further validation attempts are expected to understand the product accuracy, such as cross-comparison with other sea-ice albedo products.

2. Significant validation uncertainty at large SZA

Large SZA affects the accuracy of both VSIA and PROMICE measurements. AWS sensors suffer from the intrinsic cosine response error at large solar zenith angles, which is reported to reach a maximum of 8% at a solar zenith angle of 80° [55]. Even for solar zenith angles less than 75°, we should expect a cosine response error around 5% [56], which can even be amplified by the riming. For VSIA, the ART model for snow/ice BRDF simulation is applicable at high solar zenith angles up to 75° [57,58].

3. Thorough spatial representative investigation is desired

In this study, we did not exclude any match-ups influenced by strong ground heterogeneity since high product performance under all conditions is expected. However, it has been investigated that ground heterogeneity around the sites will amplify the uncertainty in validation at different spatial scales [59]. In this way, further investigation using higher resolution imageries as ancillary data is on schedule.

4. Further investigation on the overestimation reason

The evaluation result shows an overestimation of the VSIA albedo. However, several issues should be kept in mind before drawing this conclusion. First, the ground measurement data does not represent the "true" value considering the big scale gap between the ground point measurement and the satellite pixel retrieval. Second, the measurements from the flux instruments contain certain uncertainty [55], due to the tilting/leveling errors and the cosine-response error of the instruments. Third, the seawater component is absent in this evaluation. According to the current result, the VSIA product performs very well in the low albedo condition; however, the missing seawater component might take some unknown influence into the algorithm performance. To solve this problem, we plan an inter-comparison by introducing other coarser resolution satellite sea-ice albedo products.

5. Influence of sea-ice algae has not been considered

Algae aggregates alter the optical properties including albedo of pond areas [60,61]. The short-lived algal layer growing on the bottom of the ice in springtime resulted in an increasing of the radiance absorption and possible decreasing of surface albedo. This influence was not considered in the theory used for pond simulation in our algorithm, which could be one reason for the overestimation of the VSIA. The improvement of the model to represent the algae optical properties is expected.

6. Some assumptions of the ART model may be violated in the Arctic

The ART model performs well in reproducing the snow/ice reflectance, but still suffers from many problems. First, it is valid for weakly absorbing semi-infinite turbid media [57] so that its validity over the first-year ice still needs further evaluation, since the sample used here is small. Second, the model assumes a flat smooth surface condition. This simplification increases the retrieval uncertainty due to slope and surface roughness [47].

The sea-ice surface roughness also affects the albedo significantly. Increased roughness generally alters the incident and reflected radiation. Its influence on the albedo thus depends on the ratio of diffuse to total incoming radiation. The evaluation of such an effect deserves more extensive work, i.e., considering micro-tomography adjustment in the regression.

5. Conclusions

The surface energy balance of the polar region is mainly dependent upon the surface albedo characteristics. Ice-albedo feedback can cause substantial alteration of absorbed solar energy with even slight fluctuations in albedo; therefore, the demand for fine-resolution satellite sea-ice albedo is increasing. We worked on developing an operational VIIRS sea-ice albedo (VSIA) product and validated the product with nearly six years of in situ measurements at 30 automatic weather stations from PROMICE and GC-NET.

The direct estimation algorithm used in the VSIA product has unique advantages for satellite albedo retrieval. First, it does not need to collect the multi-angle dataset as an input for real-time processing. Second, it generated blue-sky albedo values that can be directly compared with ground measurements.

We used two sources of ground measurements as reference data for VSIA validation. The first is the long-term AWS observed albedo over Greenland. These datasets have high quality and large samples. The comparison reveals good agreement between the VSIA and PROMICE observations with a bias of 0.028 and RMSE of 0.072—a comparable result with historical validation results from the snow-specific algorithm in the Greenland area. At some large SZA conditions, the VSIA uncertainty shows a slight increasing trend, which is a result of combined uncertainties in the LUT precision, observation uncertainty, model accuracy, and spatial heterogeneity influence. The limitation of using the AWS albedo measurements is that the dataset cannot represent the surface conditions other than the optically thick sea ice. Therefore, we also used the ground measurements from previous field experiments over the sea-ice surface as the reference data. The comparison of VSIA with these datasets suffered from many issues including the limited sample size, scale difference, and strong spatial heterogeneity. Still, the VSIA retrievals match the ground measurements in magnitude and roughly reflect the evolution trend. The bias between the retrieved instantaneous albedo and measured albedo is 0.09 in the central Arctic dataset, and the bias between the retrieved daily mean albedo with the measured value is 0.077 in the Alaska dataset. Future continuous evaluation attempts are planned for fully understanding VSIA accuracy, including cross-comparison with other available sea-ice albedo products and direct-validation using more ground measurements.

With the completion of the S-NPP VIIRS granule albedo development, we now have the integration plan of a 1 km gridded global surface albedo product. The gridded albedo product will be map-projected and convenient to overlay with other data sets in applications. Considering the daily composite requirement for a gridded albedo product, we will develop a LUT directly linking TOA reflectance to the daily mean albedo of sea ice. Daily mean albedo will consider the diurnal variation pattern of sea-ice albedo so that the values from different collection times can be calculated together. Moreover, the NOAA-20 (designated as JPSS-1) was launched on 18 November 2017, with the VIIRS carried aboard, and joined the S-NPP satellite in the same orbit. The granule/gridded VSIA albedo will also be produced using NOAA-20 observations.

Author Contributions: Conceptualization, Y.Y and J.P.; Methodology, J.P.; Validation, J.P.; Formal Analysis, J.P.; Writing-Original Draft Preparation, J.P.; Writing-Review & Editing, Y.Y., P.Y., and S.L.; Funding Acquisition, Y.Y.".

Funding: This study was supported by NOAA grant NA14NES4320003 (Cooperative Institute for Climate and Satellites -CICS) at the University of Maryland/ESSIC.

Acknowledgments: This study was supported by JPSS program at NOAA/NESDIS Center for Satellite Applications and Research, for the VIIRS albedo product development. The manuscript contents are solely the opinions of the authors and do not constitute a statement of policy, decision, or position on behalf of NOAA or the U.S. Government. The authors would like to thank Dr. Xiwu Zhan and Dr. Jeffrey R. Key for helpful comments on earlier drafts of the manuscript. Many thanks to Mr. Lizhao Wang for precious suggestions and data support for the study. They are also grateful to Mr. Joshua Hrisko for language editing. We thank two anonymous reviewers for their careful reading of our manuscript and their many insightful comments and suggestions on earlier drafts of the manuscript.

Conflicts of Interest: The authors declare no conflict of interest.

References

1. Kashiwase, H.; Ohshima, K.I.; Nihashi, S.; Eicken, H. Evidence for ice-ocean albedo feedback in the Arctic Ocean shifting to a seasonal ice zone. *Sci. Rep.* **2017**, *7*, 8170. [CrossRef] [PubMed]
2. Cao, Y.; Liang, S.; Chen, X.; He, T. Assessment of sea ice albedo radiative forcing and feedback over the Northern Hemisphere from 1982 to 2009 using satellite and reanalysis data. *J. Clim.* **2015**, *28*, 1248–1259. [CrossRef]
3. Comiso, J.C.; Parkinson, C.L.; Gersten, R.; Stock, L. Accelerated decline in the Arctic sea ice cover. *Geophys. Res. Lett.* **2008**, *35*, L1703. [CrossRef]
4. Hall, A. The role of surface albedo feedback in climate. *J. Clim.* **2004**, *17*, 1550–1568. [CrossRef]
5. Laine, V.; Manninen, T.; Riihelä, A.; Andersson, K. Shortwave broadband black-sky surface albedo estimation for Arctic sea ice using passive microwave radiometer data. *J. Geophys. Res. Atmos.* **2011**, *116*. [CrossRef]
6. Xiong, X.; Stamnes, K.; Lubin, D. Surface albedo over the Arctic Ocean derived from AVHRR and its validation with SHEBA data. *J. appl. Meteorol.* **2002**, *41*, 413–425. [CrossRef]
7. Comiso, J.C. Satellite-observed variability and trend in sea-ice extent, surface temperature, albedo and clouds in the Arctic. *Ann. Glaciol.* **2001**, *33*, 457–473. [CrossRef]
8. De Abreu, R.A.; Key, J.; Maslanik, J.A.; Serreze, M.C.; LeDrew, E.F. Comparison of in situ and AVHRR-derived broadband albedo over Arctic sea ice. *Arctic* **1994**, 288–297. [CrossRef]
9. Lindsay, R.W.; Rothrock, D.A. Arctic sea ice albedo from AVHRR. *J. Clim.* **1994**, *7*, 1737–1749. [CrossRef]
10. Key, J.R.; Wang, X.; Stoeve, J.C.; Fowler, C. Estimating the cloudy-sky albedo of sea ice and snow from space. *J. Geophys. Res. Atmos.* **2001**, *106*, 12489–12497. [CrossRef]
11. Key, J.; Wang, X.; Liu, Y.; Dworak, R.; Letterly, A. The AVHRR polar pathfinder climate data records. *Remote Sens.* **2016**, *8*, 167. [CrossRef]
12. Wang, X.; Key, J.R. Arctic surface, cloud, and radiation properties based on the AVHRR Polar Pathfinder dataset. Part I: Spatial and temporal characteristics. *J. Clim.* **2005**, *18*, 2558–2574. [CrossRef]
13. Riihelä, A.; Laine, V.; Manninen, T.; Palo, T.; Vihma, T. Validation of the Climate-SAF surface broadband albedo product: Comparisons with in situ observations over Greenland and the ice-covered Arctic Ocean. *Remote Sens. Environ.* **2010**, *114*, 2779–2790. [CrossRef]
14. Stroeve, J.; Box, J.E.; Gao, F.; Liang, S.; Nolin, A.; Schaaf, C. Accuracy assessment of the MODIS 16-day albedo product for snow: comparisons with Greenland in situ measurements. *Remote Sens. Environ.* **2005**, *94*, 46–60. [CrossRef]
15. Qu, Y.; Liu, Q.; Liang, S.; Wang, L.; Liu, N.; Liu, S. Direct-estimation algorithm for mapping daily land-surface broadband albedo from MODIS data. *IEEE Trans. Geosci. Remote Sens.* **2014**, *52*, 907–919. [CrossRef]
16. Schaaf, C.B.; Gao, F.; Strahler, A.H.; Lucht, W.; Li, X.; Tsang, T.; Strugnell, N.C.; Zhang, X.; Jin, Y.; Muller, J. First operational BRDF, albedo nadir reflectance products from MODIS. *Remote Sens. Environ.* **2002**, *83*, 135–148. [CrossRef]
17. Comiso, J.C. Large decadal decline of the Arctic multiyear ice cover. *J. Clim.* **2012**, *25*, 1176–1193. [CrossRef]
18. Maslanik, J.; Stroeve, J.; Fowler, C.; Emery, W. Distribution and trends in Arctic sea ice age through spring 2011. *Geophys. Res. Lett.* **2011**, 38. [CrossRef]
19. Wen, J.; Dou, B.; You, D.; Tang, Y.; Xiao, Q.; Liu, Q.; Qinhuo, L. Forward a small-timescale BRDF/Albedo by multisensor combined brdf inversion model. *IEEE Trans. Geosci. Remote Sens.* **2017**, *55*, 683–697. [CrossRef]
20. Riihelä, A.; Manninen, T.; Key, J.; Sun, Q.; Sütterlin, M.; Lattanzio, A.; Schaaf, C. A Multisensor Approach to Global Retrievals of Land Surface Albedo. *Remote Sens.* **2018**, *10*, 848. [CrossRef]
21. Liang, S.; Stroeve, J.; Box, J.E. Mapping daily snow/ice shortwave broadband albedo from Moderate Resolution Imaging Spectroradiometer (MODIS): The improved direct retrieval algorithm and validation with Greenland in situ measurement. *J. Geophys. Res. Atmos.* **2005**, *110*. [CrossRef]
22. Wang, D.; Liang, S.; He, T.; Yu, Y. Direct estimation of land surface albedo from VIIRS data: Algorithm improvement and preliminary validation. *J. Geophys. Res. Atmos.* **2013**, *118*, 12577–12586. [CrossRef]

23. Qu, Y.; Liang, S.; Liu, Q.; Li, X.; Feng, Y.; Liu, S. Estimating Arctic sea-ice shortwave albedo from MODIS data. *Remote. Sens. Environ.* **2016**, *186*, 32–46. [CrossRef]
24. Cao, C.; Xiong, X.; Wolfe, R.; De Luccia, F.; Liu, Q.; Blonski, S.; Lin, G.; Nishihama, M.; Pogorzala, D.; Oudrari, H. *Visible Infrared Imaging Radiometer Suite (VIIRS) Sensor Data Record (SDR) User's Guide; NOAA* Technical Report; NESDIS: College Park, MD, USA, 2013.
25. Hillger, D.; Kopp, T.; Lee, T.; Lindsey, D.; Seaman, C.; Miller, S.; Solbrig, J.; Kidder, S.; Bachmeier, S.; Jasmin, T. First-light imagery from Suomi NPP VIIRS. *Bull. Am. Meteor. Soc.* **2013**, *94*, 1019–1029. [CrossRef]
26. Stamnes, K.; Hamre, B.; Stamnes, J.J.; Ryzhikov, G.; Biryulina, M.; Mahoney, R.; Hauss, B.; Sei, A. Modeling of radiation transport in coupled atmosphere-snow-ice-ocean systems. *J. Quant. Spectrosc. Radiat. Transf.* **2011**, *112*, 714–726. [CrossRef]
27. Feng, Y.; Liu, Q.; Qu, Y.; Liang, S. Estimation of the ocean water albedo from remote sensing and meteorological reanalysis data. *IEEE Trans. Geosci. Remote Sens.* **2016**, *54*, 850–868. [CrossRef]
28. Zege, E.; Malinka, A.; Katsev, I.; Prikhach, A.; Heygster, G.; Istomina, L.; Birnbaum, G.; Schwarz, P. Algorithm to retrieve the melt pond fraction and the spectral albedo of Arctic summer ice from satellite optical data. *Remote Sens. Environ.* **2015**, *163*, 153–164. [CrossRef]
29. Morassutti, M.P.; LeDrew, E.F. Albedo and depth of melt ponds on sea-ice. *Int. J. Climatol. J. Roy. Meteor. Soc.* **1996**, *16*, 817–838. [CrossRef]
30. Vermote, E.F.; Tanré, D.; Deuze, J.L.; Herman, M.; Morcette, J. Second simulation of the satellite signal in the solar spectrum, 6S: An overview. *IEEE Trans. Geosci. Remote Sens.* **1997**, *35*, 675–686. [CrossRef]
31. Gardner, A.S.; Sharp, M.J. A review of snow and ice albedo and the development of a new physically based broadband albedo parameterization. *J. Geophys. Res: Earth Surf.* **2010**, *115*. [CrossRef]
32. Key, J.R.; Mahoney, R.; Liu, Y.; Romanov, P.; Tschudi, M.; Appel, I.; Maslanik, J.; Baldwin, D.; Wang, X.; Meade, P. Snow and ice products from Suomi NPP VIIRS. *J. Geophys. Res. Atmos.* **2013**, *118*, 12–816. [CrossRef]
33. Liu, N.F.; Liu, Q.; Wang, L.Z.; Liang, S.L.; Wen, J.G.; Qu, Y.; Liu, S.H. A statistics-based temporal filter algorithm to map spatiotemporally continuous shortwave albedo from MODIS data. *Hydrol. Earth Syst. Sci.* **2013**, *17*, 2121–2129. [CrossRef]
34. Moustafa, S.E.; Rennermalm, A.K.; Román, M.O.; Wang, Z.; Schaaf, C.B.; Smith, L.C.; Koenig, L.S.; Erb, A. Evaluation of satellite remote sensing albedo retrievals over the ablation area of the southwestern Greenland ice sheet. *Remote Sens. Environ.* **2017**, *198*, 115–125. [CrossRef]
35. Ahlstrøm, A.P.; Gravesen, P.; Andersen, S.B.; Van As, D.; Citterio, M.; Fausto, R.S.; Nielsen, S.; Jepsen, H.F.; Kristensen, S.S.; Christensen, E.L. PROMICE project team. 2008. A new programme for monitoring the mass loss of the Greenland ice sheet. *Geol. Surv. Den. Greenl. Bull.* **2007**, *15*, 61–64.
36. Bøggild, C.E.; Brandt, R.E.; Brown, K.J.; Warren, S.G. The ablation zone in northeast Greenland: ice types, albedos and impurities. *J. Glaciol.* **2010**, *56*, 101–113. [CrossRef]
37. Ryan, J.C.; Hubbard, A.; Irvine Fynn, T.D.; Doyle, S.H.; Cook, J.M.; Stibal, M.; Box, J.E. How robust are in situ observations for validating satellite-derived albedo over the dark zone of the Greenland Ice Sheet? *Geophys. Res. Lett.* **2017**, *44*, 6218–6225. [CrossRef]
38. Steffen, K.; Box, J.E.; Abdalati, W. Greenland climate network: GC-Net. *US Army Cold Reg. Reattach Eng. (CRREL), CRREL Spec. Rep.* **1996**, 98–103.
39. Istomina, L.; Nicolaus, M.; Perovich, D.K. Surface spectral albedo complementary to ROV transmittance measurements at 6 ice stations during POLARSTERN cruise ARK-XXVII/3 (IceArc) in 2012, PANGAEA. 2016. [CrossRef]
40. Dou, T.; Xiao, C.; Du, Z.; Schauer, J.J.; Ren, H.; Ge, B.; Xie, A.; Tan, J.; Fu, P.; Zhang, Y. Sources, evolution and impacts of EC and OC in snow on sea ice: a measurement study in Barrow, Alaska. *Sci. Bull.* **2017**, *62*, 1547–1554. [CrossRef]
41. Hudson, S.R.; Granskog, M.A.; Karlsen, T.I.; Fossan, K. Horizontal profiles of longwave and shortwave radiation components over sea ice near Barrow, Alaska during the 2011 melt, PANGAEA. 2012. [CrossRef]
42. Wientjes, I.; Van de Wal, R.; Reichart, G.; Sluijs, A.; Oerlemans, J. Dust from the dark region in the western ablation zone of the Greenland ice sheet. *Cryosphere* **2011**, *5*, 589–601. [CrossRef]
43. Tedesco, M.; Fettweis, X.; Van den Broeke, M.R.; Van de Wal, R.; Smeets, C.; van de Berg, W.J.; Serreze, M.C.; Box, J.E. The role of albedo and accumulation in the 2010 melting record in Greenland. *Environ Res. Lett.* **2011**, *6*, 14005. [CrossRef]

44. Mernild, S.H.; Malmros, J.K.; Yde, J.C.; Knudsen, N.T. Multi-decadal marine-and land-terminating glacier recession in the Ammassalik region, southeast Greenland. *Cryosphere* **2012**, *6*, 625–639. [CrossRef]
45. Ryan, J.C.; Hubbard, A.; Stibal, M.; Irvine-Fynn, T.D.; Cook, J.; Smith, L.C.; Cameron, K.; Box, J. Dark zone of the Greenland Ice Sheet controlled by distributed biologically-active impurities. *Nat. Commun.* **2018**, *9*, 1065. [CrossRef] [PubMed]
46. Grenfell, T.C.; Perovich, D.K. Seasonal and spatial evolution of albedo in a snow-ice-land-ocean environment. *J. Geophys. Res. Atmos Oceans* **2004**, *109*. [CrossRef]
47. Dumont, M.; Arnaud, L.; Picard, G.; Libois, Q.; Lejeune, Y.; Nabat, P.; Voisin, D.; Morin, S. In situ continuous visible and near-infrared spectroscopy of an alpine snowpack. *Cryosphere* **2017**, *11*, 1091–1110. [CrossRef]
48. Wang, W.; Cao, C.; Bai, Y.; Blonski, S.; Schull, M.A. Assessment of the NOAA S-NPP VIIRS Geolocation Reprocessing Improvements. *Remote Sens.* **2017**, *9*, 974. [CrossRef]
49. Wang, X.; Zender, C.S. MODIS snow albedo bias at high solar zenith angles relative to theory and to in situ observations in Greenland. *Remote Sens. Environ.* **2010**, *114*, 563–575. [CrossRef]
50. Natural Earth. Available online: https://www.naturalearthdata.com/downloads/50m-cross-blend-hypso/50m-cross-blended-hypso-with-shaded-relief/URL (accessed on 16 November 2018).
51. Singh, M.K.; Gautam, R.; Gatebe, C.K.; Poudyal, R. PolarBRDF: A general purpose Python package for visualization and quantitative analysis of multi-angular remote sensing measurements. *Comput. Geosci.* **2016**, *96*, 173–180. [CrossRef]
52. Gardner, A.S.; Sharp, M.J. A review of snow and ice albedo and the development of a new physically based 770 broadband albedo parameterization. *J. Geophys. Res: Earth Surf.* **2010**, *115*, F1. [CrossRef]
53. Perovich, D.K.; Grenfell, T.C.; Light, B.; Hobbs, P.V. Seasonal evolution of the albedo of multiyear 772 Arctic sea ice. *J. Geophys. Res. Atmos Oceans* **2002**, *107*, SHE-20.
54. De Abreu, R.A.; Barber, D.G.; Misurak, K.; LeDrew, E.F. Spectral albedo of snow-covered first-year and multi-year sea ice during spring melt. *Ann. Glaciol.* **1995**, *21*, 337–342. [CrossRef]
55. Wang, W.; Zender, C.S.; Van As, D.; Smeets, P.; van den Broeke, M.R. A Retrospective, Iterative, Geometry-776 Based (RIGB) tilt correction method for radiation observed by Automatic Weather Stations on snow-777 covered surfaces: application to Greenland. *Cryosphere Discuss.* **2015**, *9*. [CrossRef]
56. Kipp, Z. CM3 Pyranometer instruction manual. Campbell Scientific Inc., 2002. Available online: https://s.campbellsci.com/documents/au/manuals/cm3.pdf.URL (accessed on 16 November 2018).
57. Zege, E.P.; Katsev, I.L.; Malinka, A.V.; Prikhach, A.S.; Heygster, G.; Wiebe, H. Algorithm for retrieval of the effective snow grain size and pollution amount from satellite measurements. *Remote Sens. Environ.* **2011**, *115*, 2674–2685. [CrossRef]
58. Zege, E.; Katsev, I.; Malinka, A.; Prikhach, A.; Polonsky, I. New algorithm to retrieve the effective snow grain size and pollution amount from satellite data. *Ann. Glaciol.* **2008**, *49*, 139–144. [CrossRef]
59. Peng, J.; Liu, Q.; Wen, J.; Liu, Q.; Tang, Y.; Wang, L.; Dou, B.; You, D.; Sun, C.; Zhao, X. Multi-scale validation strategy for satellite albedo products and its uncertainty analysis. *Sci. China Earth Sci.* **2014**, *58*, 573–588. [CrossRef]
60. Malinka, A.; Zege, E.; Istomina, L.; Heygster, G.; Spreen, G.; Perovich, D.; Polashenski, C. Reflective properties of melt ponds on sea ice. *Cryosphere* **2018**, *12*, 1921–1937. [CrossRef]
61. Perovich, D.K.; Roesler, C.S.; Pegau, W.S. Variability in Arctic sea ice optical properties. *J. Geophys. Res. Atmos Oceans* **1998**, *103*, 1193–1208. [CrossRef]

Article

Measuring Landscape Albedo Using Unmanned Aerial Vehicles

Chang Cao [1,*], Xuhui Lee [1,2,*], Joseph Muhlhausen [3], Laurent Bonneau [4] and Jiaping Xu [5]

[1] Yale-NUIST Center on Atmospheric Environment & Jiangsu Key Laboratory of Agriculture Meteorology, Nanjing University of Information Science & Technology, Nanjing 210044, China
[2] School of Forestry and Environmental Studies, Yale University, New Haven, CT 06511, USA
[3] WeRobotics, 1812 Bolton Street, Baltimore, MD 21217, USA; joseph@werobotics.org
[4] Center for Earth Observation, Yale University, New Haven, CT 06511, USA; laurent.bonneau@yale.edu
[5] Key Laboratory of Transportation Meteorology, China Meteorological Administration & Jiangsu Institute of Meteorological Sciences, Nanjing 210009, China; fengxuxudechui@sina.com
* Correspondence: ichangnj@sina.com (C.C.); xuhui.lee@yale.edu (X.L.); Tel.: +86-025-5869-5681 (C.C.); +1-203-432-6271 (X.L.)

Received: 11 September 2018; Accepted: 13 November 2018; Published: 15 November 2018

Abstract: Surface albedo is a critical parameter in surface energy balance, and albedo change is an important driver of changes in local climate. In this study, we developed a workflow for landscape albedo estimation using images acquired with a consumer-grade camera on board unmanned aerial vehicles (UAVs). Flight experiments were conducted at two sites in Connecticut, USA and the UAV-derived albedo was compared with the albedo obtained from a Landsat image acquired at about the same time as the UAV experiments. We find that the UAV estimate of the visibleband albedo of an urban playground (0.037 ± 0.063, mean ± standard deviation of pixel values) under clear sky conditions agrees reasonably well with the estimates based on the Landsat image (0.047 ± 0.012). However, because the cameras could only measure reflectance in three visible bands (blue, green, and red), the agreement is poor for shortwave albedo. We suggest that the deployment of a camera that is capable of detecting reflectance at a near-infrared waveband should improve the accuracy of the shortwave albedo estimation.

Keywords: Unmanned Aerial Vehicle (UAV); albedo; landscape; consumer-grade camera; radiometric calibration

---

1. Introduction

Surface albedo is a key parameter in the surface energy balance, and it therefore plays an important role in land–climate interactions. As a key biophysical property of land ecosystems, surface albedo can change throughout the season, due to changes in the vegetation morphology, and it can also be affected by sky conditions [1]. Quantification of the surface albedo at the landscape scale is still subject to many sources of uncertainty, especially over urban land [1,2].

Satellite remote sensing has been widely used for the determination of land surface albedo [3–6]. An advantage of satellite monitoring is that it provides global coverage. New satellites can provide albedo measurements at reasonably high frequencies (2–3 days in the best case for Sentinel 2) and spatial resolutions (pixel size 10 m in the case of Sentinel 2, and several cm in the case of DigitalGlobe) to provide useful information for studies on ecosystem (tens of meters) to landscape (several kilometers to tens of kilometers) scales. However, all satellite measurements are biased towards cloud-free sky conditions. In urban landscapes with heavy haze pollution, retrieval of the true surface albedo from satellite imageries must remove signal contamination caused by particle scattering. Lightweight unmanned aerial vehicles (UAVs) as an alternative for albedo monitoring may be able to overcome these limitations. UAVs can cover areas ranging from 0.01 km^2 to 100 km^2, depending on battery life and type of UAV [7]. They provide measurements at sub-decimeter spatial resolutions, and they can be used to obtain data under both clear sky and cloudy conditions [8,9]. UAV experiments can be conducted at almost any time, and at any locations [10–12]. Furthermore, the labor and financial expense of UAVs are much lower than those of aircraft [13]. Finally, UAVs can measure albedos at locations that are not accessible by ground-based instruments, such as steep rooftops in cities.

In a typical UAV experiment, consumer-grade digital cameras are utilized as multispectral sensors, similar to their counterparts on board satellites, to measure the spectral radiance reflected by ground targets, typically in the red, green, and blue wavebands and occasionally with modification to include a near-infrared (NIR) waveband. These at-camera radiance data are stored as digital numbers (DN), usually with an 8-bit resolution ranging from 0 to 255, to represent the brightness of the targeted object [14]. Vegetation indices derived from the spectral information [15] allow for the monitoring of vegetation growth status [16–18], and the estimation of crop biomass [19]. Because the UAV flies below cloud layers, cloud interference is no longer a problem. Also, because of the low flight altitude in typical UAV missions, the at-sensor radiance is a direct measure of the actual surface reflected radiance, but this is not true for satellite monitoring (Even with UAVs flying at higher flight altitudes, atmospheric interference is still much less severe than with satellite monitoring). Alternatively, a UAV can be deployed as a platform to carry pyranometers to measure albedo. For example, Levy et al. measured surface albedo over vegetation using a ground-based pyranometer paired with a pyranometer mounted on a quadcopter [20].

Although increasingly being used to produce image mosaics and to construct 3D point clouds of surface features, consumer-grade cameras on board of UAVs have rarely been used to study land surface albedo. The land surface albedo here refers to blue sky albedo, which means that it is the albedo under ambient illumination conditions. By using a Finnish Geodetic Institute Field Goniospectrometer, which includes a fisheye camera on a fixed-wing drone, Hakala et al. [21] estimated the bidirectional reflectance distribution function (BRDF) of a snow surface. Ryan et al. [22] measured snow albedo in the Arctic region with pyranometers on board a fixed-wing UAV. Since consumer-grade digital cameras are not calibrated for radiance measurements, many internal factors, such as Gamma correction, color filter array interpolation (CFA), and the vignetting effect, can contribute to radiometric instability [23]. Gamma correction is a nonlinear transformation of the electro-photo signal received by the detector at the focal plane of the camera to a DN output. The DN values and their corresponding raw data are not exactly 1-to-1 matched [22]. Because there is only one band value (e.g., green) for each pixel on a charge-coupled device (CCD) or a complementary metal-oxide-semiconductor (CMOS), the other band values (i.e., red and blue) have to be estimated by using the CFA interpolation method. The vignetting effect refers to the phenomenon whereby objects farther away from the image center will appear darker. All of these factors must be accounted for if the DN value is to be converted to true reflectance. Some researchers argue that using raw images from the camera instead of the compressed JEPG or TIFF images can avoid the Gamma correction and CFA [22]. Applying a vignetting mask on the images can effectively alleviate the vignetting effect [24].

Determination of the albedo with UAV data consists of three steps. First, accurate albedo determination requires radiometric calibration of the camera's DN values, to represent physically meaningful surface reflectance. Currently, radiometric calibration methods fall into two categories: absolute and relative. In absolute calibration, the calibration function that relates the DN value to the reflectance is based on the measurement of an accurately known, uniform radiance field [25]. Relative calibration is especially needed for sensors having more than one detector per band. By normalizing the outputs of different detectors, a uniform response can be obtained [25–28].

The next step after the radiometric correction involves the conversion of the spectral reflectance at the camera viewing angle to the hemispheric reflectance or spectral albedo [29,30], ideally using a BRDF of the target. Because the determination of the BRDF requires measurements at multiple illumination and viewing angles, a process that requires elaborate preparation and post-processing [21], it is commonly assumed that all the objects are Lambertian [22]. The albedo values that we retrieved in this study should be considered as "Lambertian-equivalent albedos". Without the consideration of the BRDF effect, spectral reflectance is essentially the same as the spectral albedo [29,30].

The third step is to convert the spectral albedo of discrete bands to the broadband albedo, that is, the total reflectance in all directions in either the visible band (wavelength 380 nm to 760 nm) or the shortwave band (250 nm to 2500 nm). Wang et al. [30] have established an empirical method to convert the spectral albedo measured by Landsat 8 to the broadband albedo. Here, we propose that the same conversion equation can be applied to UAV albedo estimation.

The objective of this study is to develop a UAV method for determining the landscape albedo. The method was tested at two sites typical of urban landscapes, and consisting of impervious and vegetation surfaces. The visible and shortwave band albedo derived from our method were compared with those of Landsat 8. This method can save labor cost, and it can be applied to the landscape albedo estimation where direct field measurement may be difficult.

2. Materials and Methods

2.1. UAV Experiments

We conducted two UAV flights along predefined routes (Table 1). One took place in a playground on the Yale University campus (41.317°N, 72.928°W) on 30 September 2015, and the other in Brooksvale Recreational Park, in Hamden, Connecticut, USA (41.453°N, 72.918°W) on 9 October 2015. The sky condition was clear on 30 September, 2015, and overcast on 9 October, 2015. For the Yale Playground, a quad-rotor drone equipped with a GARMIN VIRB-X digital camera was used for the image acquisition (Figure 1a). A fixed-wing drone designed and assembled by CielMap [31] equipped with a Sony NEX-5N camera was used in the second experiment (Figure 1b). Both cameras had fixed aperture and automatic shutter speed. The spectral sensitivity of the Sony NEX 5N camera can be found in Ryan et al. [22]. However, the spectral sensitivity of GARMIN VIRB-X was not released by the manufacturer. Although Ryan et al. [22] used raw images for retrieving ice sheet reflectance, the study by Lebourgeois [23] showed little improvement from using the raw images over JPEG images. They demonstrated that JPEG and RAW imagery data have a linear relationship for the range of DN values needed for crop monitoring. In this study, we used the compressed JPEG images.

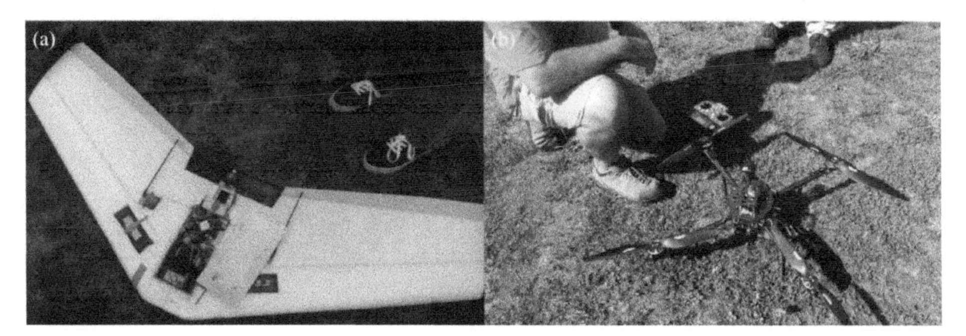

Figure 1. The unmanned aerial vehicles (UAVs) before launch. (**a**,**b**) are the fixed-wing and rotor wing UAVs for Brooksvale Park and Yale Playground, respectively.

Table 1. Information about the study sites and the drone experiments. Here, image overlap is defined as the number of photos that were sampled in the same pixel. For example, an overlap of nine means that each pixel is seen by at least nine photos.

	Brooksvale Recreation Park	Yale Playground
Location	41.453°N 72.918°W	41.317°N 72.928°W
Drone experiment date	9 October 2015	30 September 2015
Drone flight time	10:00 to 10:30	14:30 to 15:00
Sky conditions	Overcast	Clear sky
Flight duration	30 min	20 min
Flight altitude (m)	120	90
Camera	Sony NEX-5N	GARMIN VIRB-X
UAV platform	Fixed-wing	Quad-rotor
Forward overlap	80%	80%
Side overlap	60%	60%
Image overlap	>9	>9
Area (km^2)	0.065	0.014

2.2. Image Processing

Agisoft Photoscan Professional Pro software 1.1.0 (Agisoft LCC, St. Petersburg, Russia) was used to generate ortho-mosaicked images for both experiments. The software has built-in structure-from-motion (SfM) and other multiview stereo algorithms. The general workflow involves aligning photos, placing ground control points, building dense points, building textures, generating orthomosaics and the digital elevation model. All of the photos underwent image quality estimation in Photoscan, and those with quality flag values under 0.5 were rejected from processing. The vignetting effect can be avoided to some degree; due to the mosaic mode of texture generation in Photoscanwe chose [32]. Instead of calculating the average value of all pixels from individual photos that overlapped on the same point, this mosaic mode only uses the value where the pixel in interest is located within the shortest distance from the image center [32].

The Environment for Visualizing Images software (ENVI, version 5.1, Harris Corporation, Melbourne, FL, USA) was utilized to conduct a classification of the mosaicked images, in order to distinguish non-vegetation and vegetation pixel types. The supervised classification scheme used for the Brooksvale Park was maximum likelihood, and for the Yale Playground, it was the spectral angle mapper. The Yale Playground was severely affected by shadows of trees and buildings because the solar elevation angle was low at the time of the experiment. The spectral angle mapper directly compares the spectra of images to known spectra, and creates vectors in order to calculate the spectral angle between them [33]. Therefore, this classifier is not sensitive to illumination conditions. The mosaicked images (Figure 2) were then used for the determination of landscape albedo.

Figure 2. Mosaicked images of Brooksvale Park (**a**) and Yale Playground (**b**). The mosaicked images have a resolution of 4 cm per pixel.

2.3. Spectrometer Measurement of Ground Targets

A high resolution spectrometer (FieldSpec Pro FR, Malvern Panalyical Ltd., Malvern, UK) was used to record the reflectance spectra of ground targets, including non-vegetative features and ground vegetation. Before the measurement of each ground target, the spectrometer was calibrated using a white reference disc (Spectralon, Labsphere). The spectrometer field experiments were carried out under both overcast and clear sky conditions at each field site (Table 2). The UAV and spectrometer field experiments were not conducted on the same day, as the former were done in the autumn of 2015 and the latter in the spring of 2016, and therefore the vegetation conditions were different. However, the solar elevation angle did not differ much in the case of the Yale Playground. A pistol grip was used to measure the single point of each ground target five times. A typical standard deviation of the reflectance in the visible bands for each ground target was less than 0.02. The non-vegetation ground targets included a wide range of brightness, from black dustbins to white-paint markings (Figure S1). Vegetation ground targets were grass at both sites. The Lambertian assumption was adopted so that the spectral reflectance measured by the spectrometer was taken as the spectral albedo. The wavelength ranges for the red, green, and blue bands in this study were defined as 620–670 nm, 540–560 nm and 460–480 nm, respectively, coinciding with the three color wavebands of the cameras.

Table 2. Dates of the spectrometer field experiments.

Sky Conditions	Brooksvale Park	Yale Playground
Clear	28 April 2016 10:00	19 April 2016 14:30
Overcast	7 March 2016 10:00	28 April 2016 14:30

We used the spectrometer measurements for three purposes. The first purpose was to calibrate the mosaicked image. A calibration curve for each of the three wavebands was established by comparing the measured spectral albedo with the DN value of the same ground target identified in the mosaicked image. These curves were then used to convert the DN values of all the image pixels to a spectral albedo.

The second purpose was to determine the broadband (visible and shortwave) albedos of these ground targets. We used the Simple Model of Atmospheric Radiative Transfer of Sunshine program (SMART, version 2.9.5) developed by the National Renewable Energy Laboratory, United States of America Department of Energy [34], to simulate the spectral irradiance of solar radiation. In the SMART simulation, the U.S. standard atmosphere was chosen as the reference atmosphere, aerosol model was set as urban type, and the sky condition was either clear or overcast. The parameters used for the SMART calculations are given in Supplementary Table S1. The data from the SMART and spectrometer had a 1 nm spectral resolution. The broadband albedo (visible or shortwave) was computed as:

$$\alpha_* = \frac{\sum_a^b \rho(\lambda) I(\lambda)}{\sum_a^b I(\lambda)} \quad (1)$$

where α_* is broadband albedo of the ground target, $I(\lambda)$ is the solar spectral irradiance, λ is wavelength, $\rho(\lambda)$ is the spectral reflectance recorded by the spectrometer at wavelength λ, and a and b denote the range of the waveband. For the visible band, a and b are 400 nm and 760 nm, respectively, and for the shortwave, they are 400 nm and 1750 nm, respectively. These albedo values were then compared with the albedo values estimated with the satellite algorithm. It should be noticed that our shortwave band (400–1750 nm), which represents the effective range of the spectrometer measurement, is narrower than the typical shortwave definition of 400–2500 nm, and therefore, our shortwave albedo that we derived here may lose a small contribution from energy at 1750–2500 nm.

The third purpose was to determine a factor for converting the visible band albedo to the shortwave band albedo. This conversion factor is needed in order to obtain an estimate of the landscape shortwave albedo from the drone mosaicked image, because the drone image consisted of only three visible bands. From the visible and shortwave band albedos for the ground targets, we determined a mean ratio of shortwave to visible band albedo for non-vegetation features, and a mean ratio for the vegetation features.

2.4. Landscape Albedo Estimation

Figure 3 depicts the workflow of albedo estimation at the landscape scale. (i) A mosaicked image of the landscape was produced from the drone photographs using the Agisoft Photoscan software. (ii) The calibration functions based on the spectrometer measurement were used to convert the DN value of each pixel in the mosaicked image to spectral albedo in the three wavebands (red, green, and blue). (iii) The Landsat 8 visible band albedo algorithm (Equation (2) below) was validated with the visible band albedo of the ground targets. The validated Landsat 8 conversion algorithm was then used to determine the visible band albedo of each pixel in the whole image. The Landsat8 algorithm is given as [30]:

$$\alpha_{vis/Landsat8} = 0.5621\alpha_2 + 0.1479\alpha_3 + 0.2512\alpha_4 - 0.0015 \quad (2)$$

Here, α_2, α_3, and α_4 represent blue, green, and red spectral albedos calculated from (ii), respectively. (iv) Pixels in the mosaicked image was classified as vegetation and non-vegetation types. (v) The shortwave albedo of the vegetation and non-vegetation pixels was obtained by multiplying their visible band albedo with the ratio of shortwave to visible band albedo obtained with the spectrometer for the vegetation ground targets and for the non-vegetation targets, respectively. (vi) The landscape shortwave albedo was calculated as the mean value of the pixels in the drone image mosaic. The visible and shortwave band albedo values were given as mean $\pm$ 1 standard deviation of all the pixel albedo values in the mosaic.

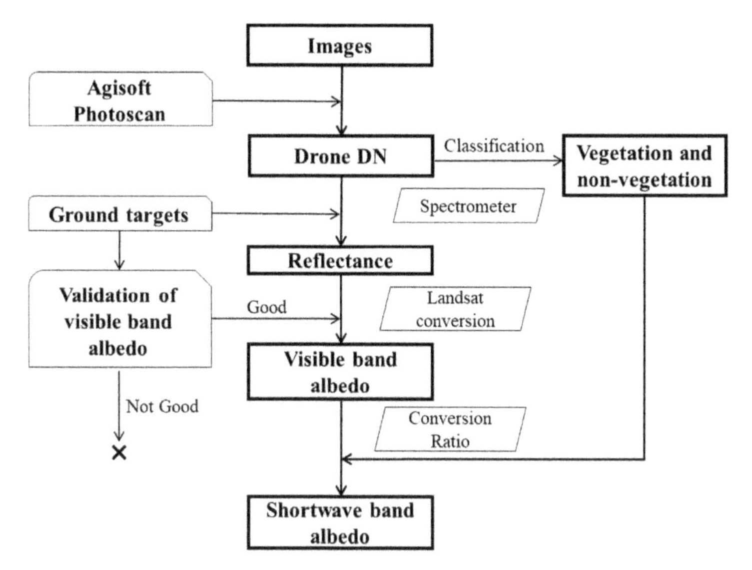

Figure 3. Workflow for estimating landscape visible and shortwave band albedo.

2.5. Retrieval of Landscape Albedo from the Landsat Satellite

Landsat 8 Operational Land Imager (OLI) surface reflectance products were used as a reference for evaluating the landscape visible and shortwave band albedo obtained with the drone images. These products have been atmospherically corrected from the Landsat 8 top of atmosphere reflectance by using the second simulation of the satellite signal in the Solar Spectrum Vectorial model [35]. They performed better than Landsat 5/7 products, by taking advantage of the new OLI coastal aerosol band (0.433–0.450 μm), which is beneficial for detecting aerosol properties [35]. The Landsat image was acquired on 6 October 2015 under clear sky condition and the WRS_PATH and Row were 13 and 31 respectively. The corresponding surface reflectance product can be ordered from https://earthexplorer.usgs.gov/. We used 72 pixels on the image that corresponded roughly to the drone image of the Brooksvale Park, and 16 pixels for the Yale Playground. The Landsat camera has a relatively small field of view (15°), and therefore, the BRDF correction is not considered in its surface reflectance product [36]. We used the Landsat 8 snow-free visible (Equation (2)) and shortwave band albedo coefficients (Equation (3)) to obtain the Landsat 8 validation values [30].

$$\alpha_{SW/Landsat8} = 0.2453\alpha_2 + 0.0508\alpha_3 + 0.1804\alpha_4 + 0.3081\alpha_5 + 0.1332\alpha_6 + 0.0521\alpha_7 + 0.0011 \quad (3)$$

Here, α_2, α_3, α_4, α_5, α_6, and α_4 represent the spectral surface reflectances of six bands of Landsat 8 (450–510 nm, 530–590 nm, 640–670 nm, 850–880 nm, 1570–1650 nm, and 2110–2290 nm, respectively) [37]. Although it is possible to use the information retrieved from MODIS to make BRDF corrections to the Landsat albedo [29,30], this correction was not performed here, to be consistent with the drone methodology, which does not account for BRDF behaviors either.

3. Results

3.1. Relationship Between the DN Values and Spectral Reflectance

A desirable fitting function of the DN values and the spectral reflectance should meet two requirements: (1) when the reflectance reaches zero, the DN value should also reach zero, and (2) the relationship should be nonlinear because of gamma correction. In the study by Lebourgeois et al. [23], the relationship between the DN values in the raw and the compressed image format is logarithmic. Inspired by their result, we adopted the following fitting function for spectral calibration:

$$y = a\,[\ln(x + 1)]^{\,b} \tag{4}$$

where x is the DN value of ground targets, y is spectral reflectance, and a and b are fitting coefficients. Equation (4) guarantees that the pixel reflectance is always positive.

This function produced a robust regression fit to the spectral reflectance of the ground targets observed in the Brooksvale Park (Figure 4) and the Yale Playground (Figure 5). The coefficients of determination (R^2) for Brooksvale Park were greater than 0.60 ($p < 0.05$) and those for the Yale Playground were greater than 0.40 ($p < 0.05$). For Brooksvale Park, all of the data points followed the fitting line closely. For the Yale Playground, there were two outliers: a yellow pavement mark, and a red brick (Figure 5). Such discrepancy may be indicative that these ground targets were not Lambertian reflectors. The view angle of the spectrometer was nadir. If the drone was relatively stable during the flight, the camera was levelled, and if only central pixels were used to form the mosaic, the camera angle would also be perfectly nadir. However, because the mosaic had pixels from other parts of the original photos, and because the camera position could deviate from the vertical, the actual camera angle viewing these targets may differ from the nadir. For this reason, a BRDF is required to correct for the non-Lamberstian behaviors, which is beyond the scope of this study.

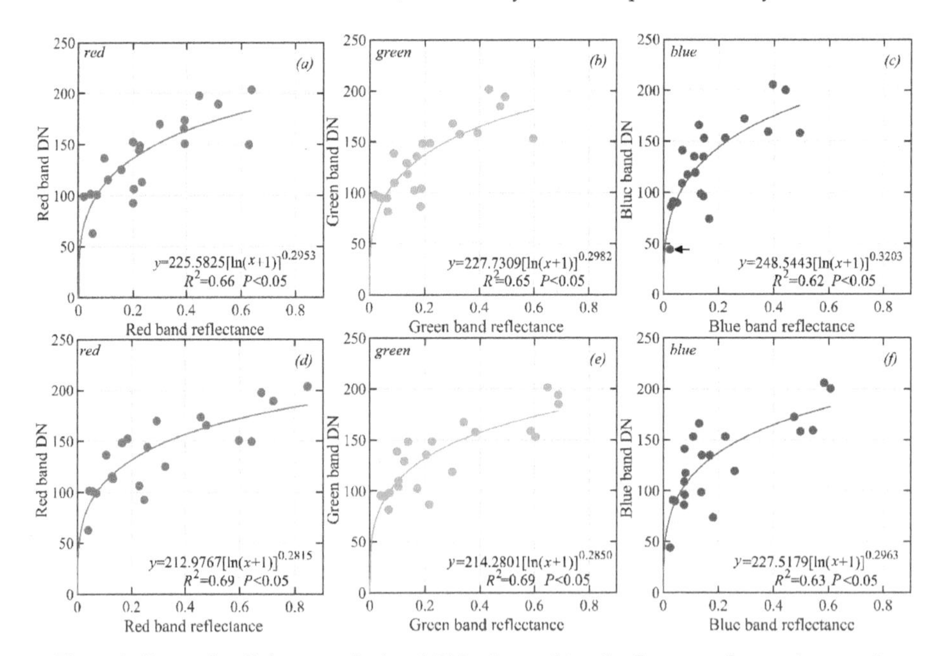

Figure 4. Regression fit between the band DN value and band reflectance of ground targets in Brooksvale Park. Panels (**a**–**c**) are for clear sky conditions, and (**d**–**f**) are for overcast sky conditions under which the spectrometer measurement took place. Also shown are the regression equation, coefficient of determination (R^2), and the confidence level (p).

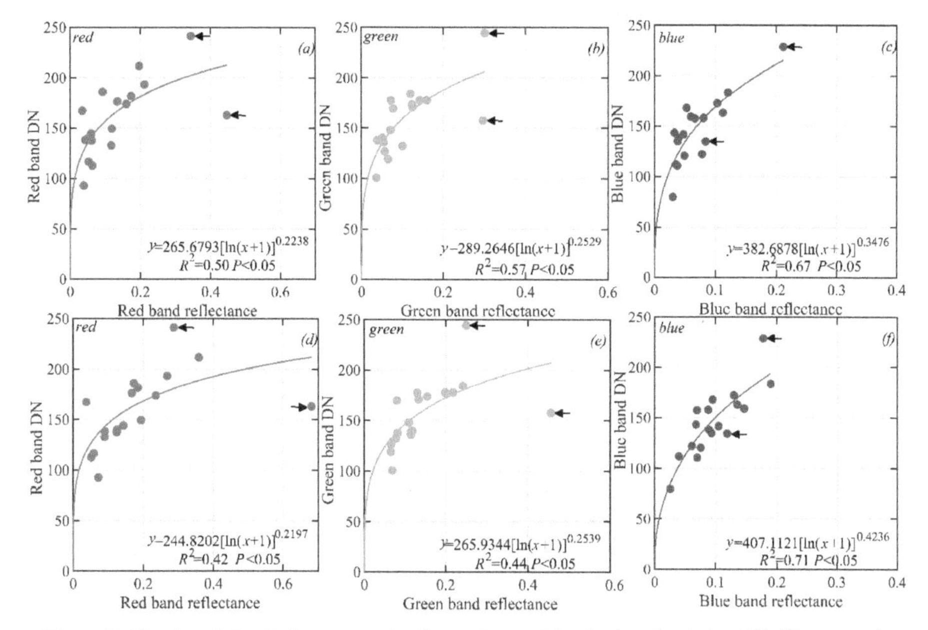

Figure 5. The description is the same as in Figure 4 except for the location being Yale Playground. Arrows indicate outliers discussed in the text.

3.2. Landscape Visible-Band and Shortwave Albedo

Applying the calibration functions (Figure 4d–f) obtained under overcast sky conditions to the pixels in Brooksvale Park, we obtained a landscape-level mean visible band albedo of 0.086 ± 0.110 (Table 3). The landscape visible band albedo of Yale Playground was 0.037 ± 0.063 according to the drone measurement. Here, the drone albedo was obtained using the clear-sky calibration functions (Figure 5a–c). The reader is reminded that the standard deviations here were computed from the albedo values of the pixels in the mosaics, and they therefore are indication of the variations across the landscapes, rather than uncertainties of our estimation.

Table 3. Comparison of drone-derived and Landsat 8 visible and shortwave band albedo under clear and overcast sky conditions for the Brooksvale Park and the Yale Playground. Refer to Supplementary Figures S3–S6 for the spatial distributions of these albedo values.

	Brooksvale Park	**Yale Playground**
Drone-derived visible band albedo	c: 0.077 ± 0.091	c: 0.037 ± 0.063
	o: 0.086 ± 0.110	o: 0.054 ± 0.090
Landsat 8 visible band albedo	0.054 ± 0.011	0.047 ± 0.012
Drone-derived shortwave band albedo	c: 0.261 ± 0.395	SN: 0.054 ± 0.074
	o: 0.332 ± 0.527	SV: 0.061 ± 0.076
Landsat 8 shortwave band albedo	0.103 ± 0.019	0.128 ± 0.013

c and o represent clear and overcast sky conditions, respectively; SN and SV represent that the shadow on the Yale Playground were taken as non-vegetation and vegetation, respectively.

The ratio of the shortwave to the visible band albedo of the non-vegetation and vegetation ground targets, are shown in Table 4. To estimate the shortwave albedo at the landscape scale, we first performed a classification of the mosaicked images. The results are illustrated in Figure S2. The vegetation and non-vegetation pixels occupied 61% and 39% of the land area in Brooksvale Park, respectively. Their visible band albedo values were multiplied by the conversion factors obtained for the vegetation and non-vegetation ground targets under overcast conditions, respectively, to obtain the shortwave albedo values. Averaging over the whole scene yielded a landscape shortwave albedo of 0.332 ± 0.527 under an overcast sky condition.

Table 4. Ratio between the shortwave and visible band albedo obtained with the spectrometer for the vegetation and non-vegetation targets of the Brooksvale Park and the Yale Playground.

Sky Condition	Brooksvale Park		Yale Playground	
	Vegetation	Non-Vegetation	Vegetation	Non-Vegetation
Clear	5.08	1.18	3.91	1.24
Overcast	6.76	1.20	5.29	1.18

At the Yale Playground, a large portion of the pixels were in shadow, with low band reflectance. The spectral information in the three visible bands was insufficient for the classifier to identify which of these pixels were vegetation, and which were non-vegetation. Therefore for the Yale Playground, the image was divided into three classes: vegetation (25%), non-vegetation (50%), and shadow (25%). For the vegetation and non-vegetation pixels, the conversion factors obtained under clear sky conditions were used to estimate their shortwave albedo. For the pixels in shadow, the conversion factors obtained under overcast sky conditions were more appropriate. Since the pixels in shadows were not identifiable, we first assumed that all of them were vegetation (grass, SV), and by applying the conversion factor for vegetation (5.29), we arrived at an estimate of the landscape albedo of 0.061 ± 0.076. We then assumed that all of the pixels in shadows were non-vegetation (SN), obtaining a landscape albedo estimate of 0.054 ± 0.074. These two estimation did not differ by much. The actual albedo value of Yale Playground should fall between these two bounds.

4. Validation

4.1. Validation of LANDSAT Visible Band Albedo Conversion Algorithm

We used the spectrometer data to validate the Landsat albedo conversion algorithm (Equation (2)). As shown in Figure 6, the visible band albedo from the Landsat algorithm was highly correlated with the spectrometer measurement, with a linear correlation coefficient greater than 0.99, and a p value of less than 0.001. The mean bias error (Landsat minus spectrometer) was 0.01 under clear sky conditions. The linear correlation and the slope of the regression for the Yale Playground was slightly lower if the Landsat algorithm was applied to the band reference values obtained under overcast sky conditions (Figure 6d), but this was not a surprise because the Landsat algorithm was intended for clear skies.

The red, green, and blue bands defined for the ground targets were 620–670 nm, 540–560 nm, and 460–480 nm, respectively, in order to match those of the camera spectral sensitivity. These bands do not correspond precisely to the Landsat bands. Figure 6 shows that despite the slight mismatches, the Landsat algorithm can be used to convert camera-acquired reflectance to albedo.

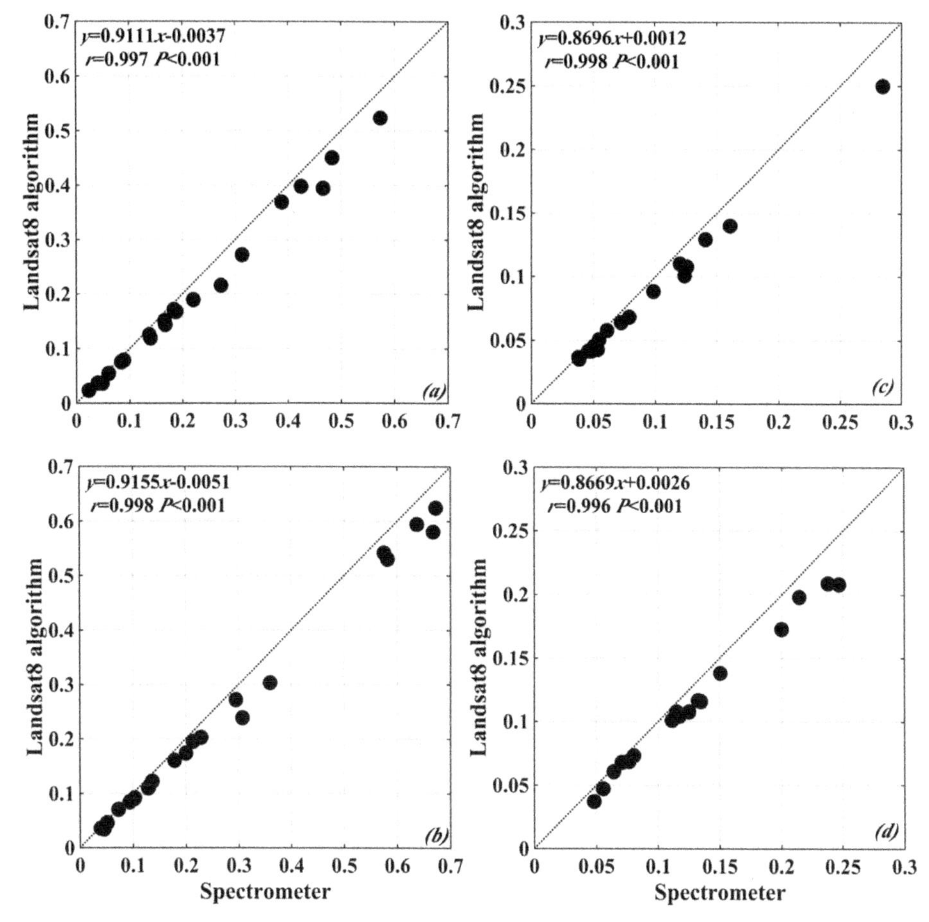

Figure 6. Comparison of visible band albedo measured with the spectrometer and derived with the Landsat conversion algorithm for the Brooksvale Park (**a**,**b**) and the Yale Playground (**c**,**d**). Panels (**a**,**c**) are for clear sky conditions and (**b**,**d**) are for overcast sky conditions.

4.2. Landscape Albedo Validation

For the Brooksvale Park, the drone-derived visible band albedo is 0.086 ± 0.110. For comparison, the Landsat visible band albedo is much lower, at 0.054 ± 0.0118. For the Yale Playground, the landscape visible band albedo was 0.037 ± 0.063 and 0.047 ± 0.012 according to the drone measurement and the Landsat measurement, respectively. Once again, because the standard deviations were computed from the individual pixel values in the scene for both the drone and the Landsat data, they indicated spatial variations of the albedo in the landscape, rather than uncertainties of estimation. The values of drone- and Landsat-derived visible band albedo for the Yale Playground were in much better agreement than those for the Brooksvale Park. We suggested that matching of sky conditions is the dominant factor for the different accuracies, as explained in the next section.

The Landsat-derived shortwave albedo is 0.103 ± 0.019 for the Brooksvale Park and 0.128 ± 0.013 for the Yale Playground. Compared with the landscape visible band albedo, the drone-derived shortwave albedo values (0.332 for Brooksvale Park and 0.054 to 0.061 for Yale Playground) are quite different from these reference values.

5. Discussion

5.1. Effect of Sky Conditions on Albedo Estimation

We conducted the spectrometer experiment under both clear sky and overcast sky conditions for the two drone field sites. The sky conditions had some effect on the regression statistics. The R^2 values for Brooksvale Park were higher if the ground measurements were made under overcast sky conditions (Figure 4), the same conditions as when the UAV mission took place, than if the measurements were made under clear sky conditions. In contrast, R^2 values for the Yale Playground were higher under clear sky conditions than under overcast sky conditions (Figure 5), keeping in mind that the UAV mission was conducted there under mostly clear sky conditions. These results indicate that the spectrometer calibration experiment should be conducted under the sky conditions that match those of the UAV experiment.

The Yale Playground flight was conducted at 14:30 on 30 September 2015 when the solar elevation angle was rather low. The long shadows due to the low elevation angle were a source of uncertainty for the albedo estimation. Choosing an appropriate flight time so that the shadow effect is minimized is very important, especially in an urban environment where tall structures are prominent features of the landscape.

The difference between the Landsat- and drone-derived visibleband albedo of the Brooksvale Park were larger than those of the Yale Playground. The main factor contributing to this discrepancy is also the sky condition. The UAV measurement was conducted under overcast conditions at Brooksvale Park, whereas the Landsat measurement was used for clear sky conditions. Generally, surface albedo is higher under cloudy skies than under clear skies [38]. At the Yale Playground, the sky conditions of the UAV experiment matched those of the Landsat observation, resulting in a much better agreement between the two albedo estimates.

5.2. Uncertainty in Landscape Shortwave Albedo

Low-cost consumer-grade cameras usually do not contain near-infrared spectral information, and therefore this may cause problems for estimating shortwave band albedo. In this study, shortwave band albedo was determined in a relatively arbitrary way. Brest and Goward [39] assigned weight factors to Landsat visible, near-infrared, and mid-infrared band reflectance, and linearly combined them to estimate the shortwave band albedo for vegetation. Similar to their method, we used the average ratio between the shortwave and visible band albedos of non-vegetation and vegetation ground targets to retrieve the landscape shortwave band albedo. The ratio for clear skies were lower than that for overcast skies for vegetation targets, suggesting that plants preferably absorb more visible radiation under clear skies than under overcast skies. For non-vegetation targets, the ratio did not differ by much between the two different sky conditions. Vegetation ground targets had higher ratios than the non-vegetation targets, due to the high near-infrared spectral reflectance of plant cell structures. The result from Brest's [40] study may provide a useful reference here: They reported that urban downtown and high-density residential neighborhoods in Hartford, Connecticut, USA, had ratios of near-infrared to visible band reflectance between 1.29 and 1.38. Their ratio for evergreen and deciduous forests was between 2.53 and 3.63. From the shortwave and visible band albedos given in their study, we infer that their non-vegetation ratio of shortwave to visible band reflectance is 1.15 to 1.19, which is in good agreement with our non-vegetation ratios (1.18 to 1.24), and their vegetation (evergreen and deciduous trees) ratio is 1.77 to 2.32, which is much lower than our ratios (3.91 to 6.67). The main reason for the difference is that grass leaves (in this study the vegetation ground targets we chose was grass) have a higher spectral reflectance in the infrared waveband than tree leaves, which can increase shortwave albedo and therefore the ratio between shortwave and visible band reflectance [40].

Three modifications to our method can potentially improve the shortwave albedo accuracy. First, the addition of a light-weight NIRcamera can give a direct reflectance measurement to every ground pixel and therefore avoid the ratio method.

Second, in the current experiments, we were only able to perform a spectrometer measurement of the short grass targets. However, a large portion of the vegetation pixels are trees (Figure 2). Applying the NIR-to-visible band ratio obtained from the grass targets to these tree pixels will cause large errors. This is especially problematic for the Brooksvale Park where some trees started to become senescent (Figure 2a). Division of the vegetation pixels into three separate categories (grass, senescent tree leaf, and green tree leaf) and establishing the NIR to visible band for each category should improve the shortwave albedo accuracy.

Third, we deliberately selected the ground targets to cover a wide range of reflectivity to establish robust calibration curves, but we did not consider variations in BRDF signatures between grasses and trees, and between sands and tarmac. Currently only two training sets (vegetation versus non-vegetation) were used for conversions to shortwave albedo. Additional training sets accounting for BRDF variations may improve our results.

5.3. Potential Applications

Using cameras on-board a drone can provide timely estimation of the landscape albedo. For example, flying drone missions at different times of the day and season and in different weather and soil moisture conditions can provide information on the dynamic variations of albedo of the landscape. Such measurements may be especially useful in situations where monitoring with conventional radiometers is not feasible. For example, white roofs are proposed as a strategy to mitigate the urban heat island in the city landscape [41]. However, white paint on building roofs can suffer from erosion and dust deposition, and thus, its albedo can quickly decrease, from the original high values of 0.7 to 0.8 to values of 0.2 to 0.3 after a few years [42]. Direct measurement of roof albedo is challenging because roof spaces are generally not accessible by micrometeorological tower instruments. Drone flights conducted at different times can help to quantify the actual albedo and inform decisions on whether the roof needs cleaning or repainting. Another advantage of the drone methodology is its fine spatial resolution, as compared to satellite monitoring. Even with Sentinel 2 with a spatial resolution of 10 m, some landscape features (such as small fish ponds and small buildings) will become mixed satellite pixels.

6. Conclusions and Future Outlook

In this paper we tested a workflow for landscape albedo determination using images acquired by drone cameras. The key findings are as follows:

(1) By adopting the method in this study, the landscape visible and shortwave band albedos of the Brooksvale Park were 0.086 and 0.332, respectively. For the Yale playground, the visible band albedo was 0.037, and shortwave albedo was between 0.054 and 0.061.

(2) The Landsat satellite algorithm for converting the satellite spectral albedo to broadband albedo can also be used to convert spectral albedo that is acquired by drones to broadband albedo.

(3) Data for spectral calibration using ground targets should be obtained under sky conditions that match those under which the drone flight take place. Because the relationship between the imagery DN value and the reflectivity is highly nonlinear, the ground targets should cover the range of reflectivity of the entire landscape.

(4) In the current configuration, the drone estimate of the visible band albedo is more satisfactory than its estimate of shortwave albedo, when compared with the Landsat-derived values. We suggest that deployment of a camera with the additional capacity of measuring reflectance in a near-infrared waveband should improve the estimate of shortwave albedo. Future cameras with the capacity to detect mid-infrared reflectance will further improve the shortwave albedo detection. The BRDF effect, which was ignored in this study, should be taken into consideration when deciding the ground calibration targets and training data in future studies.

Supplementary Materials: The following are available online at http://www.mdpi.com/2072-4292/10/11/1812/s1, Figure S1: Pictures of some of the selected ground targets for Brooksvale Park and Yale Playground, Figure S2: Classification of the mosaicked image for Brooksvale Park (a) and Yale Playground (b), Figure S3: Visible (a), shortwave band albedo for the Yale Playground when all shadow were taken as non-vegetation (b) and vegetation (c) under clear sky conditions, Figure S4: Visible (a) and shortwave band albedo (b) for the Brooksvale Park under clear sky conditions, Figure S5: Landsat visible (a) and shortwave band albedo (b) for the Yale Playground, Figure S6: Landsat visible (a) and shortwave band albedo (b) for the Brooksvale Park, Table S1: Input parameters for the SMART model.

Author Contributions: Conceptualization, X.L. and C.C.; Methodology, J.M., N.T., G.T., C.C., J.X., and L.B.; Formal Analysis, C.C.; Writing—Original Draft Preparation, C.C.; Writing—Review & Editing, X.L.

Funding: This research was funded by the Startup Foundation for Introducing Talent of Nanjing University of Information Science and Technology (grant 2017r067), Natural Science Foundation of Jiangsu Province (grants BK20180796 BK20181100), National Natural Science Foundation of China (grant 41805022), the Ministry of Education of China (grant PCSIRT), the Priority Academic Program Development of Jiangsu Higher Education Institutions (grant PAPD), and a Visiting Fellowship from China Scholarship Council (to C.C.).

Acknowledgments: We would like to thank Nina Kantcheva Tushev and Georgi Tushev for their help in conducting the drone experiment at the Yale Playground.

Conflicts of Interest: The authors declare no conflict of interest.

References

1. Liang, S. Narrowband to broadband conversions of land surface albedo I: Algorithms. *Remote Sens. Environ.* **2001**, *76*, 213–238. [CrossRef]
2. Hock, R. Glacier melt: A review of processes and their modelling. *Prog. Phys. Geogr.* **2005**, *29*, 362–391. [CrossRef]
3. Jin, Y.; Schaaf, C.B.; Gao, F.; Li, X.; Strahler, A.H.; Zeng, X.; Dickinson, R.E. How does snow impact the albedo of vegetated land surfaces as analyzed with MODIS data? *Geophys. Res. Lett.* **2002**, *29*, 1374. [CrossRef]
4. Myhre, G.; Kvalevåg, M.M.; Schaaf, C.B. Radiative forcing due to anthropogenic vegetation change based on MODIS surface albedo data. *Geophys. Res. Lett.* **2005**, *32*, L21410. [CrossRef]
5. Jin, Y.; Randerson, J.T.; Goetz, S.J.; Beck, P.S.; Loranty, M.M.; Goulden, M.L. The influence of burn severity on postfire vegetation recovery and albedo change during early succession in North American boreal forests. *J. Geophys. Res. Biogeosci.* **2012**, *117*, G01036. [CrossRef]
6. Cescatti, A.; Marcolla, B.; Vannan, S.K.; Pan, J.Y.; Román, M.O.; Yang, X.; Ciais, P.; Cook, R.B.; Law, B.E.; Matteucci, G.; et al. Intercomparison of MODIS albedo retrievals and in situ measurements across the global FLUXNET network. *Remote Sens. Environ.* **2012**, *121*, 323–334. [CrossRef]
7. Fernández, T.; Pérez, J.L.; Cardenal, J.; Gómez, J.M.; Colomo, C.; Delgado, J. Analysis of landslide evolution affecting olive groves using UAV and photogrammetric techniques. *Remote Sens.* **2016**, *8*, 837. [CrossRef]
8. Watts, A.C.; Ambrosia, V.G.; Hinkley, E.A. Unmanned aircraft systems in remote sensing and scientific research: Classification and consideration of use. *Remote Sens.* **2012**, *4*, 1671–1692. [CrossRef]
9. Salamí, E.; Barrado, C.; Pastor, E. UAV flight experiments applied to the remote sensing of vegetated areas. *Remote Sens.* **2014**, *6*, 11051–11081. [CrossRef]
10. Nex, F.; Remondino, F. UAV for 3D mapping applications: A review. *Appl. Geomat.* **2014**, *6*, 1–15. [CrossRef]
11. Laliberte, A.S.; Goforth, M.A.; Steele, C.M.; Rango, A. Multispectral remote sensing from unmanned aircraft: Image processing workflows and applications for rangeland environments. *Remote Sens.* **2011**, *3*, 2529–2551. [CrossRef]
12. Stolaroff, J.K.; Samaras, C.; O'Neill, E.R.; Lubers, A.; Mitchell, A.S.; Ceperley, D. Energy use and life cycle greenhouse gas emissions of drones for commercial package delivery. *Nat. Commun.* **2018**, *9*, 409. [CrossRef] [PubMed]
13. Yang, G.; Li, C.; Wang, Y.; Yuan, H.; Feng, H.; Xu, B.; Yang, X. The DOM generation and precise radiometric calibration of a UAV-mounted miniature snapshot hyperspectral imager. *Remote Sens.* **2017**, *9*, 642. [CrossRef]
14. Sonnentag, O.; Hufkens, K.; Teshera-Sterne, C.; Young, A.M.; Friedl, M.; Braswell, B.H.; Milliman, T.; O'Keefe, J.; Richardson, A.D. Digital repeat photography for phenological research in forest ecosystems. *Agric. For. Meteorol.* **2012**, *152*, 159–177. [CrossRef]

15. Saari, H.; Pellikka, I.; Pesonen, L.; Tuominen, S.; Heikkilä, J.; Holmlnd, C.; Mäkynen, J.; Ojala, K.; Antila, T. Unmanned Aerial Vehicle (UAV) operated spectral camera system for forest and agriculture applications. *Proc. SPIE* **2011**, *8174*, 466–471.
16. Johnson, L.F.; Herwitz, S.; Dunagan, S.; Lobitz, B.; Sullivan, D.; Slye, R. Collection of ultra high spatial and spectral resolution image data over California vineyards with a small UAV. In Proceedings of the International Symposium on Remote Sensing of Environment, Beijing, China, 22–26 April 2003.
17. Berni, J.A.J.; Suárez, L.; Fereres, E. Remote Sensing of Vegetation from UAV Platforms Using Lightweight Multispectral and Thermal Imaging Sensors. Available online: http://www.isprs.org/proceedings/XXXVIII/1_4_7-W5/paper/Jimenez_Berni-155.pdf (accessed on 21 March 2016).
18. Zahawi, R.A.; Dandois, J.P.; Holl, K.D.; Nadwodny, D.; Reid, J.L.; Ellis, E.C. Using lightweight unmanned aerial vehicles to monitor tropical forest recovery. *Biol. Conserv.* **2015**, *186*, 287–295. [CrossRef]
19. Bendig, J.; Yu, K.; Aasen, H.; Bolten, A.; Bennertz, S.; Broscheit, J.; Gnyp, M.L.; Bareth, G. Combining UAV-based plant height from crop surface models, visible, and near infrared vegetation indices for biomass monitoring in barley. *Int. J. Appl. Earth Obs.* **2015**, *39*, 79–87. [CrossRef]
20. Levy, C.; Burakowski, E.; Richardson, A. Novel measurements of fine-scale albedo: Using a commercial quadcopter to measure radiation fluxes. *Remote Sens.* **2018**, *10*, 1303. [CrossRef]
21. Hakala, T.; Suomalainen, J.; Peltoniemi, J.I. Acquisition of Bidirectional Reflectance Factor Dataset Using a Micro Unmanned Aerial Vehicle and a Consumer Camera. *Remote Sens.* **2010**, *2*, 819–832. [CrossRef]
22. Ryan, J.; Hubbard, A.; Box, J.E.; Brough, S.; Cameron, K.; Cook, J.; Cooper, M.; Doyle, S.H.; Edwards, A.; Holt, T.; et al. Derivation of High Spatial Resolution Albedo from UAV Digital Imagery: Application over the Greenland Ice Sheet. *Front. Earth Sci.* **2017**, *5*. [CrossRef]
23. Lebourgeois, V.; Bégué, A.; Labbé, S.; Mallavan, L.; Prévost, B.R. Can Commercial Digital Cameras Be Used as Multispectral Sensors? A Crop Monitoring Test. *Sensors* **2008**, *8*, 7300–7322. [CrossRef] [PubMed]
24. Lelong, C.C.D.; Burger, P.; Jubelin, G.; Roux, B.; Labbé, S.; Baret, F. Assessment of Unmanned Aerial Vehicles Imagery for Quantitative Monitoring of Wheat Crop in Small Plots. *Sensors* **2008**, *8*, 3557–3585. [CrossRef] [PubMed]
25. Markelin, L.; Honkavaara, E.; Peltoniemi, J.; Ahokas, E.; Kuittinen, R.; Hyyppä, J.; Suomalainen, J.; Kukko, A. Radiometric Calibration and Characterization of Large-format Digital Photogrammetric Sensors in a Test Field. *Photogramm. Eng. Remote Sens.* **2008**, *74*, 1487–1500. [CrossRef]
26. Honkavaara, E.; Arbiol, R.; Markelin, L.; Martinez, L.; Cramer, M.; Bovet, S.; Chandelier, L.; Ilves, R.; Klonus, S.; Marshal, P.; et al. Digital Airborne Photogrammetry—A New Tool for quantitative remote sensing? A State-of-Art review on Radiometric Aspects of Digital Photogrammetric Images. *Remote Sens.* **2009**, *1*, 577–605. [CrossRef]
27. Du, M.; Noguchi, N. Monitoring of Wheat Growth Status and Mapping of Wheat Yield's within-Field Spatial Variations Using Color Images Acquired from UAV-camera System. *Remote Sens.* **2017**, *9*, 289. [CrossRef]
28. Wang, M.; Chen, C.; Pan, J.; Zhu, Y.; Chang, X. A Relative Radiometric Calibration Method Based on the Histogram of Side-Slither Data for High-Resolution Optical Satellite Imagery. *Remote Sens.* **2018**, *10*, 381. [CrossRef]
29. Shuai, Y.; Masek, J.; Gao, F.; Schaaf, C. An algorithm for the retrieval of 30-m snow-free albedo from Landsat surface reflectance and MODIS BRDF. *Remote Sens. Environ.* **2011**, *115*, 2204–2216. [CrossRef]
30. Wang, Z.; Erb, A.M.; Schaaf, C.B.; Sun, Q.; Liu, Y.; Yang, Y.; Shuai, Y.; Casey, K.A.; Roman, M.O. Early spring post-fire snow albedo dynamics in high latitude boreal forests using Landsat-8 OLI data. *Remote Sens. Environ.* **2016**, *185*, 71–83. [CrossRef] [PubMed]
31. Available online: http://www.cielmap.com/cielmap/ (accessed on 28 June 2016).
32. Agisoft Photoscan User Manual. Available online: http://www.agisoft.com/pdf/photoscan-pro_1_2_en.pdf/ (accessed on 18 July 2016).
33. Kruse, F.A.; Lefkoff, A.B.; Boardman, J.W.; Heidebrecht, K.B.; Shapiro, A.T.; Barloon, P.J.; Goetz, A.F.H. The spectral image processing system (SIPS)—Interactive visualization and analysis of imaging spectrometer data. *Remote Sens. Environ.* **1993**, *44*, 145–163. [CrossRef]
34. Gueymard, C.A. The sun's total and spectral irradiance for solar energy applications and solar radiation models. *Sol. Energy* **2004**, *76*, 423–453. [CrossRef]

35. He, T.; Liang, S.; Wang, D.; Cao, Y.; Gao, F.; Yu, Y.; Feng, M. Evaluating land surface albedo estimation from Landsat MSS, TM, ETM+, and OLI data based on the unified direct estimation approach. *Remote Sens. Environ.* **2018**, *204*, 181–196. [CrossRef]
36. Landsat8 Surface Reflectance Code (LaSRC) Product Guide. Available online: https://landsat.usgs.gov/sites/default/files/documents/lasrc_product_guide.pdf (accessed on 14 February 2018).
37. Vermote, E.; Justice, C.; Claverie, M.; Franch, B. Preliminary analysis of the performance of the Landsat 8/OLI land surface reflectance product. *Remote Sens. Environ.* **2016**, *185*, 46–56. [CrossRef]
38. Payne, R.E. Albedo of the sea surface. *J. Atmos. Sci.* **1972**, *29*, 959–970. [CrossRef]
39. Brest, C.; Goward, S. Deriving surface albedo measurements from narrow band satellite data. *Int. J. Remote Sens.* **1987**, *8*, 351–367. [CrossRef]
40. Brest, C. Seasonal albedo of an urban/rural landscape from satellite observations. *J. Clim. Appl. Meteorol.* **1987**, *26*, 1169–1187. [CrossRef]
41. Zhao, L.; Lee, X.; Schultz, M. A wedge strategy for mitigation of urban warming in future climate scenarios. *Atmos. Chem. Phys.* **2017**, *17*, 9067–9080. [CrossRef]
42. Akbari, H.; Kolokotsa, D. Three decades of urban heat islands and mitigation technologies research. *Energy Build.* **2016**, *133*, 834–842. [CrossRef]

Article

The Role of Climate and Land Use in the Changes in Surface Albedo Prior to Snow Melt and the Timing of Melt Season of Seasonal Snow in Northern Land Areas of 40°N–80°N during 1982–2015

Kati Anttila *, Terhikki Manninen, Emmihenna Jääskeläinen, Aku Riihelä and Panu Lahtinen

Finnish Meteorological Institute, Meteorological Research Unit, FI-00101 Helsinki, Finland; terhikki.manninen@fmi.fi (T.M.); emmihenna.jaaskelainen@fmi.fi (E.J.); aku.riihela@fmi.fi (A.R.); panu.lahtinen@fmi.fi (P.L.)

* Correspondence: kati.anttila@fmi.fi; Tel.: +358-50-4412298

Received: 14 August 2018; Accepted: 9 October 2018; Published: 11 October 2018

Abstract: The rapid warming of the Northern Hemisphere high latitudes and the observed changes in boreal forest areas affect the global surface albedo and climate. This study looks at the trends in the timing of the snow melt season as well as the albedo levels before and after the melt season in Northern Hemisphere land areas between 40°N and 80°N over the years 1982 to 2015. The analysis is based on optical satellite data from the Advanced Very High Resolution Radiometer (AVHRR). The results show that the changes in surface albedo already begin before the start of the melt season. These albedo changes are significant (the mean of absolute change is 4.4 albedo percentage units per 34 years). The largest absolute changes in pre-melt-season albedo are concentrated in areas of the boreal forest, while the pre-melt albedo of tundra remains unchanged. Trends in melt season timing are consistent over large areas. The mean of absolute change of start date of melt season is 11.2 days per 34 years, 10.6 days for end date of melt season and 14.8 days for length of melt season. The changes result in longer and shorter melt seasons, as well as changed timing of the melt, depending on the area. The albedo levels preceding the onset of melt and start of the melt season correlate with climatic parameters (air temperature, precipitation, wind speed). The changes in albedo are more closely linked to changes in vegetation, whereas the changes in melt season timing are linked to changes in climate.

Keywords: snow; albedo; climate

1. Introduction

Climate change over the Northern Hemisphere high latitudes and boreal forest zone has affected the snow and vegetation cover and, thus, the surface albedo [1–12]. Previous studies have shown that the duration and timing of the snow melt season has changed differently in different areas [13–17]. The snow cover extent has decreased especially significantly in the spring [3,5,7], and the surface albedo during the melt season months has also decreased, largely due to a decline in the area covered by snow [10]. These changes affect the local and global energy budgets [6]. The changing climate also changes the vegetation. The increased size of vegetation decreases wintertime albedo by covering the land surface and casting larger shadows on snow-covered surfaces. This is particularly relevant in the late winter. Changes in vegetation affect the local climate and the scattering properties of the forest (with larger shadows and increased multiple scattering due to increased forest height or density) [11]. The changes in the vegetation are also linked to changes in the spatial coverage of permafrost.

Changes in the timing and duration of the melt season, as well as in the surface albedo, are important parameters for climate models [18,19]. They can be used as comparison data for model parameters during the run. More accurate information on these parameters is needed to improve the models. In particular, the albedo of vegetated surfaces overlain by snow, such as boreal forests, introduce uncertainty to the model outputs [20–22].

Changes in the albedo of snow-covered surfaces can be caused by changes in a number of variables, such as climate, impurities in snow, vegetation, permafrost and changes in the properties of the snow surface [23–27]. Some of these factors also interact with each other. For example, changes in air temperature, precipitation, and wind speed affect both the snow cover extent and snow surface properties, which in turn change the snow surface albedo [10,24,27]. The optical scattering properties of a snow surface are most heavily determined by grain size and shape [28], with grain size being the main physical factor responsible for snow-albedo variations [24]. These, together with climatic factors such as wind, air temperature, and the existence of vegetation affect snow surface roughness, which also affects the brightness of the surface [29,30].

The vast area covered by seasonal snow provides a reason to study it at global scale. This can be characterized using satellite remote sensing. Snow cover and surface albedo have been analyzed from satellite data for decades [31–33]. The first studies covered small areas and short periods of time, and the available data was limited in coverage and resolution. These studies form the basis for the new satellite-derived data records, which offer better spatial and temporal coverage and, thus, enable climatic studies of various key parameters. The timing of the melt season has typically been estimated using passive microwave satellite data [14,15], which are sensitive to the presence and amount of liquid water. Therefore, microwave data are good at detecting changes in the snow moisture content, but do not react to thin snow covers, which affect the surface albedo significantly, but have very low liquid water content. Moreover, microwave data cannot easily differentiate between wet snow and wet ground. Surface albedo data have also been used to determine the timing of melt season [34], but the spatial and temporal coverage of these studies has been limited. There are many different definitions for the start and end of melt [35]. One way to define these dates is to use the time when the open areas are less than half covered with snow [35]. The choice of definition depends on the intended application of the data.

Previous studies on snow cover have shown a decrease in the area covered by snow, as in the melt season albedo [2,3,5,10], but whether or not climate change has caused the albedo of the snow surface to change prior to the onset of melt has so far been unclear. Studies of changes in snow season surface albedo have typically been based on either specific calendar months assumed to represent the melt season [10], or on the maximum albedo of the snow season [36]. This paper presents a study of the changes in the surface albedo of the land areas of the Northern Hemisphere between latitudes 40°N and 80°N prior to the melt season. This has an effect on the global energy budget, as well as the length of the melt season and the surface albedo during the melt season. The study also investigates the changes in the timing of the melt season. The analysis utilizes 5-day mean surface albedo data, derived from optical satellite data for the years 1982 to 2015, to determine the start and end dates of the melt season and the corresponding surface albedo levels. The trends of these parameters over the 34-year period and their relationships to land use and trends in climatic parameters are also investigated.

2. Data

This study is based on the 5-day mean surface albedo data of the Satellite Application Facility for Climate Monitoring (CM SAF, funded by EUMETSAT) CLouds, Albedo and RAdiation second release Surface ALbedo (CLARA-A2 SAL) data record [37,38], which is constructed using Advanced Very High Resolution Radiometer (AVHRR) data. The albedo is defined as the broadband shortwave directional-hemispherical reflectance, i.e., the black-sky albedo. The retrieved albedo corresponds to the wavelength range 0.25–2.5 μm and the observations are averaged to a $0.25° \times 0.25°$ grid, which is also the resolution of the final product. The albedo values are given in the range 0–100%. At the time of the analysis this was the longest available homogeneous data record of surface albedo.

The basis of the derivation of the 5-day mean albedo product used here is similar to CLARA-A1 SAL [39]. The albedo values for a five day period are first determined by observation and then averaged. After cloud masking the satellite data, the effect of topography and inclined slope and related shading on location and reflectance of the satellite data is corrected. The land pixels are then corrected for atmospheric effects on the radiation. The atmospheric correction utilizes a dynamic aerosol optical depth (AOD) time series [40] as input. The AOD time series was constructed using the total ozone mapping spectrometer (TOMS) and ozone monitoring instrument (OMI) aerosol index data [40]. The scattering properties of the surface are described by bidirectional reflectance distribution functions (BRDF) for different land-use types. The land-use classes are derived from four different land-use classifications, using always the classification which has been constructed from data that is temporally closest to the observation in question. Finally, the 0.6 and 0.8 micrometer (AVHRR channels 1 and 2) albedos are converted into broadband albedo. The reflectance characteristics of snow surfaces vary between different snow types [41]. Therefore the albedo of snow- and ice-covered areas is derived by averaging the broadband bidirectional reflectance values of the AVHRR overpasses into pentad and monthly means. The albedo of open water, such as oceans, is constructed using solar zenith angle and wind speed. The existence of sea ice is verified using the Ocean and Sea Ice Satellite Application Facility (OSI SAF) sea ice extent data [42].

The data record has been validated against in situ data and compared with the Moderate Resolution Imaging Spectroradiometer (MODIS) MCD43C3 edition 5 data set [43,44]. The mean relative retrieval error of CLARA-A2 SAL is −0.6%, the mean root mean square error (RMSE) is 0.075 and the decadal relative stability (over Greenland Summit) is 8.5%. Larger differences between the in situ measurements and the satellite-based albedo value are mostly related to the heterogeneity of the land surface within CLARA-A2 SAL pixels [45]. A comparison between CLARA-A2 SAL and MODIS MCD43C3 showed that the two products are in good agreement. The relative difference between the two products is typically between −10% and 10%, with the global mean CLARA-A2 SAL surface albedo being 2–3% higher than the MCD43C3 global mean surface albedo for some periods. One has to take into account that the SAL product includes a topographic correction in mountainous areas, whereas the MODIS product does not [43], and that mountains typically cause underestimation of albedo due to shadowing [37]. The water areas are excluded from the comparison as the MODIS product is not defined over water areas or sea ice.

Our study utilizes the global version of the CLARA-A2 SAL products and covers the land areas between latitudes 40°N and 80°N, and the years from 1982 to 2015. Using a 5-day mean albedo limits the role of possible individual low-quality albedo values with large retrieval errors (due to observation geometry, cloud contamination or geolocation error). Furthermore, using sigmoid fitting (described in Section 3) for the analysis limits the effect of possible erroneous individual mean albedo values.

The influence of climatic parameters on melt season and albedo was studied using ERA-Interim reanalysis data [46] for 14 day period before the previously defined date for the onset of melt. The parameters extracted from the data were air temperature (2 m), wind speed (10 m above ground), accumulated precipitation, amount of snow fall (giving also the accumulated rain) and the number of days on which the maximum temperature during that period was above 0 °C, −4 °C and −10 °C. These parameters were chosen due to their possible effect on snow reflectance, metamorphism and albedo. The air temperature affects the amount of liquid water and heat flux within the snow pack. Wind can affect the surface albedo by affecting the mechanical breaking of the surface crystals, by producing wind related surface structures such as ripples and ablation and accumulation areas. It can also affect the amount of vegetation visible above the snow surface and fraction of bare ground by removing snow partly or altogether from some areas. It also typically affects the amount of snow on trees. In the case of evergreen trees, this can have a significant effect on surface albedo. Precipitation can affect the surface albedo through adding fresh snow crystals on the surface and on vegetation and by affecting the snow depth. In the case of rain-on-snow, this can bring heat into the snow pack thus affecting the melt processes. The three temperature thresholds were chosen based on the relationship between air temperature, snow metamorphism and albedo. The 0 °C was chosen since it is the melting point for snow in normal conditions. The −4 °C was used to take into account the fact that snow metamorphism starts already at sub-zero temperatures. In the wide variety of snow albedo models, the simplest parameterizations presume a steady albedo for colder temperatures, and then a linear decline in snow surface albedo for air temperatures from −5 °C to 0 °C. At −4 °C the heating of the sun can already affect the snow surface crystals; −10 °C was chosen to represent a temperature at which the air temperature does not considerably affect the snow surface crystals, so if the maximum temperature of the day stays colder than this it can be presumed there is no change in the snow surface due to the temperature. The data was originally in 6-hourly temporal resolution (for snow fall 12 hourly) from which it was further processed to daily values. The resolution of the ERA-Interim data was 0.25°.

The role of land use in the trends in melt season albedo and timing is assessed using data from GlobCover2009 [47]. The data was coarsened to the same resolution as the melt season data (0.25°) by choosing the most common land-use class within the melt season grid cell. Figure 1 shows the GlobCover data at CLARA-A2 SAL resolution. The GlobCover land-use classes present in the study area are listed in Table 1.

Table 1. GlobCover2009 classes found to be most common within one surface albedo (SAL) resolution unit in the study area and the number of occurrences of each class as the most common land-use class in one resolution unit of melt season data.

LUC Class	Label	Number of Occurrence
11	Post-flooding or irrigated croplands (or aquatic)	420
14	Rainfed croplands	5137
20	Mosaic cropland (50–70%)/vegetation (grassland/shrubland/forest) (20–50%)	5771
30	Mosaic vegetation (grassland/shrubland/forest) (50–70%)/cropland (20–50%)	2996
50	Closed (>40%) broadleaved deciduous forest (>5 m)	7775
70	Closed (>40%) needleleaved evergreen forest (>5 m)	2472
90	Open (15–40%) needleleaved deciduous or evergreen forest (>5 m)	31415
100	Closed to open (>15%) mixed broadleaved and needleleaved forest (>5 m)	3605
110	Mosaic forest or shrubland (50–70%)/grassland (20–50%)	2299
120	Mosaic grassland (50–70%)/forest or shrubland (20–50%)	1883
130	Closed to open (>15%) (broadleaved or needleleaved, evergreen or deciduous) shrubland (<5 m)	1371
140	Closed to open (>15%) herbaceous vegetation (grassland, savannas or lichens/mosses)	3869
150	Sparse (<15%) vegetation	28741
180	Closed to open (>15%) grassland or woody vegetation on regularly flooded or waterlogged soil–Fresh, brackish or saline water	1639
190	Artificial surfaces and associated areas (urban areas >50%)	103
200	Bare areas	7523
210	Water bodies	115239
220	Permanent snow and ice	11022

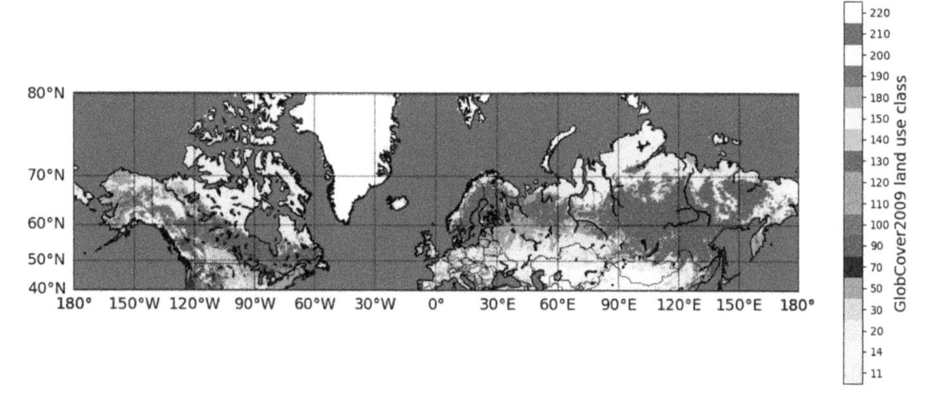

Figure 1. GlobCover2009 land-use classes coarsened to melt season data resolution.

3. Methods

The melt season parameters were determined by fitting sigmoids to CLARA-A2 SAL pentad (5 day) mean albedo values using non-linear regression [48]. For each grid cell and year the 5-day mean albedo values from the end of January until the end of August were used for the sigmoid fitting. Figure 2 shows two examples of sigmoid fitting. To include all changes in albedo during the melt season, the dates of snow melt at the onset was taken to be the date at which the sigmoid reached 99% of its variation range (i.e., a change of 1% (relative) from the pre-melt albedo level). Likewise the end of the snow melt season was defined to be the date at which the sigmoid reached 1% of its variation range. These thresholds were chosen in order to include the whole dynamic change in surface albedo during melt season. The length of the melt season was then the difference between the start and end date of melt. The albedo values corresponding to the dates of the onset and end of melt were used as the representative albedo values preceding and following the melt season.

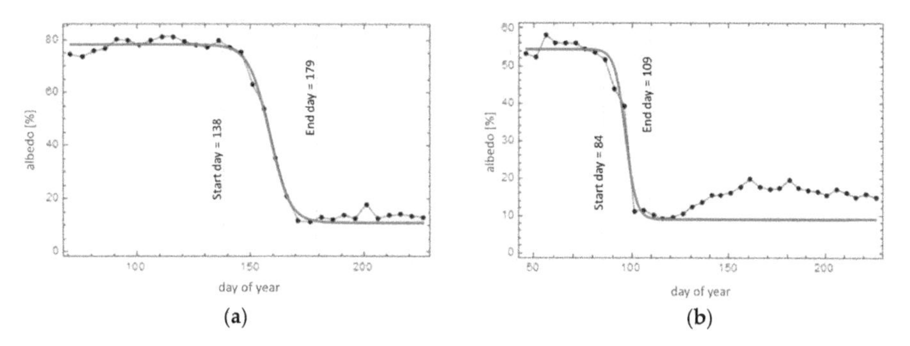

Figure 2. Sigmoid fitting for 5-day mean albedo data for (**a**) location 55.375°N, 47.625°E for year 2006 and (**b**) location 68.125°N, 120.125°E for the year 2007. The growth of vegetation after snow melt is manifested in (**b**) where the albedo values after melt season increase slightly as the vegetation starts to produce leaves.

In some cases the snow melt onset could not be determined, because the melt had already started before the first cloud-free albedo pentad of the year was available for the grid cell in question. The final analyses included only the grid cells for which (1) both the snow melt onset and end days were retrieved successfully; (2) the albedo difference between the start and end date of melt was larger than 5% absolute albedo units; and (3) data meeting these two criteria were available for at least 10 years. The mean values of R^2 and RMSE of the final sigmoid regressions were 0.989 and 5.55 (albedo percentage), respectively. The corresponding median values were 0.993 and 4.75. In the analyses only the grid cells for which $R^2 > 0.95$ and RMSE < 20 (for the sigmoid fitting) were included. This lead to discarding about 2% of the data. The final dataset consisted of 2.46 million grid cell level melt seasons. Figure 3 shows the number of years with successful melt season retrievals per resolution unit.

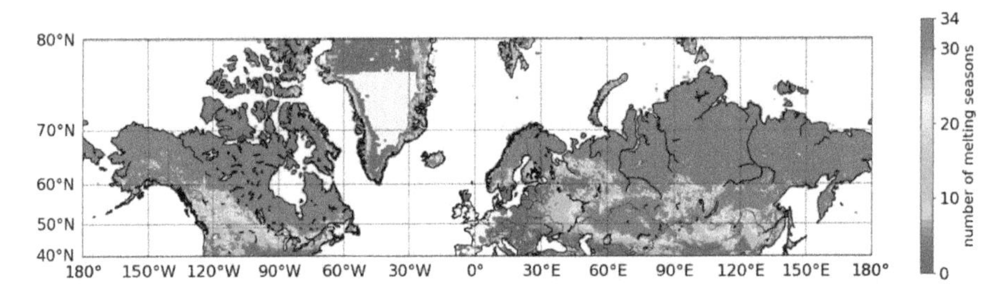

Figure 3. Number of successful retrievals of melt season per resolution unit during the 34 years.

The effect of random error in the albedo data on the derivation of the melt season timing was estimated by using 6 different sigmoids with different levels of albedo prior to melt. After constructing artificial data around these sigmoids, the data was modified by introducing relative random error of +12.5% to −12.5% to it. This error is larger than the level of typical variation of albedo values. These data with random error were then used to produce sigmoids and to extract the melt season parameters. The analysis was repeated for 100 different cases for each of the six chosen pre-melt levels with random error. The effect of relative random error in the albedo data on the start date of melt was 1.3 days (standard deviation) and 1.1 days for the end dates of melt.

The trends for the melt season parameters (start and end times of melt, the length of the melt season and albedo levels before and after the melt season) for each grid cell over the 34 years were detected using linear regression. The trends were determined using rolling 5-year means. Only 5-year means with data from at least 3 years were included in the trend fitting, and the fitting was carried out only for grid cells that had at least 20 mean values during the 34 years. This gave 72092 grid cell level estimates of trends in pre melt season albedo. From these 30% had R^2 larger than 0.5.

The climatic dependencies of the melt season parameters were studied using ERA-Interim re-analysis data [46] for air temperature at 2 m, precipitation and wind speed prior to melt. The time interval used in the analysis was 14 days prior to melt. Correlation analysis between ERA-Interim climate data and surface albedo or day of onset of melt were carried out only for the grid cells for which data were available for at least 20 years. The analysis was performed by looking at the linear correlation coefficient between the melt season parameter in question and the climatic parameters.

The GlobCover2009 land-use data set (Figure 1) was coarsened to the same resolution as the melt season data (CLARA-A2 resolution of 0.25°). The area of the boreal forest was determined by looking at GlobCover classes 70, 90 and 100. The tundra areas were determined as the grid cells with land-use class 140, 150 and 200 that are north of 55°N in North America and 65°N in Europe.

4. Results

4.1. Albedo Before and After the Melt Season

The change in albedo of Northern Hemisphere land areas between 40°N and 80°N before the melt season shows large spatial variations (Figure 4). The changes are concentrated in large areas with homogeneous change characteristics. These areas are listed in Table 2. The areas are shown on a map together with place names in Figure 5. For the observations with clear trends ($R^2 > 0.5$, 30% of all observations) the pre-melt albedo decreased by 2.4 absolute albedo percentage units on average over the whole study area over the 34 years of record. The statistically reliable trends in pre-melt season albedo are concentrated in the region of the boreal forest zone. This can be seen in Figure 4b, which shows the land-use classes for the resolution units with albedo trends with $R^2 > 0.5$. The trends in the boreal forest zone show decreasing pre-melt albedo in the southern half of the Central Siberian Plain and Scandinavia, and increasing pre-melt albedo in the north of Mongolia and China. In North America the direction of changes varies over short distances, whereas in Eurasia there are larger areas with similar direction of change in albedo.

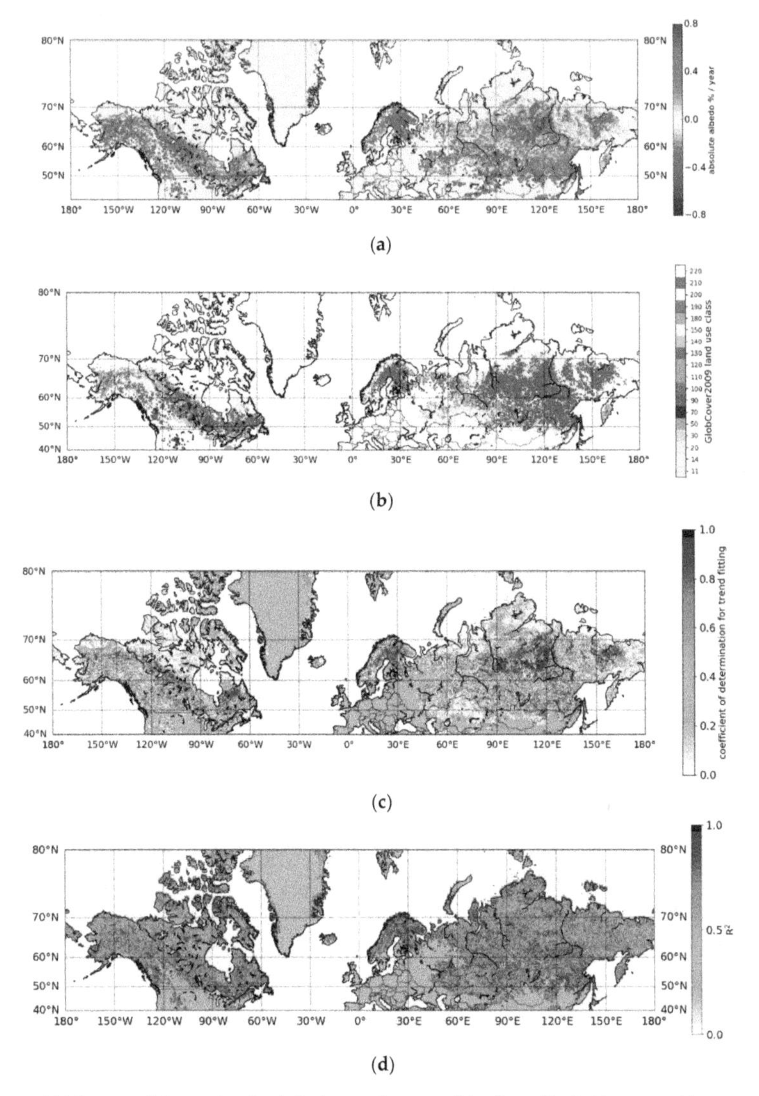

Figure 4. (**a**) The rate of change in albedo before melt season (absolute albedo % per year) between 1982 and 2015 using 5 year rolling mean albedo (showing cases for which R^2 of the fitting was larger than 0.5). The positive rates of change mean higher albedo and negative lower albedo values. (**b**) The GlobCover land-use class for the areas with clear trends in pre-melt albedo. (**c**) The coefficient of determination (R^2) value for trends in albedo before the onset of melt. (**d**) R^2 value for multiple variable linear correlation between albedo before melt season and 6 climatic parameters for 14 days prior onset of melt. The climatic parameters used are the mean air temperature, number of days with maximum temperature above 0 °C, −4 °C and −10 °C, accumulated precipitation and mean wind speed.

Table 2. The mean trends ($R^2 > 0.5$) for all the melt season and climatic parameters in areas with observed consistently homogenous change. The trends for snowfall and rain were not included in the table due to the fact that the trends for the pre-defined regions presented in the table were so weak that they were less than the yearly variation of the parameter. All climate parameters are derived for 14 days preceding the onset of snow melt. The number of observations in the area are described in the parenthesis. The area "RCM" refers to the area around the border of Russia, China and Mongolia. The negative trends for the melt season timing mean earlier onset or end of melt and shorter melt season and the positive trend directions mean later onset and end of melt and longer melt seasons.

	Pre-Melt Albedo	Post-Melt Albedo	Start Day of Melt (Days Per Year)	End Day of Melt (Days Per Year)	Melt Season Length (Days Per Year)	Mean Air Temp. (K/Year)	No. Days Above 0 °C (Days Per Year)	No. Days Above > −4 °C (Days Per Year)	No. Days Above > −10 °C (Days Per Year)	Accum. Precip (mm/Year)	Wind Speed (m/s Per Year)
Cent. Sib. Plain 7809	−0.25 (4100)	−0.08 (1765)	−0.61 (2486)	−0.50 (2243)	0.46 (317)	−0.12 (370)	−0.06 (359)	−0.11 (373)	−0.11 (399)	−0.28 (169)	0.01 (112)
RCM 7191	0.15 (2505)	−0.09 (1872)	−0.95 (1030)	0.71 (577)	1.51 (925)	−0.25 (720)	−0.09 (573)	−0.14 (667)	−0.18 (604)	−0.58 (477)	0.01 (378)
Labrador 2640	−0.09 (1111)	−0.06 (603)	0.28 (95)	−0.59 (289)	−0.94 (228)	0.21 (380)	0.03 (113)	0.16 (193)	0.19 (499)	0.33 (84)	0.00 (142)
Rocky Mnts 2840	−0.05 (449)	−0.07 (626)	−1.18 (170)	1.17 (303)	2.13 (280)	−0.21 (254)	−0.12 (197)	−0.19 (243)	−0.18 (197)	−0.12 (127)	0.01 (104)
Alaska 4866	−0.15 (1121)	−0.08 (1024)	−0.43 (156)	−0.22 (204)	0.37 (267)	−0.08 (176)	−0.05 (214)	−0.16 (268)	−0.07 (161)	−0.09 (130)	0.01 (101)
Europ. Arctic 3526	−0.29 (1233)	−0.01 (227)	−0.79 (510)	−0.39 (102)	0.88 (303)	−0.12 (276)	−0.06 (217)	−0.06 (376)	−0.09 (252)	−0.09 (102)	0.02 (121)
Canad. archip. 1796	−0.11 (222)	−0.26 (432)	−0.61 (79)	0.03 (79)	0.81 (95)	0.01 (32)	−0.06 (20)	−0.1 (30)	−0.04 (29)	0.33 (27)	−0.03 (83)
The Alps 299	−0.17 (22)	−0.12 (34)	−1.48 (28)	1.53 (19)	2.69 (30)	−0.18 (5)	−0.27 (5)	−0.18 (4)	−0.04 (3)	−0.00 (8)	−0.02 (7)

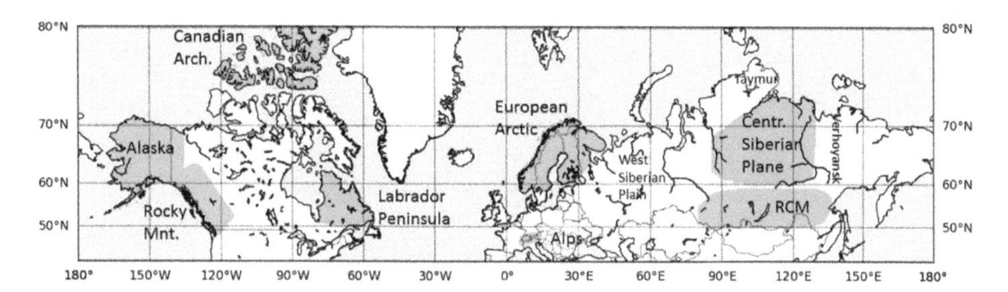

Figure 5. The areas listed in Table 2.

Most of the tundra areas show no significant change in albedo prior to melt. This can be seen in Figure 4c, which shows the R^2 values for all albedo trend retrievals. For all the areas with an R^2 value, the retrieval of melt season parameters was successful, but the data show no reliable trends (having R^2 values lower than 0.5). The slope values of these excluded trends are typically close to 0. In many areas the annual variability of the pre-melt albedo values is so large that even 34 years is not long enough to determine a small trend in the albedo value. In southern Eurasian tundra the pre-melt albedo shows weak negative trends, but no trends in the higher latitudes. The Kola Peninsula and northern Finnish Lapland show strong negative trends (−0.29 albedo percentage units per year).

The role of vegetation in the observed pre-melt albedo changes can be estimated by looking at the albedo levels right after the snow has melted, before the vegetation has started greening. Figure 6 shows the trends for albedo after melt as well as the corresponding R^2 values for the trend fitting. The trends in the level of albedo after melt are much weaker than those in the pre-melt season albedo before melt season, which can be expected since the differences in the albedo between different biomes are much smaller than the changes in the snow cover. The post-melt albedo of the northern Eurasian tundra decreased over the study period. In more southerly areas of tundra, however, there are no clear trends. In the boreal forest zone the trends are towards lower albedo values in most of the area, except for the area west of the River Ob in Russia. One potential reason for the darkening of the boreal forest zone after the melt season could be the increased size of trees and denser forests, causing more shadowing of the surface and increased multiple scattering. The darkening of the southern tundra prior to melt could be explained by the reported shrubification of tundra [49].

The role of climate change in altering the pre-melt season albedo was studied using the linear fitting between the melt season data and the ERA-Interim reanalysis data [45] on air temperature, precipitation, wind speed and the number of days with maximum temperature above 0 °C, −4 °C and −10 °C for 14 days prior to onset of melt. Figure 7 shows the 34-year trends in air temperature, accumulated precipitation and wind speed. Changes in all these climatic parameters contributed to the changes in the pre-melt albedo (mean R^2 for the whole area being 0.64 and the 80th percentile being 0.79) (Figure 4d). In the area around the borders of China, Mongolia and Russia (Figure 5) the climatic parameters explained almost all of the albedo change. The mean air temperature was the dominant influence (mean R^2 = 0.51 for the whole area). It was the largest explanatory factor in particular in Yablonovyy and Verhoyansk Mountain Ranges, Northern West Siberian Plain, Kola Peninsula, Baffin Island and Central Siberian Plain.

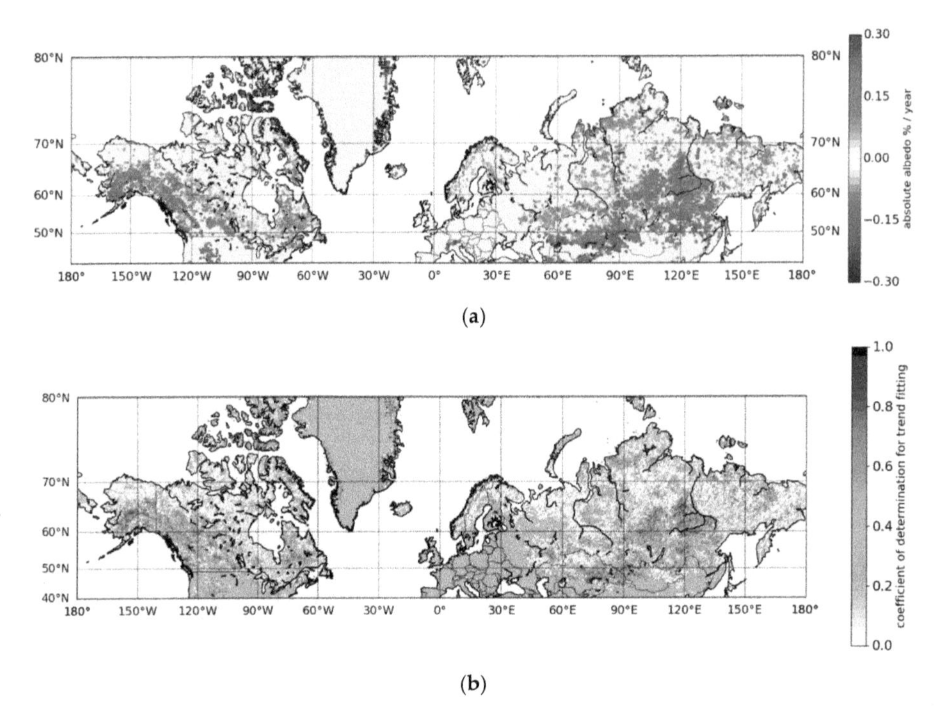

Figure 6. (**a**) The rates of change for the albedo after melt season (absolute albedo % per year) between 1982 and 2015 using 5-year rolling mean albedo (showing cases for which R^2 of the fit was larger than 0.5). (**b**) The coefficient of determination (R^2) value for trends in albedo after the end of melt.

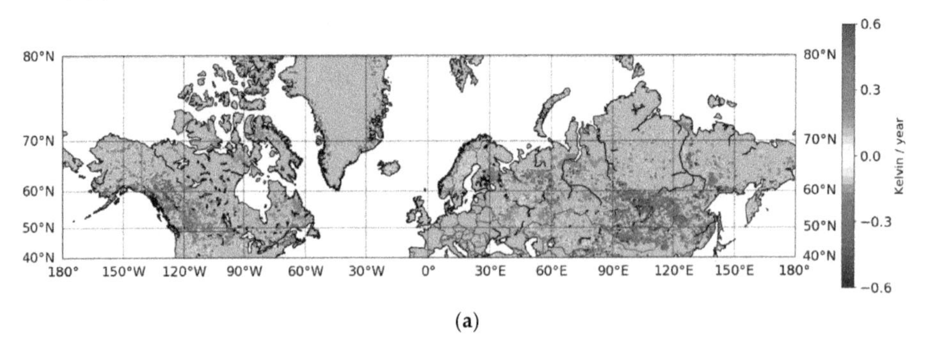

(**a**)

Figure 7. *Cont.*

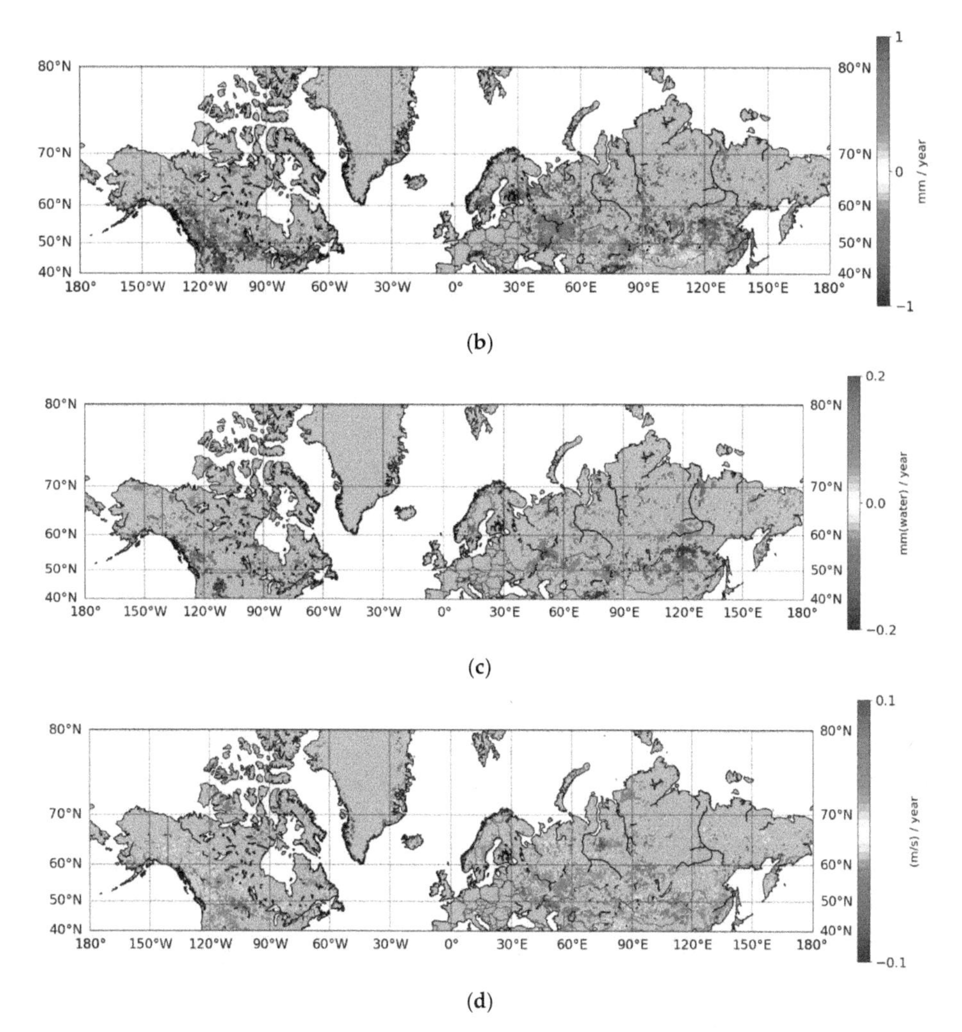

Figure 7. ERA-Interim trends for 1982–2015 for (**a**) mean air temperature (**b**) accumulated precipitation (**c**) snowfall (units are given as amount of snow converted into liquid water) (**d**) wind speed for 14 days prior to melt onset. The maps show trends for which the coefficient of determination was larger than 0.5.

4.2. Melt Season Timing

In addition to the changes in albedo, the timing of the melt season has also changed (Figures 8–10 and Table 2). The changes are, as with albedo, significant and spatially consistent but they also vary within the study area. The observations with high values for coefficient of determination ($R^2 > 0.5$) are concentrated in large distinct areas. The changes were in general towards longer melt seasons and earlier onset of melt. The mean start date of melt season in the pixels for which melt season data are available for the whole 34 years, became 6.1 days earlier over the 34 years. Similarly, the melt ended on average 5.2 days earlier and the melt season, therefore, became 1 day longer on average. The majority of the observations showed no clear reliable trends (Figure 11), but in many areas the changes were significant (Figures 8–10). In many areas the inter-annual variation in the start and/or end dates of the melt season were so large that it was not possible to detect a statistically significant trend. In Eurasia all the parameters showed changes over large homogenous areas, but in North America the trends are typically more localized and variable.

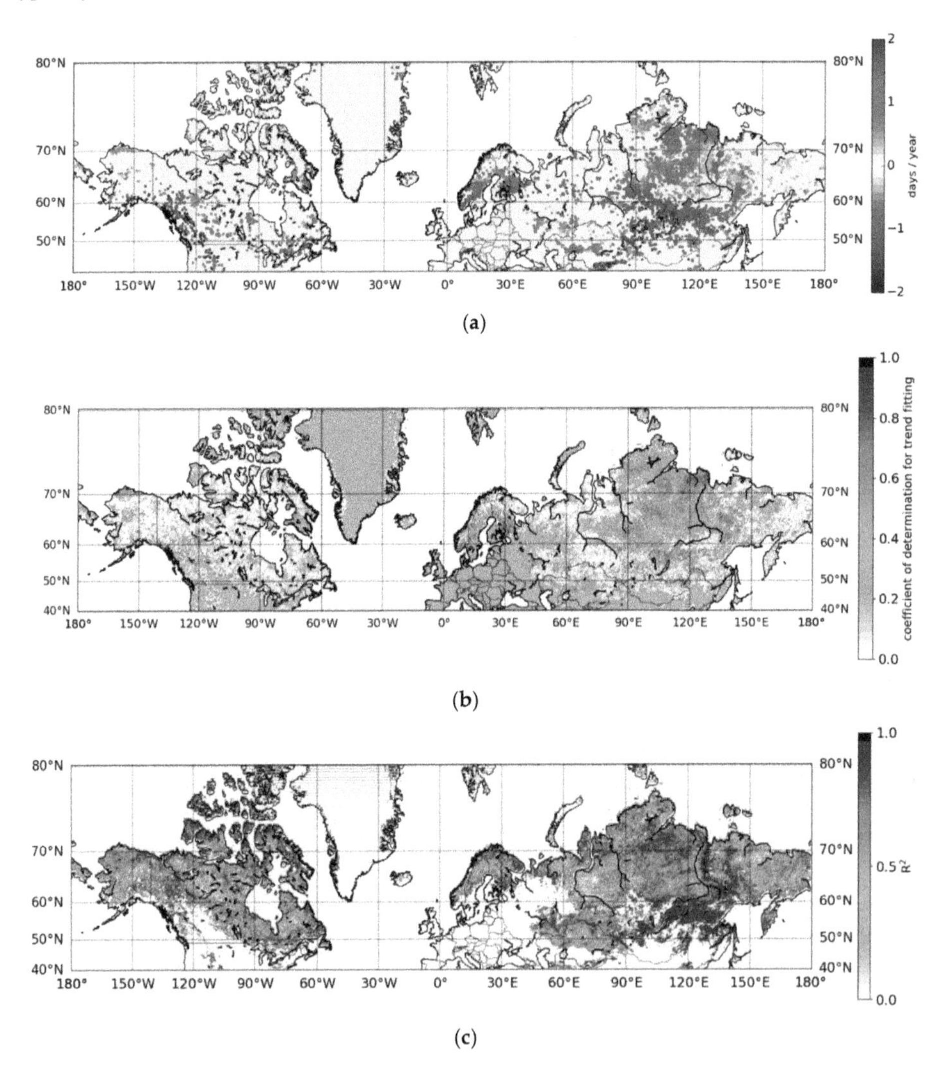

Figure 8. (**a**) The rates of change for the start day of melt between 1982 and 2015 using 5-year rolling mean albedo (showing cases for which R^2 of the fit was larger than 0.5). The negative rates of change mean earlier onset of melt and the positive rates of change mean later dates of onset of melt. (**b**) The coefficient of determination (R^2) values for the trend fitting for start day of melt. (**c**) The multiple variable correlation (R^2) between ERA-Interim climate data and start day of melt. R^2 value for linear correlation between the time that melt season started (day of year) and 3 climatic parameters for 14 days prior to the onset of melt. The climatic parameters used are the mean air temperature, accumulated precipitation and mean wind speed.

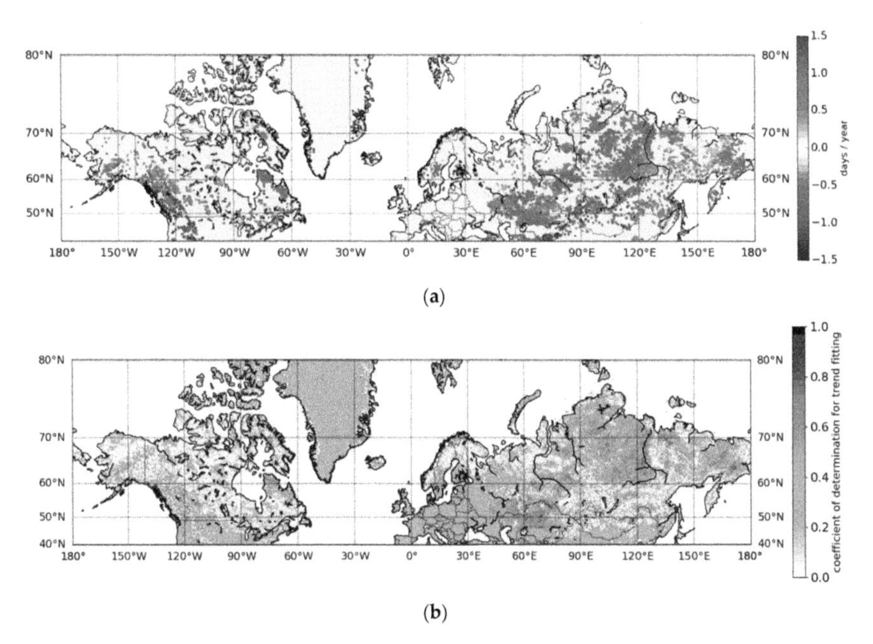

Figure 9. (**a**) The rates of change for the end date of melt between 1982 and 2015 using 5-year rolling mean (showing cases for which R^2 of the fit was larger than 0.5). The negative rates of change mean earlier end of melt and the positive rates of change mean later dates of end of melt. (**b**) The coefficient of determination (R^2) values for the trend fitting for end day of melt.

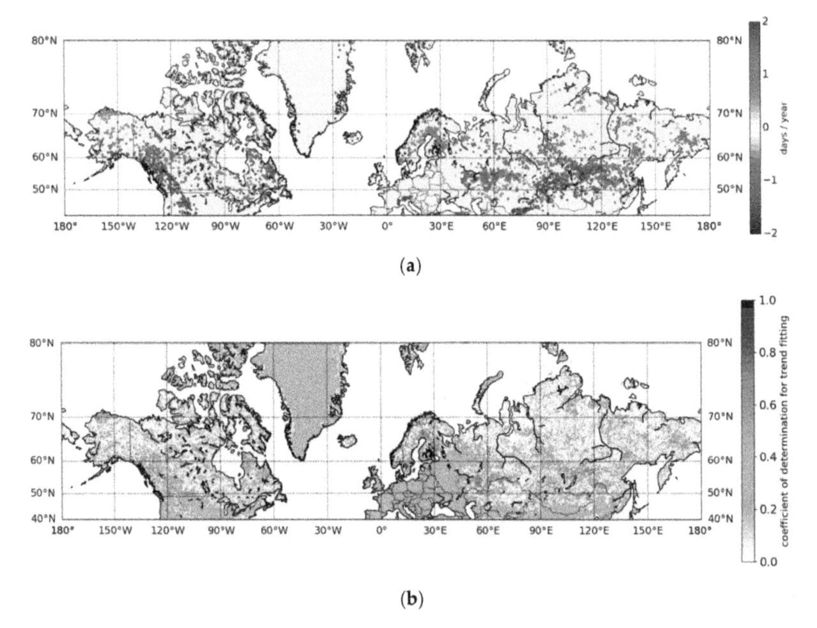

Figure 10. (**a**) The rates of change for the length of melt season between 1982 and 2015 using 5-year rolling mean (showing cases for which R^2 of the fit was larger than 0.5). The negative rates of change mean shorter melt seasons and the positive rates of change mean longer melt season. (**b**) The coefficient of determination (R^2) values for the trend fitting for length of melt season.

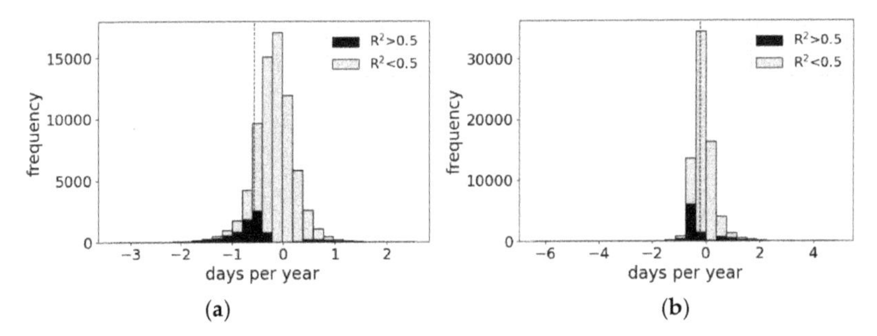

Figure 11. The rate of change observations for (**a**) start and (**b**) end day of melt. The dashed line shows the mean value of the observations for which R^2 is larger than 0.5.

The distribution of all the computed trends for start and end day of melt can be seen in Figure 11. The majority of the trends had R^2 values lower than 0.5. and the rate of change of these were typically close to zero but slightly towards earlier onset and end of melt.

The trends in melt season timing showed no significant dependency on the land use. In the Central Siberian Plain the melt started and ended earlier (Table 2), resulting in earlier melt seasons across the whole region regardless of the vegetation type (both boreal and tundra). The decreasing trends for mean air temperature before the melt onset in the Central Siberian Plain (Figure 7a, Table 2) show similar spatial patterns as the melt season timing parameters. This can be explained by the fact that the air temperatures prior to melt are not derived from the same time of year, but change together with the start date of melt. With earlier onset of melt, the air temperatures are also derived from an earlier period. In the mid-winter the air temperature is more heavily influenced by the lack of heating from the Sun, whereas later in the spring other climatic factors start to affect the air temperature more significantly. Earlier onset of melt can be associated with colder air temperatures prior to melt onset and more rapid change in the air temperature from cold mid-winter values to melting conditions.

In the area around the borders of Russia, Mongolia and China the melt starts earlier and ends later (Table 2), resulting in longer melt seasons. This is also the case for the Canadian Rocky Mountains. In North America, the northern parts of Labrador Peninsula, which are tundra, also show trends towards a shorter melt season and earlier end of melt.

Using the climatic data from ERA-Interim, three parameters (mean air temperature, mean wind speed and accumulated precipitation) are required to explain the changes in the start date of melt, giving a mean R^2 value of 0.65 for the whole study area (Figure 8). In some regions, the mean wind speed and accumulated precipitation (for 14 days prior to melt) are strongly correlated with the starting time of the melt, while the mean air temperature is not (Figure 12), thus supporting the multivariate explanation. For example, wind speed affects the start of melt more than air temperature in the Southern West Siberian Plain and the Southern Byrranga Mountains. Changes in wind conditions affect snow surface scattering by affecting sublimation, mechanical metamorphism of the surface crystals, and distribution of the snow, thus affecting the albedo and depth of snow, and the length of the melt season. The distribution of impurities, such as litter from vegetation on the snow cover, can change due to wind conditions. Impurities increase both the absorption of solar energy into the snow pack and the melt rate. Precipitation (together with air temperature) correlates with the start of melt particularly in the Yablonovyy Range and south of Taymur Peninsula. The trends in the 5-year rolling mean of climatic parameters in individual grid cells show similar spatial patterns as the melt season parameters. The trends in these areas are summarized in Table 2.

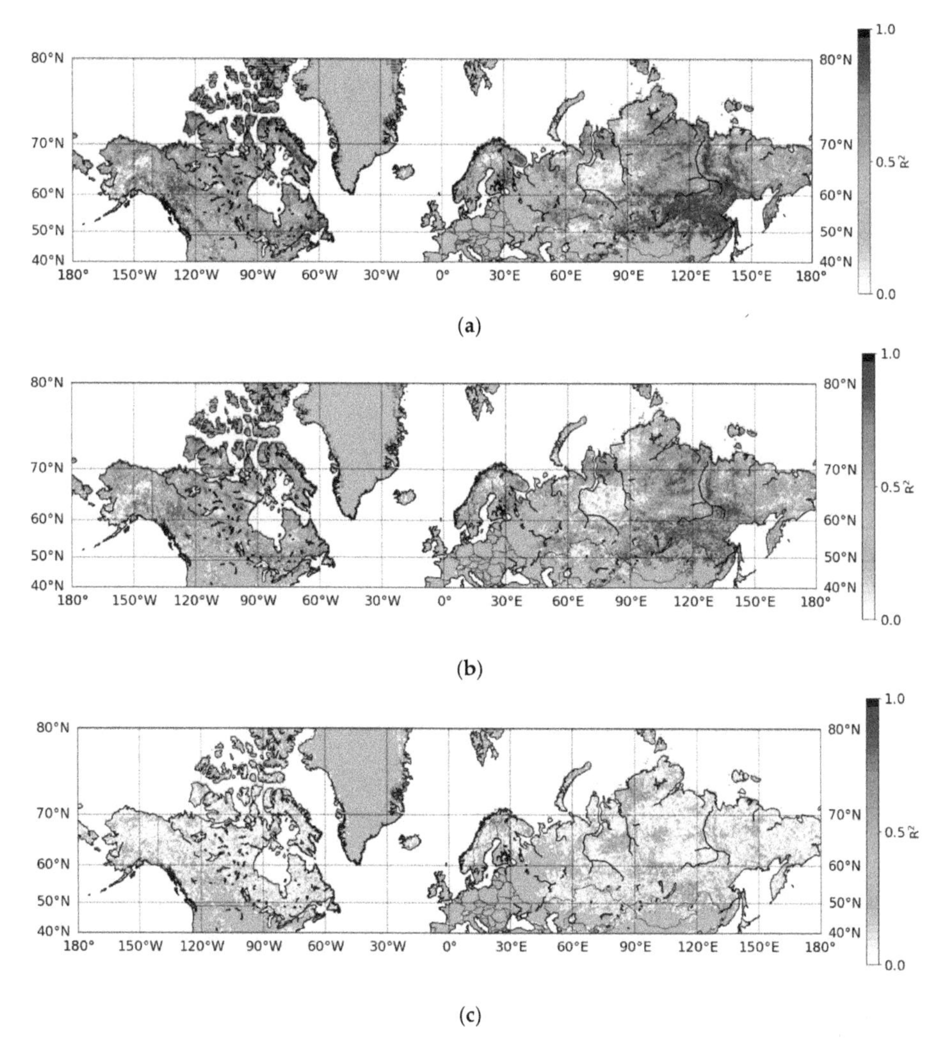

Figure 12. The correlation (R^2) between start day of melt and (**a**) mean air temperature (**b**) total precipitation and (**c**) wind speed for 14 days prior to the onset of melt.

5. Discussion

The magnitude of changes in albedo prior to melt season seem to be linked strongly to vegetation characteristics, whereas the changes in melt season timing seem to be more strongly linked to climatic factors. The influence of vegetation on surface albedo prior to melt can be understood by considering the difference between the albedo of vegetation and the albedo of snow. Even a small increase in the vegetation cover can alter the surface albedo by several absolute albedo percentage units [49,50], whereas changes in climatic factors prior to melt affect the snow depth and/or the surface crystal structure, and have smaller (but still significant) impact on the surface albedo. Changes in the number of days with snow on the trees are an obvious exception to this, putting emphasis on the influence of climatic parameters on surface albedo. Increased precipitation in vegetated areas would mean increased albedo while, in open areas, such as tundra, more precipitation would more strongly affect the depth and surface crystal structure of the snow pack, with a weaker effect on the surface albedo.

The start date of melt is linked to the air temperature. Increasing temperatures would result in earlier melt onset. However, in this study the focus was on the temperature prior to melt. That means that the time period in question changes from year to year. An increase in air temperatures prior to melt onset would mean that at the start of melt the snow pack is already warmer and the surface crystals may have been affected by melt and sublimation. Conversely, lower air temperatures, such as in the area around the borders of Russia, Mongolia and China, would result in a colder snowpack, and, in the case of a thin snow cover, resulting from low rates of precipitation, colder ground. This would mean slower melt. Both the end date of melt and the length of the melt season are affected by the air temperature, snow depth, snowpack characteristics, and the ground temperature. These are all regulated by climatic factors. For changes in vegetation to cause significant changes in the melt season timing, such as observed in this study, the vegetation would need to change considerably, for example from bare ground to tree cover. At larger spatial scales, such changes take more than 34 years.

The relationship between vegetation and climate is not straightforward. Existing studies of changes in vegetation, permafrost and impurities in snow [26,51–55] reveal similar spatial patterns of consistent change as do the melt season parameters, but the effect is not the same in all areas. None of the aforementioned factors alone are able to explain the changes over the whole study area. In fact, they can have opposite effects on the albedo in different regions.

The inconsistency in the effects of vegetation on winter time albedo can be partly explained by the different response of vegetation to the changing climate. According to Xu et al. [26], in Eurasia the normalized difference vegetation index (NDVI) is positively correlated with warming temperature, whereas in North America the effect varies in different regions and patterns of greening and browning in North America are fragmented [26]. It is noteworthy that while the NDVI data describe growing season conditions and are thus not directly translatable to winter conditions, they should be considered an indicator of changes that may also be visible in winter, such as growth in shrub size and coverage. The observed changes in vegetation cannot be directly translated to changes in surface albedo. According to Myers-Smith et al. [56] shrub growth is responsive to different drivers in different regions. Sturm et al. [50] found that if shrubs protrude above the snow and cover 10% of the surface, the albedo will decrease by 30%. The spatial coverage of continuous permafrost shows similar spatial patterns with many of the trends for the melt season parameters. For example, at the southern edge of Central Siberian Plain and in Labrador Peninsula continuous permafrost ends in the same area as the trends for melt season change. The difference in trends for melt season parameters can be due to changes in permafrost coverage causing changes in land use, such as vegetation or formation of melt ponds, or by different response of vegetation and snow cover to climatic changes in areas with and without permafrost.

6. Conclusions

The surface albedo before the onset of melt shows clear but spatially varying trends. The clearest trends are observed in the boreal forest zone, and can be as large as 9 absolute albedo percentage units over the 34-year-long study period. This is about 10–15% of the albedo of clean snow cover with no vegetation protruding above the snow surface. At grid cell level, both the albedo prior to melt onset and the start date of the melt season are responsive to changes in air temperature, wind speed and precipitation amount, with air temperature being the most significant driving factor. At most latitudes the mean albedo before the onset of melt has decreased over the 34 years.

The timing of the melt season shows strong rates of change in localized areas. These changes are better explained by climatic factors than land-use changes. Areas with consistent homogenous changes are larger in Eurasia than in North America. In the Central Siberian Plain the melt season takes place earlier than before. In the Canadian Rocky Mountains and the area around the borders of Russia, China and Mongolia, the melt season starts earlier, ends later, and lasts longer than before.

All in all, both the timing of the melt season and the albedo prior to the melt season changed in large areas between 1982 and 2015. In most areas, both the start and end dates of the melt season have advanced, and the albedo prior to melt onset has decreased, indicating a darkening of winter snow surfaces.

Author Contributions: Data curation, K.A., T.M., E.J., A.R. and P.L.; Formal analysis, K.A.; Funding acquisition, T.M.; Investigation, K.A.; Methodology, T.M.; Project administration, K.A.; Supervision, T.M.; Writing—original draft, K.A.

Funding: This research was financially supported by the CM SAF project funded by The European Organisation for the Exploitation of Meteorological Satellites (EUMETSAT).

Acknowledgments: The authors would like to thank the CM SAF team, Vesa Laine and Kaj Andersson for their co-operation in producing the CLARA-A2 SAL data record, and Outi Meinander for supporting discussions. EUMETSAT supported financially the generation of the CLARA-A2 SAL, which is available at www.cmsaf.eu.

Conflicts of Interest: The authors declare no conflict of interest.

References

1. Hall, A. The role of surface albedo feedback in climate. *J. Clim.* **2004**, *17*, 1550–1568. [CrossRef]
2. Smith, N.; Saatchi, S.; Randerson, J. Trends in high northern latitude soil freeze and thaw cycles from 1988 to 2002. *J. Geophys. Res.* **2004**, *109*, D12101. [CrossRef]
3. Déry, S.; Brown, R. Recent Northern Hemisphere snow cover extent trends and implications for the snow-albedo feedback. *Geophys. Res. Lett.* **2007**, *34*, L22504. [CrossRef]
4. Solomon, S.; Qin, D.; Manning, M.; Averyt, K.; Marquis, M.; Averyt, K.B.; Tignor, M.; Miller, H.L. *Climate Change 2007: The Physical Science Basis; Working Group I Contribution to the Fourth Assessment Report of the IPCC*; Cambridge University Press: Cambridge, UK; New York, NY, USA, 2007; Volume 4.
5. Brown, R.D.; Robinson, D.A. Northern Hemisphere spring snow cover variability and change over 1922–2010 including an assessment of uncertainty. *Cryosphere* **2011**, *5*, 219–229. [CrossRef]
6. Flanner, M.; Shell, K.; Barlage, M.; Perovich, D.; Tschudi, M. Radiative forcing and albedo feedback from the Northern Hemisphere cryosphere between 1979 and 2008. *Nat. Geosci.* **2011**, *4*, 151–155. [CrossRef]
7. Derksen, C.; Brown, R. Spring snow cover extent reductions in the 2008–2012 period exceeding climate model projections. *Geophys. Res. Lett.* **2012**, *39*. [CrossRef]
8. Foster, J.L.; Cohen, J.; Robinson, D.A.; Estilow, T.W. A look at the date of snowmelt and correlation with the Arctic Oscillation. *Ann. Glaciol.* **2013**, *54*, 196–204. [CrossRef]
9. IPCC. *Climate Change 2013: The Physical Science Basis*; Contribution of Working Group I to the Fifth Assessment Report of the Intergovernmental Panel on Climate Change; Stocker, T.F., Qin, D., Plattner, G.-K., Tignor, M., Allen, S.K., Boschung, J., Nauels, A., Xia, Y., Bex, V., Midgley, P.M., Eds.; Cambridge University Press: Cambridge, UK; New York, NY, USA, 2013; p. 1535.
10. Atlaskina, K.; Berninger, F.; Leeuw, G. Satellite observations of changes in snow-covered land surface albedo during spring in the Northern Hemisphere. *Cryosphere* **2015**, *9*, 1879–1893. [CrossRef]
11. Manninen, T.; Stenberg, P. Simulation of the effect of snow covered forest floor on the total forest albedo. *Agric. For. Meteorol.* **2009**, *149*, 303–319. [CrossRef]
12. Fassnacht, S.; Cherry, M.; Venable, N.; Saavedra, F. Snow and albedo climate change impacts across the United States Northern Great Plains. *Cryosphere* **2016**, *10*, 329–339. [CrossRef]
13. Li, Q.; Ma, M.; Wu, X.; Yang, H. Snow Cover and Vegetation-Induced Decrease in Global Albedo from 2002 to 2016. *J. Geophys. Res. Atmos.* **2018**, *123*, 124–138. [CrossRef]
14. Markus, T.; Stroeve, J.C.; Miller, J. Recent changes in Arctic sea ice melt onset, freezeup, and melt season length. *J. Geophys. Res.* **2009**, *114*, C12024. [CrossRef]
15. Wang, L.; Derksen, C.; Brown, R.; Markus, T. Recent changes in pan-Arctic melt onset from satellite passive microwave measurements. *Geophys. Res. Lett.* **2013**, *40*, 1–7. [CrossRef]
16. Chen, X.; Liang, S.; Cao, Y.; Cao, T.; Wang, D. Observed contrast changes in snow cover phenology in northern middle and high latitudes from 2001–2014. *Sci. Rep.* **2015**, *5*. [CrossRef] [PubMed]
17. Malnes, E.; Karlsen, R.S.; Johansen, B.; Bjerke, J.W.; Tømmervik, H. Snow season variability in a boreal-Arctic transition area monitored by MODIS data. *Environ. Res. Lett.* **2016**, *11*, 125005. [CrossRef]
18. Rautiainen, K.; Parkkinen, T.; Lemmetyinen, J.; Schwank, M.; Wiesmann, A.; Ikonen, J.; Derksend, C.; Davydove, S.; Davydovae, A.; Boike, J.; et al. SMOS prototype algorithm for detecting autumn soil freezing. *Remote Sens. Environ.* **2016**, *180*, 346–360. [CrossRef]

19. Bhatt, U.; Walker, D.A.; Rauynolds, M.K.; Bienek, P.A.; Epstein, H.E.; Comiso, J.C.; Pinzon, J.E.; Tucker, C.J.; Steele, M.; Ermold, W.; et al. Changing seasonality of panarctic tundra vegetation in relationship to climatic variable. *Environ. Res. Lett.* **2017**, *12*, 055003. [CrossRef]
20. Essery, R. Large-scale simulations of snow albedo masking by forests. *Geophys. Res. Lett.* **2013**, *40*. [CrossRef]
21. Thackeray, C.W.; Fletcher, C.G.; Derksen, C. Quantifying the skill of CMIP5 models in simulating seasonal albedo and snow cover evolution. *J. Geophys. Res. Atmos.* **2015**, *120*, 5831–5849. [CrossRef]
22. Abe, M.; Takata, K.; Kawamiya, M.; Watanabe, S. Vegetation masking effect on future warming and snow albedo feedback in a boreal forest region of northern Eurasia according to MIROC-ESM. *J. Geophys. Res. Atmos.* **2017**, *122*. [CrossRef]
23. Warren, S. Impurities in snow: Effects on albedo and snowmelt (review). *Ann. Glaciol.* **1984**, *5*, 177–179. [CrossRef]
24. Domine, F.; Salvatori, R.; Legagneux, L.; Salzano, R.; Fily, M.; Casacchia, R. Correlation between the specific surface area and the short wave infrared (SWIR) reflectance of snow. *Cold Reg. Sci. Technol.* **2006**, *46*, 60–68. [CrossRef]
25. Ménégoz, M.; Krinner, G.; Balkanski, Y.; Cozic, A.; Boucher, O.; Ciais, P. Boreal and temperate snow cover variations induced by black carbon emissions in the middle of the 21st century. *Cryosphere* **2013**, *7*, 537–554. [CrossRef]
26. Xu, L.; Myneni, R.; Chapin, F., III; Callaghan, T.; Pinzon, J.; Tucker, C.; Zhu, Z.; Bi, J.; Ciais, P.; Tømmervik, H.; et al. Temperature and vegetation seasonality diminishment over northern lands. *Nat. Clim. Chang.* **2013**, *3*, 581–586. [CrossRef]
27. Wiscombe, W.J.; Warren, S.G. A model for the spectral albedo of snow. I: Pure snow. *J. Atmos. Sci.* **1980**, *37*, 2712–2733. [CrossRef]
28. Shi, J.; Dozier, J. Estimation of Snow Water Equivalence Using SIR-C/X-SAR, Part II: Inferring Snow Depth and Particle Size. *IEEE Trans. Geosci. Remote Sens.* **2000**, *38*, 2475–2488.
29. Warren, S.; Brandt, R.; Hinton, P. Effect of surface roughness on bidirectional reflectance of Antarctic snow. *J. Geophys. Res.* **1998**, *103*, 25789–25807. [CrossRef]
30. Nagler, T.; Rott, H. Retrieval of wet snow by means of multitemporal SAR data. *Trans. Geosci. Remote Sens.* **2000**, *38*, 754–765. [CrossRef]
31. Robinson, D.A.; Kukla, G. Albedo of a Dissipating Snow Cover. *J. Clim. Appl. Meteorol.* **1984**, *23*, 1626–1634. [CrossRef]
32. Robinson, D.A.; Kukla, G. Maximum Surface Albedo of Seasonally Snow-Covered Lands in the Northern Hemisphere. *J. Clim. Appl. Meteorol.* **1985**, *24*, 402–411. [CrossRef]
33. Kuittinen, R. Determination of areal snow-water equivalent values using satellite imagery and aircraft gamma-ray spectrometry. In *Hydrologic Applications of Space Techology: Proceedings of an International Workshop on Hydrologic Applications of Space Technology, Held in Cocoa Beach, FL, USA, 19–23 August 1985*; IAHS Press: Oxfordshire, UK; Institute of Hydrology: Wallingford, UK, 1986; Volume 160, pp. 181–189.
34. Rinne, J.; Aurela, M.; Manninen, T. A Simple Method to determine the timing of snow melt by remote sensing with application to the CO_2 balances of northern mire and heath ecosystems. *Remote Sens.* **1986**, *1*, 1097–1107. [CrossRef]
35. Solantie, R.; Drebs, A.; Hellsten, E.; Saurio, P. *Lumipeitteen tuo-, lähtö-ja Kestoajoista Suomessatalvina 1960/1961–1992/1993*; Finnish Meteorological Institute, English Summary; Meteorological publications: Helsinki, Finland, 1996; Volume 34, 159p.
36. Barlage, M.; Zeng, X.; Wei, H.; Mitchell, K.E. A global 0.05° maximum albedo dataset of snow-covered land based on MODIS observations. *Geophys. Res. Lett.* **2005**, *32*, L17405. [CrossRef]
37. Anttila, K.; Jääskeläinen, E.; Riihelä, A.; Manninen, T.; Andersson, K.; Hollman, R. Algorithm Theoretical Basis Document: CM SAF Cloud, Albedo, Radiation Data Record Ed. 2—Surface Albedo. 2016. Available online: https://icdc.cen.uni-hamburg.de/fileadmin/user_upload/icdc_Dokumente/EUMETSAT-CMSAF/SAF_CM_FMI_ATBD_GAC_SAL_2_3.pdf (accessed on 14 August 2018).
38. Karlsson, K.-G.; Anttila, K.; Trentmann, J.; Stengel, M.; Meirink, J.F.; Devastale, A.; Hanschmann, T.; Kothe, S.; Jääskeläinen, E.; Sedlar, J.; et al. CLARA-A2: The second edition of the CM SAF cloud and radiation data record from 34 years of global AVHRR data. *Atmos. Chem. Phys.* **2017**, *17*, 5809–5828. [CrossRef]
39. Riihelä, A.; Manninen, T.; Laine, V.; Andersson, K.; Kaspar, F. CLARA-SAL: A global 28 yr timeseries of Earth's black-sky surface albedo. *Atmos. Chem. Phys.* **2013**, *13*, 3743–3762. [CrossRef]

40. Jääskeläinen, E.; Manninen, T.; Tamminen, J.; Laine, M. The Aerosol Index and Land Cover Class Based Atmospheric Correction Aerosol Optical Depth Time Series 1982–2014 for the SMAC Algorithm. *Remote Sens.* **2017**, *9*, 1095. [CrossRef]
41. Peltoniemi, J.I.; Suomalainen, J.; Hakala, T.; Puttonen, E.; Näränen, J.; Kaasalainen, S.; Torppa, J.; Hirschmugl, M. Reflectance of various snow types: Measurements, modelling and potential for snow melt monitoring. In *Light Scattering Reviews 5: Single Light Scattering and Radiative Transfer*; Springer Praxis Books: Berlin/Heidelberg, Germany, 2010; Chapter 9; pp. 393–450. [CrossRef]
42. Eastwood, S. Sea Ice Product User's Manual OSI-401-a, OSI-402-a, OSI-403-a, Version 3.11. 2014. Available online: http://osisaf.met.no/docs/osisaf_ss2_pum_ice-conc-edge-type_v3p11.pdf (accessed on 14 August 2018).
43. Schaaf, C.B.; Gao, F.; Strahler, A.H.; Lucht, W.; Li, X.; Tsang, T.; Strugnell, N.C.; Zhang, X.; Jin, Y.; Muller, J.-P.; et al. First operational BRDF, albedo nadir reflectance products from MODIS. *Remote Sens. Environ.* **2002**, *83*, 135–148. [CrossRef]
44. Anttila, K.; Jääskeläinen, E.; Riihelä, A.; Manninen, T.; Andersson, K.; Hollman, R. Validation Report: CM SAF Cloud, Albedo, Radiation Data Record Ed. 2—Surface Albedo. 2016. Available online: https://icdc.cen.uni-hamburg.de/fileadmin/user_upload/icdc_Dokumente/EUMETSAT-CMSAF/SAF_CM_FMI_ATBD_GAC_SAL_2_3.pdf (accessed on 14 August 2018).
45. Riihelä, A.; Laine, V.; Manninen, T.; Palo, T.; Vihma, T. Validation of the Climate-SAF surface broadband albedo product: Comparisons with in situ observations over Greenland and the ice-covered Arctic Ocean. *Remote Sens. Environ.* **2010**, *114*, 2779–2790. [CrossRef]
46. Dee, D.P.; Uppala, S.; Simmons, A.; Berrisford, P.; Poli, P.; Kobayashi, S.; Andrae, U.; Alonso-Balmaseda, M.; Balsamo, G.; Bauer, P.; et al. The ERA–Interim reanalysis: Configuration and performance of the data assimilation system. *Q. J. R. Meteorol. Soc.* **2011**, *137*, 553–597. [CrossRef]
47. Arino, O.; Ramos, J.; Kalogirou, V.; Defourny, P.; Achard, F. GlobCover 2009. In Proceedings of the Living Planet Symposium, Bergen, Norway, 28 June–2 July 2010.
48. Böttcher, K.; Aurela, M.; Kervinen, M.; Markkanen, T.; Mattila, O.P.; Kolari, P.; Metsämäki, S.; Aalto, T.; Arslan, A.N.; Pulliainen, J. MODIS tile-series-derived indicators for the beginning of the growing season in boreal coniferous forest—A comparison with the CO_2 flux measurements and phenological observations in Finland. *Remote Sens. Environ.* **2014**, *140*, 625–638. [CrossRef]
49. Sturm, M.; Douglas, T.; Racine, C.; Liston, G. Changing snow and shrub conditions affect albedo with global implications. *J. Geophys. Res.-Biogeosci.* **2005**, *110*, G01004. [CrossRef]
50. Bonan, G.B.; Pollard, D.; Thompson, S.L. Effects of boreal forest vegetation on global climate. *Nature* **1992**, *359*, 716. [CrossRef]
51. Rigina, O. Environmental impact assessment of the mining and concentration activities in the Kola Peninsula, Russia by multidate remote sensing. *Environ. Monit. Assess.* **2002**, *75*, 11–31. [CrossRef] [PubMed]
52. Piao, S.; Wang, X.; Ciais, P.; Zhu, B.; Wang, T. Changes in satellite-derived vegetation growth trend in temperate and boreal Eurasia from 1982 to 2006. *Glob. Chang. Biol.* **2011**, *17*, 3228–3239. [CrossRef]
53. Buitenwerf, R.; Rose, L.; Higgins, S. Three decades of multi-dimensional change in global leaf phenology. *Nat. Clim. Chang.* **2015**, *5*, 364–368. [CrossRef]
54. Bullard, J.; Baddock, M.; Bradwell, T.; Crusius, J.; Darlington, E.; Gaiero, D.; Gassó, S.; Gisladottir, G.; Hodgkins, R.; McCulloch, R.; et al. High-latitude dust in Earth system. *Rev. Geophys.* **2016**, *54*, 447–485. [CrossRef]
55. Helbig, M.; Wischnewski, K.; Kljun, N.; Chasmer, L.E.; Quinton, W.L.; Detto, M.; Sonnentag, O. Regional atmospheric cooling and wetting effect of permafrost thaw-induced boreal forest loss. *Glob. Chang. Biol.* **2016**, *22*, 4048–4066. [CrossRef] [PubMed]
56. Myers-Smith, I.H.; Elmerdorf, S.; Becl, P.; Wilmking, M.; Hallinger, M.; Blok, D.; Tape, K.D.; Rayback, S.A.; Macias-Fauria, M.; Forbes, B.C.; et al. Climate sensitivity of shrub growth across the tundra biome. *Nat. Clim. Chang.* **2015**, *5*, 887–891. [CrossRef]

Article

Novel Measurements of Fine-Scale Albedo: Using a Commercial Quadcopter to Measure Radiation Fluxes

Charlotte R. Levy [1,*], Elizabeth Burakowski [2] and Andrew D. Richardson [3,4]

[1] Department of Ecology and Evolutionary Biology, Cornell University, Ithaca, NY 14850, USA
[2] Institute for the Study of Earth, Oceans, and Space, University of New Hampshire, Durham, NH 03824, USA; elizabeth.burakowski@unh.edu
[3] School of Informatics, Computing and Cyber Systems, Northern Arizona University, Flagstaff, AZ 86011, USA; andrew.richardson@nau.edu
[4] Center for Ecosystem Science and Society, Northern Arizona University, Flagstaff, AZ 86011, USA
* Correspondence: crl222@cornell.edu; Tel.: +1-857-636-9396

Received: 19 June 2018; Accepted: 15 August 2018; Published: 18 August 2018

Abstract: Remote sensing of radiative indices must balance spatially and temporally coarse satellite measurements with finer-scale, but geographically limited, in-situ surface measurements. Instruments mounted upon an Unmanned Aerial Vehicle (UAV) can provide small-scale, mobile remote measurements that fill this resolution gap. Here we present and validate a novel method of obtaining albedo values using an unmodified quadcopter at a deciduous northern hardwood forest. We validate this method by comparing simultaneous albedo estimates by UAV and a fixed tower at the same site. We found that UAV provided stable albedo measurements across multiple flights, with results that were well within the range of tower-estimated albedo at similar forested sites. Our results indicate that in-situ albedo measurements (tower and UAV) capture more site-to-site variation in albedo than satellite measurements. Overall, we show that UAVs produce reliable, consistent albedo measurements that can capture crucial surface heterogeneity, clearly distinguishing between different land uses. Future application of this approach can provide detailed measurements of albedo and potentially other vegetation indices to enhance global research and modeling efforts.

Keywords: albedo; land use; remote sensing; Unmanned Aerial Vehicles; vegetation indices

---

1. Introduction

Over the past few decades, the simultaneous rise of remote sensing technologies and earth system models has generated a broad, cross-disciplinary need for radiometric datasets with both global extent and fine-scale parameterization. Radiometric indices are used to estimate global primary productivity, vegetative cover, energy fluxes, and many more properties essential to understanding present and future climate and ecosystem functioning [1,2]. An uneven or too sparse global distribution of sites will bias estimates and cause these ecosystem properties to be poorly represented by global climate models [3,4]. At the same time, local disturbances (forest fires, drought, plowing, thinning, snow aging) [5–7] can have outsize effects on regional and global climate [5,8–11], yet be poorly captured by coarse global measurements or too underrepresented to be well modeled by earth system models [12,13]. To understand current and future trends in ecosystem functioning and climatic change, we must be able to capture both global extent and fine-scale variation in remotely-sensed, radiometric datasets [6,14–20].

Patterns at the global scale are generally derived from broadband satellite products [3,16,21,22], that are far-reaching but coarse-scaled. The most commonly used albedo dataset, the MODIS data products, are scaled as 500 m sinusoidal grid resolutions, limiting their ability to register small-scale land use and management strategies [4,23,24]. Development of a well-validated LANDSAT albedo product is ongoing and will provide a 30 m product at 16-day intervals, significantly improving the spatial resolution of the remotely-sensed albedo measurements; however, fine-scale in-situ estimates will still be needed to continue to validate this product [7,25,26]. In-situ measurements can corroborate satellite data but have their own limitations. Fixed towers are immobile, few in number, and have physical limitations on maximum height that limit their spatial range. Thus, scattered point measurements from towers may not accurately represent variation across larger landscapes [3,24]. Portable spectroradiometers have been used to quantify radiation fluxes in fields and the understory, and are generally very effective for evaluating effects of snow depth [27], snow age, grain size, and layer structure [28,29]. However these tools are limited in their application above canopy [27]. Airborne high-resolution hyperspectral sensors mounted on planes or helicopters have permitted quantification of radiation fluxes across broader regions, but tend to be extremely costly and logistically complex. They can capture only single time point measurements along the flight path and are subject to technical issues caused by the scattering of light by aerosols and water vapor at higher altitudes between the sensor and the land surface [27].

Unmanned aerial vehicles (UAVs) can increase both the flexibility and affordability of fine-scale measurements, providing an essential bridge between ground-truthing and global satellite data [30,31]. UAVs can move freely over tree canopies, allowing measurement over entire forest stands rather than just single points. UAVs can adjust to a range of canopy heights, giving them more flexibility to achieve optimal observation heights [24]. UAV flights are more affordable than piloted airborne missions; moreover, in the United States recent adjustments to Federal Aviation Administration regulations have made UAV technology more accessible for researchers [32]. Several caveats must be considered: flights are limited in range and flight time by the strength of the radio signal, the battery life, the payload, and the angle of view of the observer. Standards for accommodating any position or height instability must still be developed. Finally, adaptation of UAVs for measurement of radiative indices requiring both incoming and reflected radiation measurements has been technically difficult to make by UAVs due to issues of payload weight and balance. Albedo is the ratio between down-welling shortwave broadband solar radiation and reflected, up-welling shortwave broadband solar radiation; it is typically measured using paired (one upward facing, one downward facing) pyranometers. However, standard UAVs are generally designed to lift objects with a center of gravity beneath the vehicle, such that mounting an upward-facing pyranometer on top or on an extended boom off of an UAV requires extensive customization and technical adjustment to ensure flight stability. In addition, the weight of two sensors imposes a significant energy cost, greatly reducing flight time. Two previous studies measuring albedo via UAV (fixed-wing craft over the Indian Ocean [33], fixed-wing craft over Greenland [34]) have required custom modifications not swiftly replicable by most research labs. The simple method of measuring albedo proposed here allows use of unmodified quadcopters such as have been widely adopted by many labs for other forms of aerial imaging while minimizing payload and maximizing flight time.

Here we employ a novel measurement method to investigate albedo over a mixed hardwood forest in central New York. UAV measurements were tested for consistency across flights and for comparability to conventional forest albedo measurements made by tower and satellite. We verify the validity of our technique through side-by-side tower and UAV comparison over a field of shrub willow. Finally, we examine albedo across three land uses and seven flights, comparing within flight variability to variability across land uses. In testing this novel method, which minimizes UAV payload and permits use of uncustomized quadcopters, we hope to expand the capacity for scientists to validate satellite estimates using fine-scale radiometric measurements.

2. Materials and Methods

2.1. A Novel Method of Measuring Albedo by UAV

In the method presented here, albedo was calculated as the ratio between reflected shortwave radiation, as measured from a downward-facing pyranometer mounted under a UAV, and incoming shortwave radiation, as measured from a separate upward-facing pyranometer mounted to a pole in an immediately adjacent open area (Figure 1). The UAV-mounted downward-facing pyranometer was a Kipp and Zonen CMP3 pyranometer (spectral range: 300–2800 nm). It was secured underneath a four-rotor Spyder 850 (Sky Hero, Pearland, TX, USA) UAV and leveled using a motorized Gaui Crane gimbal (Figure 2).

The UAV was only modified to the extent of having the carbon-fiber support legs lengthened, to provide additional clearance for the pyranometer during take-off and landing. The UAV pyranometer was paired with an upward-facing Kipp and Zonen CMP6 pyranometer (spectral range: 285–2800 nm) mounted on a pneumatic telescoping pole (Total Mast Solutions, CP56-08) and secured to a portable tripod. The pyranometer was fixed on a 30 cm leveled boom, oriented to the south, at a height of 9.09 m [35]. To obtain reference albedo measurements for validation flights, a second Kipp and Zonen CMP6 pyranometer was fixed and leveled below the first, to determine reflected radiation from beneath the tower. All instruments had a sensitivity of 5 to 20 $\mu V/W/m^2$, a response time of 18 sec or less (95%), and an effective half field of view of 81°. The thirty-second averages of up-welling and down-welling shortwave radiation from both pyranometers were recorded by an attached Kipp and Zonen METEON datalogger. The internal clocks of the two dataloggers were synchronized by a common laptop computer an hour prior to the experiment start. For each individual flight, the sum of all reflected radiation values was divided by the sum of incoming radiation values to get a flux-weighted albedo value for that flight. The viewing area of the pyranometer was calculated as the area from which 99% of sensor input came. This area was calculated based on Kipp and Zonen (2016) recommendations:

$$\text{Footprint diameter} = 2 * \text{height} * \tan(\text{effective half field of view}) \quad (1)$$

Figure 1. Diagram of flight design depicting the UAV with downward-facing pyranometer (**left**) and the fixed pole with the upward-facing pyranometer (**right**).

Figure 2. Preparation of the UAV for flight. The gimbal (**A**) is visible underneath the UAV, equipped with the downward-facing pyranometer secured beneath (**B**).

2.2. Experimental Design and Study Area

Albedo measurements consisted of targeted forest measurements by UAV (Section 2.2.1), along with comparative measurements of similar forests by tower and satellite (Section 2.2.2), validating measurements over a local willow field by UAV and tower (Section 2.2.3), and a final comparison of UAV measurements of forest, field, and coniferous forest (Section 2.2.4). The targeted forest measurements demonstrated the internal consistency of UAV measurements, while tower and satellite measurements showed the comparability of UAV albedo to ground and satellite measurements at similar sites. Validation flights compared simultaneous UAV and tower albedo. Finally, the comparison of deciduous, coniferous, and willow sites contrasted the variability across flights with variability across land uses.

Targeted UAV surveys over mixed hardwood forest took place in Tully, NY, USA at a closed-canopy mixed northern hardwood forest stand (Figure 3b; Table 1). Comparative tower and satellite-based measurements from other mixed hardwood sites were obtained from three sites with existing long-term tower albedo measurements, in Bartlett, NH; Durham, NH; and Petersham, MA (Figure 3a; Table 1). All three sites represented a temperate climate and mixed northern hardwood forest land cover. Albedo at each comparative site was obtained from a fixed-point tower and from MODIS satellite data. Validation UAV flights took place in Geneva, NY, USA over a cropped willow field (July 2017), where low height of vegetation allowed both tools to be used simultaneously (Figure 3c; Table 1). Finally, additional UAV flights at a Norway spruce monoculture stand (July 2017) were combined with 2017 forest and willow data for a comparison of different land uses (Table 1).

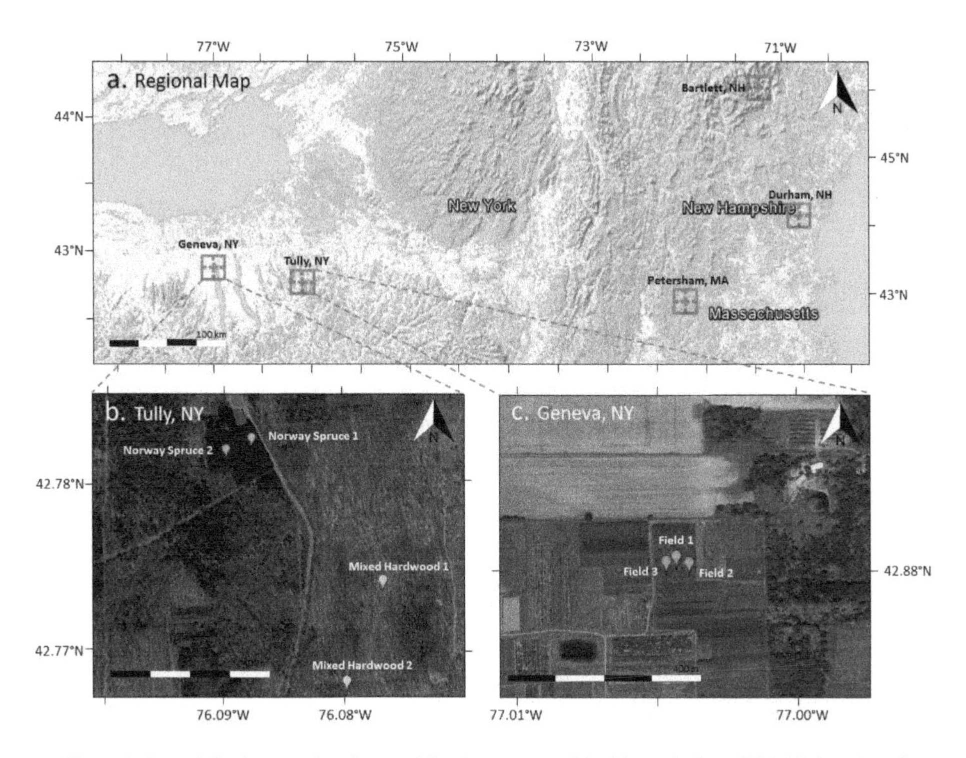

Figure 3. Inset A depicts a regional map of the sites measured in this study. Inset B highlights sites of UAV flights in Tully, NY. Inset C highlights sites of UAV flights in Geneva, NY. Inset A is sourced from 2018 NOAA Imagery. Insets B and C are sourced from Google Earth satellite imagery, April and July 1995 respectively.

Table 1. The locations, typical July temperatures and precipitation, and dominant vegetation of the five study sites.

Objective	Site	Lat (°)	Lon (°)	Canopy Height (avg, m)	Dominant Land Cover
Targeted Flights *UAV, Satellite*	Tully, NY *Mixed Forest*	42.733	−76.081	23	*Acer saccharum, Fagus grandifolia*
Comparative *Tower, Satellite*	Durham, NH † *Mixed Forest*	43.111	−70.955	17	*Quercus rubra, Pinus strobus, Acer rubrum, Carya ovata, Quercus alba*
Comparative *Tower, Satellite*	Bartlett, NH ‡ *Mixed Forest*	44.065	−71.289	21	*F. grandifolia, Picea rubens, A. rubrum, Abies balsamea, Tsuga canadensis, A. saccharum, Betula alleghaniensis*
Comparative, *Tower, Satellite*	Petersham, MA § *Mixed Forest*	42.535	−72.190	16	*Q. rubra, P. strobus, A. rubrum, T. canadensis*
Validation, *UAV/Tower*	Geneva, NY *Cropped Willow*	42.883	−77.004	3	*Salix* spp.
Land Use Flight, *UAV*	Tully, NY *Spruce Stand*	42.733	−76.081	23	*Picea abies*

† [27]; ‡ [36]; § [37].

2.2.1. Targeted UAV Measurements over Mixed Hardwood Forest

The UAV made five flights at Tully, NY over deciduous hardwood forest, one at local solar noon, two flights one and two hours prior to local solar noon, and two flights one and two hours after local solar noon on 27 July 2016 (Supplementary Table S1). In each flight, the UAV followed a pre-programmed course to the designated coordinates and altitude in approximately one minute. The UAV then held its position until the battery was nearly exhausted, approximately ten minutes, before returning to the staging area. Conditions on 27 July 2016 were clear, with minimal cloud cover moving in around local solar noon, and local air quality index less than 50 for both particulate matter and ozone [38].

2.2.2. Comparative Tower and Satellite Measurements over Mixed Hardwood Forest

Tower albedo measurements for the three other mixed hardwood forest sites used here were made in July 2014 and 2015 between 20 July and 24 July. Only measurements taken between 2.5 h prior to and 2.5 h post solar noon were used, to better match UAV data. Readings at Durham were taken every 30 s, Bartlett readings were taken every 5 s, and Petersham measurements were taken every 1 s. Half-hour averages of these measurements were used. For each individual day, the sum of all half-hourly reflected radiation values was divided by the sum of incoming radiation values to get a flux-weighted albedo value for the day.

Durham, NH, USA albedo was measured by a Kipp and Zonen CMA6 (effective half field of view = 81°) placed on a 4.5 m leveling boom extended from 25 m up a 30 m tower [27]. Albedo at Bartlett, NH, USA was collected using two Kipp and Zonen CMP3 pyranometers (effective half field of view = 81°) placed 23.8 m and 25 m up a 30 m tower, facing downwards on a 3 m leveling boom and upwards on a 1 m boom respectively [36]. Albedo values at Petersham, MA, USA were taken using a CNR-4 Kipp & Zonen 4-channel net radiometer mounted on a 3 m boom extending south from a 40 m tower (Effective Field of View 81°) [39,40].

Satellite albedo measurements for Tully mixed hardwood forest and the three comparative forest sites were extracted from the MODIS bidirectional reflectance distribution function albedo product (MCD43A3: MODIS/Terra and Aqua Albedo Daily L3 Global 500 m SIN Grid V006) [41], for DOY 201–215, from 2014, 2015, and 2016. Pixels marked as low-quality in the MODIS quality control data were removed from the analysis. Due to these conditions, only data from 2015 and 2016 was available for Bartlett, NH and Durham, NH. Satellite albedo at the UAV flight site at Tully were extracted from four pixels, a square half kilometer each (Supplementary Table S2). Satellite albedo for the tower sites were pulled from single pixels (Supplementary Table S3). Satellite shortwave albedo at solar noon were converted from black-sky and white-sky albedo to blue-sky using a standard conversion formula [42,43]. Aerosol optical depth (AOD; unitless) was assumed to be 0.2, although a realistic range of environmental depths from 0.1–0.5 was also examined to test sensitivity (Supplementary Table S4). The sensitivity analysis showed that the low and high estimates were not significantly different from 0.2 for any of the examined satellite datasets, and so the 0.2 AOD value was used for the final comparison (NASA, 2016).

2.2.3. Validation Measurements Comparing Simultaneous UAV and Tower Data

Validation flights were conducted on 31 July 2017 over a cropped willow field in Geneva, NY to compare UAV-measured albedo to tower-based measurements. Fixed tower data was collected from a mounted Kipp and Zonen CMA6 albedometer fixed at 8 m on a 30 cm boom. The UAV was first positioned one meter due west of the mounted albedometer, maintaining a height of 8 m (Supplementary Table S1). This first flight took place 30 min prior to local solar noon; two subsequent flights took place 15 min prior, and 30 min post local solar noon. For the second and third flight the UAV was positioned at the same height, 24 m west and 29 m east of the tower, which remained fixed at the center point. All flights took place on a clear day with the local air quality index less than 50 for ozone and below 100 for particulate matter (unitless) [38]. Partial cloud cover appeared towards the end of the third validation flight.

2.2.4. UAV Measurements Comparing Albedo across Multiple Scenarios of Land Use

Follow-up flights took place a year later, on 30 July 2017. First and second flights on 30 July were made at Tully, NY over a Norway spruce (*Picea abies*) plantation, over two neighboring locations within the same spruce plantation, at 0.5 and 1 h post local solar noon, respectively. A second and third flight revisited the same deciduous hardwood forest site as was measured above, as well as a second deciduous hardwood site within the same forest block, at 1.5 h post solar noon and 2 h post local solar noon. 30 July 2017 was completely clear with no clouds; local air quality index was less than 50 for both particulate matter and ozone [38]. Finally, willow albedo data as collected above was used alongside the spruce and deciduous forest data to compare albedo over three different land uses.

2.3. Data Processing and Analysis

Outliers were removed where measured incoming solar radiation was less than 60% of predicted solar insolation.

$$\text{Predicted Solar Insolation} = \text{Solar Constant} * \cos\left(\frac{\text{Zenith Angle} * \pi}{180}\right) \quad (2)$$

This removed values representing 18%, 48%, 57%, and 46% of the original data, at Tully, Durham, Bartlett, and Petersham, respectively (the multiple day measurements at the last three sites resulted in there having been more clouded days to remove). We also examined albedo at solar noon, as solar noon measurements are more comparable to solar noon-approximated satellite values. For both UAV and tower data, albedo at solar noon was defined as all measurements within one hour of solar noon at that site on the day of the measurement.

Data were analyzed in R version 3.2.1 [44]. We conducted a Type II ANOVA [45] and the Tukey HSD test from stats v3.4.1., to compare site level differences across both in-situ and satellite measurements; Anova residuals were normally distributed. We used the R *t*-test from stats v3.4.1. to conduct a Student *t*-test to compare in-situ and satellite measurements; albedo was transformed with a negative reciprocal 7th power transformation.

3. Results

3.1. Targeted UAV Measurements over Mixed Hardwood Forest

We examined the temporal consistency of albedo estimates across flights and years in a series of flights over a mixed hardwood forest in Tully, NY. The series of five flights spaced hourly around local solar noon measured a summer forest albedo of 0.145 $\pm$ 0.005 SD, $n = 5$, and ranged from 0.140 to 0.146. The mixed hardwood albedo at solar noon, 0.145, was the same as the mean and was consistent with albedo values recorded over the course of the day.

3.2. Comparative Tower and Satellite Measurements over Mixed Hardwood Forest

UAV albedo values were compared to tower measurements from three similar mixed northern hardwood forests (Figure 4).

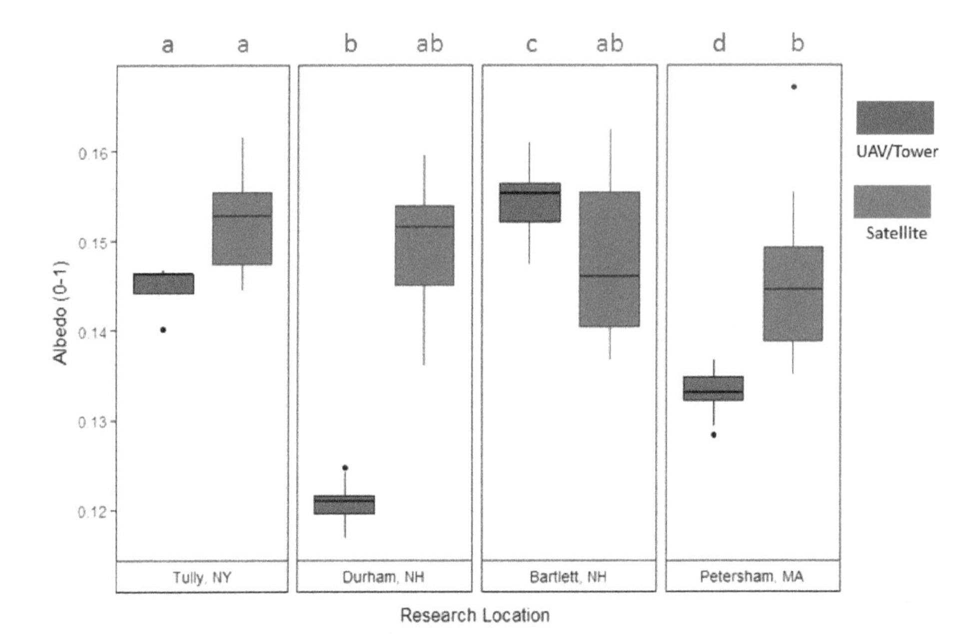

Figure 4. Boxplots showing albedo over mixed temperate forest as measured by UAV (red) and MODIS satellite data (blue) in (**a**) Tully, NY and by fixed towers (red, 0.145 ± 0.003 SD, n = 5) and MODIS satellite data (blue, 0.152 ± 0.005 SD, n = 70) at (**b**) Durham (red, 0.121 ± 0.015 SD, n = 12; blue, 0.151 ± 0.006 SD, n = 23), (**c**) Bartlett (red, 0.155 ± 0.008 SD, n = 13; blue, 0.148 ± 0.008 SD, n = 18), and (**d**) Petersham (red, 0.133 ± 0.009 SD, n = 12; blue, 0.145 ± 0.008 SD, n = 18). Box and whisker plots show medians, data quartiles, and outliers. Letters over tower and satellites represent significant differences within UAV and tower measurements at Tully, Durham, Bartlett, and Petersham (calculated by ANOVA, Tukey HSD), and within satellite measurements at each of the same (calculated by ANOVA, Tukey HSD).

The average summer albedo at the Durham, NH tower was lower than UAV measurements by 0.02, while albedo at the Bartlett tower was higher by 0.02. Albedo at the Petersham, MA tower was lower by 0.01. While differences between each of the four sites were small, they were statistically significant, showing clear across-site heterogeneity (Type II Anova, df = 3, F value = 271, $p < 0.001$). At all sites, albedo at solar noon was within 0.01 units of the five hour albedo, and was not significantly different from the full albedo. Overall, all sites fell within the needed accuracy of 0.02–0.05 albedo units of each other. In comparison, satellite albedo data varied little across sites, with only Tully and Petersham showing a significant difference of less than 0.01 units, showing very little across site heterogeneity.

We then compared the in-situ UAV and tower measurements to albedo measurements made by satellite. At the Tully site, MODIS average albedo was slightly higher than in-situ, UAV-measured albedo (Student t-test: df = 8.05, $t = -6.34$, $p < 0.001$). In Durham, satellite albedo was also significantly greater than in-situ, tower-measured albedo values (Student t-test: df = −15.3, $t = 23.1$, $p < 0.001$). Likewise, Petersham satellite albedo was significantly greater than the in-situ measurements by tower (Student t-test: df = 21.2, $t = -6.03$, $p < 0.001$). However, at Bartlett, the average albedo value measured by satellite was significantly less than the average tower albedo (Student t-test: df = 18.0, $t = -4.60$, $p < 0.001$). Both UAV showed similar consistency (0.01 albedo units lower) with satellite measurements as tower measurements had with respective satellite measurements (Durham: 0.03 lower, Bartlett: 0.01 higher, Petersham: 0.01 higher). Overall, the coefficient of variation across all summer UAV albedo measurements made over mixed hardwood forest at Tully (2.1%) was similar to the variability observed at the in-situ estimates made by towers (Durham: 1.9%, Bartlett: 2.7%, Petersham: 1.9%).

3.3. Validation Measurements comparing Simultaneous UAV and Tower Data

We compared albedo measurements taken by tower and UAV approaches, over the same field of willow biofuels. Initial side by side flights by both methods produced closely matched albedo estimates (Figure 5).

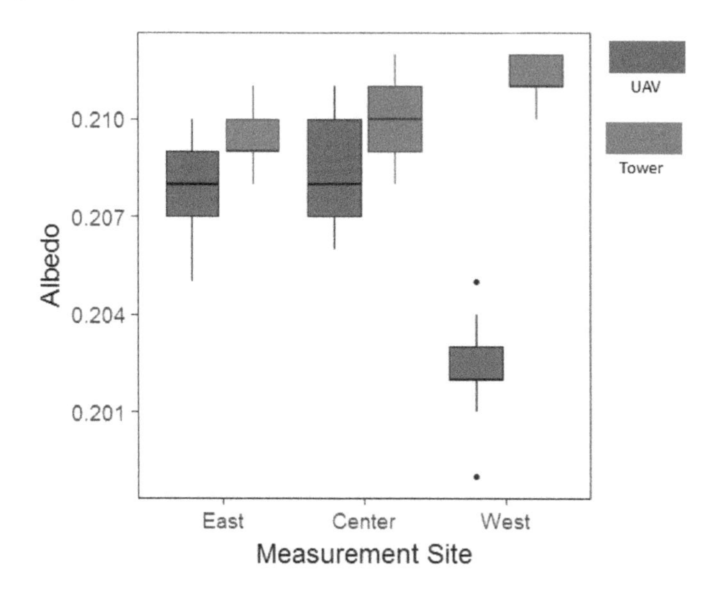

Figure 5. Albedo was measured by UAV over a willow biofuel field at three sites: immediately adjacent to a fixed tower (0.208 ± 0.001 SD, n = 19), 24 meters east of the tower (0.208 ± 0.002 SD, n = 23), and 24 m west of the tower (0.202 ± 0.002 SD, n = 21). Tower measurements were taken simultaneously with each flight from a fixed location at center (0.209 ± 0.001 SD, n = 19; 0.210 ± 0.001 SD, n = 23; 0.211 ± 0.001 SD, n = 21). UAV measurements are depicted in red, while paired, simultaneous fixed tower measurements are depicted in blue. Box and whisker plots show medians, data quartiles, and outliers.

Measurements were repeated over the same willow field, using the UAV in flights 24 m west and then 29 m east of the tower. All UAV-derived willow albedo measurements were well within ±0.01 of each other, but there was slightly greater correspondence between the side-by-side measurements and measurements made over willow a distance from the tower. UAV albedo tended to have greater variability than tower albedo, regardless of the sub-site, although overall variance was low.

3.4. UAV Measurements Demonstrating Albedo across Multiple Scenarios of Land Use

Three land use types were examined in parallel: surveys of monoculture of Norway spruce (Figure 6) measured an albedo of 0.0743 and 0.0824. Two flights resurveyed the same forest site and an adjacent point, as described above, with albedo of 0.149 and 0.154. Three flights over a cropped willow field, as described above, were measured at 0.208, 0.208, and 0.202 albedo. The variation across sites was an order of magnitude less than the variation across the different land uses.

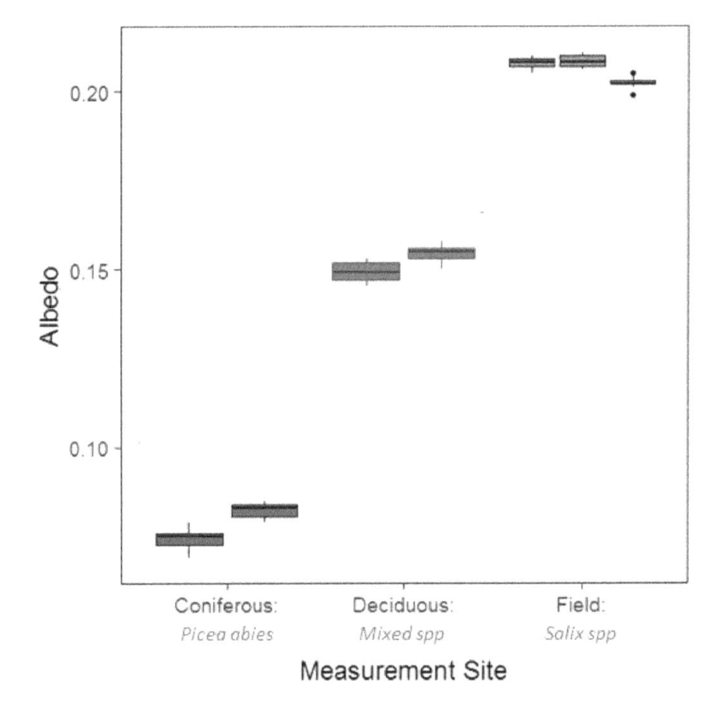

Figure 6. Albedo measured by UAV for three different land use types: two adjacent coniferous forest sites at Tully, NY (0.074 ± 0.002, n = 24; 0.082 ± 0.002, n = 23) (red), two adjacent mixed northern hardwood forest sites at Tully, NY (0.149 ± 0.002, n = 22; 0.154 ± 0.002, n = 23) (blue), and three adjacent cropped willow sites at Geneva, NY (0.208 ± 0.001, n = 19; 0.208 ± 0.002, n = 23; 0.202 ± 0.002, n = 21) (gray). Error bars represent the variance, reported as standard error, within each flight.

4. Discussion

Evaluation of earth system models requires surface albedo estimates with an absolute accuracy of between 0.02-0.05 units of albedo [24,42,46]. In this study, UAV flights over a mixed northern hardwood forest produced measurements within 0.01 units of each other across five flights and five hours. This albedo fell well within the range of site variation in albedo for fixed tower measurements of mixed forest albedo in Bartlett, NH; Durham, NH; and Petersham, MA, differing by no more than 0.02 units despite small cross-site differences in field of view, topography, and species composition. The tight grouping of UAV and tower estimates over a common, well–surveyed land use type supports the idea that this methodology will be able to provide accurate albedo measurements over land use types that have been poorly surveyed by conventional methods. Measurements made over a Norway spruce monoculture and cropped willow field show that the UAV easily distinguishes between land use types across multiple flights. These flights demonstrate how UAV may be used to sample sites that are not well captured by either satellite or tower measurements due to their small footprint or high rate of turnover for harvesting.

Although ground measurements (UAV and tower) were tightly grouped within a small range of albedo, each site was significantly different from the others, indicating site to site variation likely caused by species, land use, and topographic differences. This variability was poorly captured by satellite measurements, which were more tightly grouped and generally not significantly different from each other. Due to this, satellite data poorly estimated variation on the ground; UAV and tower data were generally lower than satellite data by 6 to 16%, although Bartlett forest reported 9% higher albedo than the satellite. UAV data did not differ from tower data in this way. This finding provides additional support for the use of ground measurements, such as can be obtained by UAV, to validate satellite measurements and refine model predictions. Satellites have a larger and differently distributed field of view which may be confounded by different land use types at the pixel boundaries [4,24]. UAV can capture albedo with greater precision, and with the flexibility to take fine-scale measurements across the entire landscape.

Certain caveats of this method should be considered; uneven cloud cover means that a homogenous down-welling flux of incoming solar radiation cannot be assumed over the UAV's entire flight path. All flights in this study had to be recorded on days with no to very minimal cloud cover. UAVs must be flown such that unaided line of site is maintained, which can limit the range and altitude possible for flights over high canopy (Supplementary Table S5). Finally, continuous albedo measurements over the course of an entire day would be dependent either on the capacity to make many serial flights over the time horizon desired.

Satellites are often insufficient to capture the fine-scale landscape heterogeneity caused by local variation in canopy density, vegetative community, terrain, and other local scale properties [4,24,42]. Fine-scale point measurements, however, often lack the range needed to assess these properties on a global scale [3]. It is our belief that using UAVs to measure albedo will improve our ability to determine sources of variability in albedo measurements. Payload mass reduction through a split upwards and downward-facing sensor widens the range of UAVs available for these types of measurements and maximizes flight time (Table 2). The method described here provides a simple method of albedo assessment accessible to the typical researcher.

Table 2. Payload capacity and expected payload for common UAVs and commercially available albedo equipment.

UAV	Payload Capacity (g)	Payload	Mass (g)
DJI Phantom 3	300	CMP3	300
Sky Hero Spyder 700	1600	Datalogger	200
Freefly Systems ALTA	6800	Gimbal *	200–600
DJI Spreading Wings S900 Professional Hexacopter	8200		

* Based on MK HiSight SLR1 and Gaui Crane Gimbal.

5. Conclusions

UAVs offer an opportunity to make flexible, efficient radiation measurements at many locations and scales. Here, we validated a method of measuring albedo by UAV while minimizing payload and technical requirements. We found that UAVs provided stable albedo measurements across multiple flights over a mixed hardwood forest, with results that were well within the range of tower-estimated albedo at similar forested sites. Simultaneous albedo estimates by UAV and a fixed tower at the same site showed that the two methods produced near identical results. Finally, we demonstrated that in-situ albedo measurements (tower and UAV) capture more site-to-site variation in albedo than satellite measurements. Overall, we show that UAVs produce reliable, consistent albedo measurements that can capture crucial surface heterogeneity, clearly distinguishing between different land uses.

Supplementary Materials: The following are available online at http://www.mdpi.com/2072-4292/10/8/1303/s1, Table S1: Dates, times, and locations of described UAV flights, Table S2: Latitude and longitude of UAV satellite pixels Table S3: Latitude and longitude of tower satellite pixels, Table S4: Blue sky albedo estimates for sites in Tully, NY. Table S5: Current FAA requirements for UAV research within the United States.

Author Contributions: Conceptualization, C.R.L.; UAV and Satellite Dataset, C.R.L.; Harvard and Bartlett Forest Dataset, A.D.R.; Durham Forest Dataset, E.B.; Validation and Analysis, C.R.L.; Writing-Original Draft Preparation, C.R.L.; Writing-Review & Editing, C.R.L., E.B., and A.D.R.

Funding: This research was funded by the David R Atkinson Center for a Sustainable Future, through the Academic Venture Fund program. Harvard and Bartlett Forest measurements were funded by the National Science Foundation through the LTER program, which has supported research at Harvard Forest (DEB-1237491) and Bartlett Experimental Forest (DEB-1114804).

Acknowledgments: Timothy Fahey was the primary investigator on this grant, and is particularly thanked for his support. The authors would also like to extend sincere thanks and appreciation to Martin LaChance of UAV Imaging Systems for his expertise and hard work in realizing this project.

Conflicts of Interest: The authors declare no conflict of interest.

References

1. Latifi, H.; Galos, B. Remote sensing-supported vegetation parameters for regional climate models: A brief review. *IForest* **2010**, *3*, 98–101. [CrossRef]
2. Pfeifer, M.; Disney, M.; Quaife, T.; Marchant, R. Terrestrial ecosystems from space: A review of earth observation products for macroecology applications. *Glob. Ecol. Biogeogr.* **2012**, *21*, 603–624. [CrossRef]
3. Pan, S.; Tian, H.; Dangal, S.R.S.; Ouyang, Z.; Tao, B.; Ren, W.; Lu, C.; Running, S. Modeling and monitoring terrestrial primary production in a changing global environment: Toward a multiscale synthesis of observation and simulation. *Adv. Meteorol.* **2014**, *2014*. [CrossRef]
4. Cescatti, A.; Marcolla, B.; Santhana Vannan, S.K.; Pan, J.Y.; Roman, M.O.; Yang, X.; Ciais, P.; Cook, R.B.; Law, B.E.; Matteucci, G.; et al. Intercomparison of MODIS albedo retrievals and in situ measurements across the global FLUXNET network. *Remote Sens. Environ.* **2012**, *121*, 323–334. [CrossRef]
5. Randerson, J.T.; Liu, H.; Flanner, M.G.; Chambers, S.D.; Jin, Y.; Hess, P.G.; Pfister, G.; Mack, M.C.; Treseder, K.K.; Welp, L.R.; et al. The impact of boreal forest fire on climate warming. *Science* **2006**, *314*, 1130–1132. [CrossRef] [PubMed]
6. Otto, J.; Berveiller, D.; Bréon, F.M.; Delpierre, N.; Geppert, G.; Granier, A.; Jans, W.; Knohl, A.; Kuusk, A.; Longdoz, B.; et al. Forest summer albedo is sensitive to species and thinning: How should we account for this in Earth system models? *Biogeosciences* **2014**, *11*, 2411–2427. [CrossRef]
7. Wang, Z.; Erb, A.M.; Schaaf, C.B.; Sun, Q.; Liu, Y.; Yang, Y.; Shuai, Y.; Casey, K.A.; Román, M.O. Early spring post-fire snow albedo dynamics in high latitude boreal forests using Landsat-8 OLI data. *Remote Sens. Environ.* **2016**, *185*, 71–83. [CrossRef] [PubMed]
8. Jackson, R.B.; Randerson, J.T.; Canadell, J.G.; Anderson, R.G.; Avissar, R.; Baldocchi, D.D.; Bonan, G.B.; Caldeira, K.; Diffenbaugh, N.S.; Field, C.B.; et al. Protecting climate with forests. *Environ. Res. Lett.* **2008**, *3*, 1–5. [CrossRef]
9. Kirschbaum, M.U.F.; Whitehead, D.; Dean, S.M.; Beets, P.N.; Shepherd, J.D.; Ausseil, A.-G.E. Implications of albedo changes following afforestation on the benefits of forests as carbon sinks. *Biogeosciences* **2011**, *8*, 3687–3696. [CrossRef]
10. Betts, R.; Falloon, P.D.; Goldewijk, K.K.; Ramankutty, N. Biogeophysical effects of land use on climate: Model simulations of radiative forcing and large-scale temperature change. *Agric. For. Meteorol.* **2007**, *142*, 216–233. [CrossRef]
11. Betts, R. Offset of the potential carbon sink from boreal forestation by decreases in surface albedo. *Nature* **2000**, *408*, 187–190. [CrossRef] [PubMed]
12. Wang, Z.; Zeng, X. Evaluation of snow albedo in land models for weather and climate studies. *J. Appl. Meteorol. Climatol.* **2010**, *49*, 363–380. [CrossRef]
13. Flanner, M.G.; Zender, C.S. Linking snowpack microphysics and albedo evolution. *J. Geophys. Res.* **2006**, *111*, 2156–2202. [CrossRef]
14. Bonan, G.B. Forests and climate change: Forcings, feedbacks, and the climate benefits of forests. *Science* **2008**, *320*, 1444–1449. [CrossRef] [PubMed]
15. Bright, R.M. Metrics for biogeophysical climate forcings from land use and land cover changes (LULCC) and their inclusion in Life Cycle Assessment (LCA): A critical review. *Environ. Sci. Technol.* **2015**, *49*, 3291–3303. [CrossRef] [PubMed]
16. Zhao, K.; Jackson, R.B. Biophysical forcings of land-use changes from potential forestry activities in North America. *Ecol. Monogr.* **2014**, *84*, 329–353. [CrossRef]
17. Chen, L.; Dirmeyer, P.A. Adapting observationally based metrics of biogeophysical feedbacks from land cover/land use change to climate modeling. *Environ. Res. Lett.* **2016**, *11*. [CrossRef]

18. Burakowski, E.; Tawfik, A.; Ouimette, A.; Lepine, L.; Novick, K.; Ollinger, S.; Zarzycki, C.; Bonan, G. The role of surface roughness, albedo, and Bowen ratio on ecosystem energy balance in the Eastern United States. *Agric. For. Meteorol.* **2018**, *249*, 367–376. [CrossRef]
19. Lee, X.; Goulden, M.L.; Hollinger, D.Y.; Barr, A.; Black, T.A.; Bohrer, G.; Bracho, R.; Drake, B.; Goldstein, A.; Gu, L.; et al. Observed increase in local cooling effect of deforestation at higher latitudes. *Nature* **2011**, *479*, 384–387. [CrossRef] [PubMed]
20. Lutz, D.A.; Burakowski, E.A.; Murphy, M.B.; Borsuk, M.E.; Niemiec, R.M.; Howarth, R.B. Tradeoffs between three forest ecosystem services across the state of New Hampshire, USA: Timber, carbon, and albedo. *Ecol. Appl.* **2016**, *26*, 146–161. [CrossRef] [PubMed]
21. Barnes, C.A.; Roy, D.P. Radiative forcing over the conterminous United States due to contemporary land cover land use albedo change. *Geophys. Res. Lett.* **2008**, *35*, 1–6. [CrossRef]
22. Montenegro, A.; Eby, M.; Mu, Q.; Mulligan, M.; Weaver, A.J.; Wiebe, E.C.; Zhao, M. The net carbon drawdown of small scale afforestation from satellite observations. *Glob. Planet. Chang.* **2009**, *69*, 195–204. [CrossRef]
23. Campagnolo, M.L.; Sun, Q.; Liu, Y.; Schaaf, C.; Wang, Z.; Román, M.O. Estimating the effective spatial resolution of the operational BRDF, albedo, and nadir reflectance products from MODIS and VIIRS. *Remote Sens. Environ.* **2016**, *175*, 52–64. [CrossRef]
24. Roman, M.O.; Schaaf, C.; Woodcock, C.E.; Strahler, A.H.; Yang, X.; Braswell, R.H.; Curtis, P.S.; Davis, K.J.; Dragoni, D.; Goulden, M.L.; et al. The MODIS (Collection V005) BRDF/albedo product: Assessment of spatial representativeness over forested landscapes. *Remote Sens. Environ.* **2009**, *113*, 2476–2498. [CrossRef]
25. Franch, B.; Vermote, E.F.; Claverie, M. Intercomparison of Landsat albedo retrieval techniques and evaluation against in situ measurements across the US SURFRAD network. *Remote Sens. Environ.* **2014**, *152*, 627–637. [CrossRef]
26. Roy, D.P.; Zhang, H.K.; Ju, J.; Gomez-Dans, J.L.; Lewis, P.E.; Schaaf, C.; Sun, Q.; Li, J.; Huang, H.; Kovalskyy, V. A general method to normalize Landsat reflectance data to nadir BRDF adjusted reflectance. *Remote Sens. Environ.* **2016**, *176*, 255–271. [CrossRef]
27. Burakowski, E.; Ollinger, S.V.; Lepine, L.; Schaaf, C.; Wang, Z.; Dibb, J.E.; Hollinger, D.Y.; Kim, J.; Erb, A.; Martin, M. Spatial scaling of reflectance and surface albedo over a mixed-use, temperate forest landscape during snow-covered periods. *Remote Sens. Environ.* **2015**, *158*, 465–477. [CrossRef]
28. Adolph, A.C.; Albert, M.R.; Lazarcik, J.; Dibb, J.E.; Amante, J.M.; Price, A. Dominance of grain size impacts on seasonal snow albedo at open sites in New Hampshire. *J. Geophys. Res.* **2017**, *122*, 121–139. [CrossRef]
29. Warren, S.G.; Wiscombe, W.J. A Model for the Spectral Albedo of Snow. II: Snow Containing Atmospheric Aerosols. *J. Atmos. Sci.* **1980**, *37*, 2734–2745. [CrossRef]
30. Anderson, K.; Gaston, K.J. Lightweight unmanned aerial vehicles will revolutionize spatial ecology. *Front. Ecol. Environ.* **2013**, *11*, 138–146. [CrossRef]
31. Cruzan, M.B.; Weinstein, B.G.; Grasty, M.R.; Kohrn, B.F.; Hendrickson, E.C.; Arredondo, T.M.; Thompson, P.G. Small Unmanned Aerial Vehicles (Micro-UAVs, Drones) in Plant Ecology. *Appl. Plant Sci.* **2016**, *4*. [CrossRef] [PubMed]
32. U.S. Department of Transportation Federal Aviation Administration (FAA). *Air Traffic Organization Policy, Order JO 7110.65V*; FAA: Washington, DC, USA, 2017.
33. Ramana, M.V.; Ramanathan, V.; Kim, D.; Roberts, G.C.; Corrigan, C.E. Albedo, atmospheric solar absorption and heating rate measurements with stacked UAVs. *Q. J. R. Meteorol. Soc.* **2007**, *133*, 937–948. [CrossRef]
34. Schneider, C.; Truffer, M.; Michael Shea, J.; Hubbard, A.; Ryan, J.C.; Box, J.E.; Brough, S.; Cameron, K.; Cook, J.M.; Cooper, M.; et al. Derivation of High Spatial Resolution Albedo from UAV Digital Imagery: Application over the Greenland Ice Sheet. *Front. Earth Sci.* **2017**, *5*, 1–13. [CrossRef]
35. Weiser, U.; Olefs, M.; Schöner, W.; Weyss, G.; Hynek, B. Correction of broadband snow albedo measurements affected by unknown slope and sensor tilts. *Cryosphere* **2016**, *10*, 775–790. [CrossRef]
36. Jenkins, J.P.; Richardson, A.D.; Braswell, B.H.; Ollinger, S.V.; Hollinger, D.Y.; Smith, M.L. Refining light-use efficiency calculations for a deciduous forest canopy using simultaneous tower-based carbon flux and radiometric measurements. *Agric. For. Meteorol.* **2007**, *143*, 64–79. [CrossRef]
37. Wang, Z.; Schaaf, C.; Strahler, A.H.; Chopping, M.J.; Román, M.O.; Shuai, Y.; Woodcock, C.E.; Hollinger, D.Y.; Fitzjarrald, D.R. Evaluation of MODIS albedo product (MCD43A) over grassland, agriculture and forest surface types during dormant and snow-covered periods. *Remote Sens. Environ.* **2014**, *140*, 60–77. [CrossRef]

38. Environmental Protection Agency AirNow. Available online: https://www.airnow.gov/index.cfm?action=airnow.mapsarchivecalendar (accessed on 25 October 2016).
39. Richardson, A.D. Radiometric and Meteorological Data from Harvard Forest Barn Tower Since 2011. Available online: http://harvardforest.fas.harvard.edu:8080/exist/apps/datasets/showData.html?id=hf249 (accessed on 25 October 2016).
40. Aubrecht, D.M.; Helliker, B.R.; Goulden, M.L.; Roberts, D.A.; Still, C.J.; Richardson, A.D. Continuous, long-term, high-frequency thermal imaging of vegetation: Uncertainties and recommended best practices. *Agric. For. Meteorol.* **2016**, *228–229*, 315–326. [CrossRef]
41. Schaaf, C. *MCD43A3 MODIS/Terra + Aqua BRDF/Albedo Daily L3 Global-500 m V006*; NASA: Washington, DC, USA, 2015.
42. Lucht, W.; Hyman, A.H.; Strahler, A.H.; Barnsley, M.J.; Hobson, P.; Muller, J.P. A comparison of satellite-derived spectral albedos to ground-based broadband albedo measurements modeled to satellite spatial scale for a semidesert landscape. *Remote Sens. Environ.* **2000**, *74*, 85–98. [CrossRef]
43. Lewis, P.; Barnsley, M. Influence of the sky radiance distribution on various formulations of the earth surface albedo. *Proc. Conf. Phys. Meas. Signatures Remote Sens.* **1994**, 707–715.
44. R Development Core Team. *R: A Language and Environment for Statistical Computing*; R Foundation for Statistical Computing: Vienna, Austria, 2011.
45. Fox, J.; Weisberg, S. *An {R} Companion to Applied Regression*; Second Edition; Sage Group: Thousand Oaks, CA, USA, 2011.
46. Liu, J.; Schaaf, C.; Strahler, A.; Jiao, Z.; Shuai, Y.; Zhang, Q.; Roman, M.; Augustine, J.A.; Dutton, E.G. Validation of moderate resolution imaging spectroradiometer (MODIS) albedo retrieval algorithm: Dependence of albedo on solar zenith angle. *J. Geophys. Res. Atmos.* **2009**, *114*, 1–11. [CrossRef]

Article

Assessing the Impacts of Urbanization on Albedo in Jing-Jin-Ji Region of China

Rongyun Tang [1,2], Xiang Zhao [1,2,*], Tao Zhou [3,4], Bo Jiang [1,2], Donghai Wu [5] and Bijian Tang [6]

1 State Key Laboratory of Remote Sensing Science, Jointly Sponsored by Beijing Normal University and Institute of Remote Sensing and Digital Earth of Chinese Academy of Sciences, Beijing 100875, China; rongyun_geo@mail.bnu.edu.cn (R.T.); bojiang@bnu.edu.cn (B.J.)
2 Beijing Engineering Research Center for Global Land Remote Sensing Products, Institute of Remote Sensing Science and Engineering, Faculty of Geographical Science, Beijing Normal University, Beijing 100875, China
3 Key Laboratory of Environmental Change and Natural Disaster of Ministry of Education, Academy of Disaster Reduction and Emergency Management, Faculty of Geographical Science, Beijing Normal University, Beijing 100875, China; tzhou@bnu.edu.cn
4 State Key Laboratory of Earth Surface Processes and Resource Ecology, Beijing Normal University, Beijing 100875, China
5 College of Urban and Environmental Sciences, Peking University, Beijing 100871, China; donghai.wu@pku.edu.cn
6 Division of Environment and Sustainability, The Hong Kong University of Science and Technology, Kowloon, Hong Kong, China; btangac@ust.hk
* Correspondence: zhaoxiang@bnu.edu.cn; Tel.: +86-10-5880-0152

Received: 4 May 2018; Accepted: 5 July 2018; Published: 10 July 2018

Abstract: As an indicative parameter that represents the ability of the Earth's surface to reflect solar radiation, albedo determines the allocation of solar energy between the Earth's surface and the atmosphere, which plays an important role in both global and local climate change. Urbanization is a complicated progress that greatly affects urban albedo via land cover change, human heat, aerosol, and other human activities. Although many studies have been conducted to identify the effects of these various factors on albedo separately, there are few studies that have quantitatively determined the combined effects of urbanization on albedo. In this study, based on a partial derivative method, vegetation index data and nighttime light data were used to quantitatively calculate the natural climate change and human activities' contributions to albedo variations in the Jing-Jin-Ji region, during its highest population growth period from 2001 to 2011. The results show that (1) 2005 is the year when urbanization starts accelerating in the Jing-Jin-Ji region; (2) albedo trends are equal to 0.0065 year^{-1} before urbanization and 0.0012 year^{-1} after urbanization, which is a reduction of 4/5; and (3) the contribution rate of urbanization increases from 15% to 48.4%, which leads to a decrease in albedo of approximately 0.05. Understanding the contribution of urbanization to variations in urban albedo is significant for future studies on urban climate change via energy balance and can provide scientific data for energy conservation policymaking.

Keywords: surface albedo; urbanization; vegetation variation; climate change; DMSP

---

1. Introduction

Surface albedo is represented by the ratio of reflected shortwave solar radiation in all directions from the Earth's surface to the total incoming solar radiation [1]. Albedo is an indicator that characterizes the reflective ability of the Earth's surface via solar radiation and determines the allocation of radiative energy between the Earth's surface and the atmosphere [2–4], making it an imperative parameter that also affects the Earth's climate [5]. The increase in albedo can reduce the absorption of solar radiation at the Earth's surface; this lowers the surface temperature and has an equivalent effect on the reduction in CO_2 emissions, which mitigates greenhouse effects [6–8]. In recent years, there have been many studies focusing on the effects of variations in surface albedo on climate change in both global and local areas [9–13], and some have suggested that the effects that variations in albedo have on climate change are comparable to those of fossil fuel combustion [14,15]. Therefore, identifying changes in albedo is of great significance for further exploring climate change.

The influential factors and causes for change in albedo have been extensively studied. Surface albedo was found to decrease as the surface irregularity increases, and albedo increases with an increasing solar zenith angle, leading to a minimum albedo at noon during diurnal variation [16]. Soil moisture was also found to be an important factor. Some research has revealed that surface albedo decreased with an increase in soil moisture, indicating a typical exponential relationship between them [17]. Furthermore, much research [18–20] has revealed that meteorological factors, such as aerosol optical depth, temperature, rainfall and snowfall et al., also contribute to changes in surface albedo. Reflectivity measurements from 61 real-world surfaces in Dana's study indicated that albedo varies with surface roughness, as well as viewing and illumination directions [21]. Roughness has been well studied by many researchers [22–26], and their results showed that the increase of roughness will make the surface albedo decrease, which was explained by the fact that surfaces with greater roughness or irregularity offer more spaces and cracks where the incident light is trapped [27]. The aforementioned studies mostly focused on a single factor that might affect surface albedo. However, surface albedo is often affected by multiple factors in reality, which complicates the reasons for variations in albedo.

Although the changes in global land surface albedo have been widely studied, the impact of urbanization that human activities induce on albedo is not well understood. As we all know, urbanization is one of the most important aspects of human activities in the terrestrial ecosystem [28,29], and it has a significant impact on regional climate change [30,31]. Climate change in urban areas has received substantial public attention, especially regarding urban heat islands (UHIs) [32,33]; there have been many comprehensive studies on the UHI, including its morphological structure [34,35] and change process [36–38]. As a parameter that affects the distribution of solar radiation, surface albedo in urban areas influences surface temperatures in cities to some extent. However, due to the coexisting influences of land cover changes, industrial pollutants, aerosols, and vegetation growth during urbanization [39], changes in urban surface albedo have various uncertainties. Therefore, it is still very difficult to quantitatively analyze the processes of energy distribution and conversion in urban areas. Although existing studies have shown that the decrease in surface albedo is one of the most important causes of urban warming [8,40], the main factor affecting changes in albedo before and after urbanization is still unknown.

Therefore, this study quantitatively calculates the contributions of vegetation and urbanization to surface albedo in the Jing-Jin-Ji region and distinguishes the main driving factors behind its spatiotemporal changes during the most rapid population growth period (2001–2011). Based on the Moderate Resolution Imaging Spectroradiometer (MODIS) global land surface albedo product [41], we used a shift linear regression method [42] to detect the breakpoint year in the albedo time series, and we also analyzed the temporal and spatial patterns of albedo in the Jing-Jin-Ji region. With the nighttime light data from the U.S Air Force Defense Meteorological Satellites Program Operational Linescan System (DMSP/OLS), we used the Digital Number (DN) to represent the intensity of urbanization [43,44]. Combined with the MODIS product of the vegetation index data [45], we quantitatively calculated the contributions of urbanization and vegetation to variations in albedo via a partial derivative-based calculation method to determine the main controlling factors. The aim of this paper is to determine the contributions of urbanization and vegetation to albedo, which influence the urban climate, to provide a case basis for developing urban energy conservation programs.

2. Study Area and Data

2.1. Study Area

The Jing-Jin-Ji region is located in the North China Plain, with the Bohai Sea to the east and the Taihang Mountains to the west. The altitude is higher to the northwest and lower to the southeast, with a total area of 185,000 km^2. The area has a typical temperate monsoon climate that is characterized by rainy summers with high temperatures and cold and dry winters. The fastest population growth period in the Jing-Jin-Ji region is from 2000 to 2010 [46]. The Jing-Jin-Ji urban agglomeration is one of the three major urban agglomerations along the eastern coast of China due to active economic activities. The fast-growing population and rapid urbanization make it a hotspot for urbanization-related scientific research in China [47,48]. To select representative cities, Beijing (BJ), Tianjin (TJ), Shijiazhuang (SJZ), Handan (HD), Tangshan (TS), and Baoding (BD) are chosen, as they have populations greater than one million based on the 2012 China City Statistical Yearbook, in order to identify the differences in albedo for each city during urbanization. The study area is shown in Figure 1, and the GlobeLand30 data for 2000 and 2010 are used for the statistics in this area (Table 1). The results (Table 1) show that the largest decreasing areas were characterized by cultivated land (−6.33%, from 2000 to 2010), while the largest growth areas were characterized by shrub lands (5.42%), followed by artificial land surfaces (1.31%).

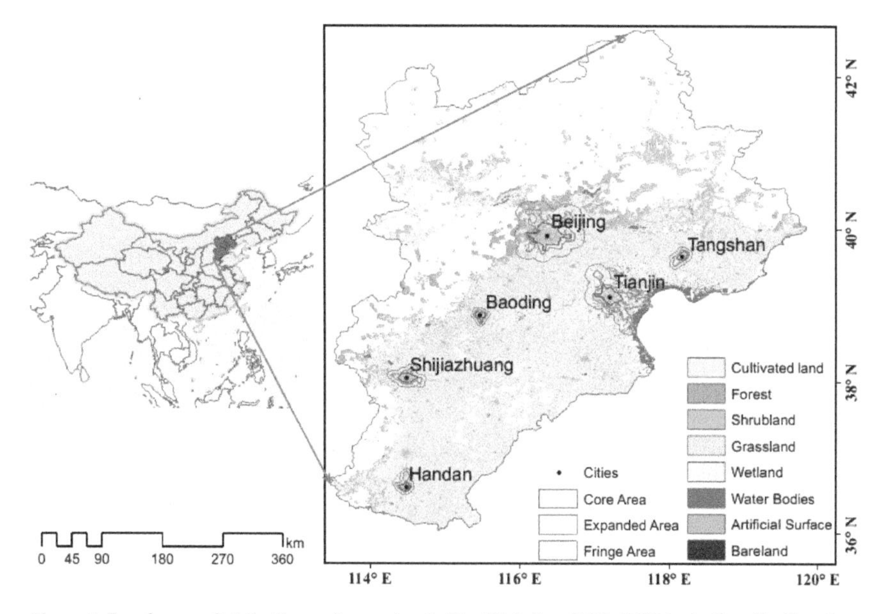

Figure 1. Land cover distribution pattern extracted by GlobeLand30 in 2000 in the Jing-Jin-Ji region.

Besides, we made a statistic about the typical albedo value for each type of land cover based on the MODIS albedo products from its multiyear average value (Table 1).

Although the albedo mean values exhibit little difference between 2010 and 2000 due to the coarse resolution of albedo (compared to GlobeLand30), we can still see differences among land cover types. So, we just list the typical albedo values in 2010 in the table above, where shrublands have the largest albedo mean value (0.137), and artificial surfaces have a relatively small albedo value which is just bigger than wetland and water bodies.

Table 1. Typical albedo values of individual land covers in 2010, and the statistic percentage of each land cover type in 2000 and 2010. Mean is the spatial mean value of the multiyear average albedo in study areas, std is the standard deviation, and variation is the percentage difference between 2010 and 2000.

Land Cover Types	Mean	Std	Percentage in 2000	Percentage in 2010	Variation
cultivated lands	0.118	0.018	77.32%	70.99%	−6.33%
forests	0.115	0.020	6.88%	6.64%	−0.23%
grasslands	0.122	0.022	0.03%	0.03%	0.00%
shrublands	0.137	0.011	0.69%	6.11%	5.42%
wetlands	0.101	0.026	0.46%	0.45%	−0.01%
water bodies	0.104	0.025	2.55%	2.39%	−0.15%
artificial surfaces	0.113	0.023	12.06%	13.37%	1.31%
barelands	0.127	0.016	0.02%	0.02%	0.00%

2.2. *Surface Albedo Data Set*

The MODIS 16-day 1 km albedo products (MCD43B3, collection5) from 2001 to 2011 were used in this study. The product contains the black-sky albedo (BSA) and the white-sky albedo (WSA), which can be used to calculate the actual (blue-sky) albedo based on the fraction of diffuse skylight [49,50]. Considering the small difference and high correlation between BSA and WSA [44,51–53], WSA was used as the index of albedo in this study. MODIS surface albedo products have been validated on a global scale, and the accuracy has been demonstrated to be suitable for studies on climate change [54,55]. We synthesized the 8-d intervals for albedo into yearly scales for the contribution analysis. The corresponding albedo quality data (MCD43B2, collection5) were also used to avoid the effects of snow cover.

2.3. *Vegetation Index Data*

The monthly MODIS Enhanced Vegetation Index (EVI) product (MOD13A3) with a spatial resolution of 1 km was used. This product was generated based on atmosphere-corrected bidirectional surface reflectance, where the atmospheric effects of water, clouds, and aerosols were removed [56,57]. This product has been widely used in studies regarding global vegetation monitoring, land cover changes, and climate researches [58,59].

2.4. *Nighttime Light Data*

Nighttime light signals detected by remote sensing satellites derive from the Defense Meteorological Satellite Program (DMSP), specifically from its visible and near infrared sensors named Operational Linescan System (OLS). DMSP/OLS nighttime light data are widely used in research regarding urban areas, such as estimating urban population [60], extracting urban extent [61], measuring urban expansion [62], and exploring human activities and its impacts on the environment in urban areas [43] etc. DMSP/OLS nighttime light data (Version 4) were used in this study, whose spatial resolution is 1 km. We excluded pixels without light (DN = 0) to ensure that there were human activities in every part of our study areas. In addition, an invariant target area method [63–65] for image correction was used to perform continuous and saturation corrections on the data. Using this method, we gained a nighttime light data time series with comparable DN values, which have been used to identify the urbanization [44].

2.5. *GlobeLand30 Landcover Data*

GlobeLand30 is one of the global land cover map products at a 30-m resolution, which was produced with a pixel-object-knowledge (POK)-based operational mapping approach [66]. The classification system includes ten land cover types, namely cultivated lands, forests, shrublands, grasslands, wetlands, water bodies, tundra, artificial surfaces, permanent snow and ice, and barren lands for the years 2000 and 2010. The overall classification accuracy is over 80% [67], and it has been widely validated in many other researches [68,69].

3. Methods

3.1. Urban Area Extraction

Based on DMSP/OLS data, the urban areas in 2000 and 2010 were extracted using a clustering algorithm [61]. Urban core area, fringe area, and rural area are characterized by metropolitan morphology, and urban fringe is a transition zone from urban to rural areas [70–72]. In this study, the core areas and the fringe areas are studied in urban expansion. Core Area is extracted by DMSP/OLS in 2000, and it is the place where the central part of the city is located. The area where the fringe area in 2000 turned into the core area in 2010 was named the Expanded Area. In order to identify this urban sprawl process of each city, the Expanded Area was defined as the urban area extracted in 2010 (excluding the Core Area). The Fringe Area in this study was defined as the buffer zone whose areas were equal to the urban areas extracted in 2010 (Figure 1). The Core Area, Expanded Area, and Fringe Area can represent not only the old urban area, the new urban area, and the suburbs, but also the initial stage, middle acceleration stage, and final stage of the urbanization process, respectively.

3.2. Breakpoint Analysis

A shift linear analysis method [42,73] was used to identify the breakpoint in the albedo time series. The idea behind this method is determining the point where the slope changes significantly in the time series before and after the point. The calculation method is as follows:

$$A_i = \begin{cases} b_1 + k_1 t_i, & t_i < b_3 \\ b_2 + k_2 t_i, & t_i \geq b_3 \end{cases} \tag{1}$$

where A_i denotes the albedo in the ith year; t_i denotes the ith year; and b_1, b_2, b_3, k_1 and k_2 are the fitted parameters. Among them, k_1 and k_2 represent the slopes of the fitted line, b_1 and b_2 represent the fitted intercepts, and b_3 represents the breakpoint position.

The Chow test which is generally used to detect changes in time series was applied to test the significance in this study and the formula is expressed as follows:

$$F = \frac{S_1 - S_2 - S_3}{S_2 + S_3} \cdot \frac{N_1 + N_2 - 2c}{c} \tag{2}$$

where $S_1 = \sum_{i=1}^{n} (A_i - \hat{A}_i)^2$ indicates the sum of square errors of the time series. For the former N_1 number of the time series, $S_2 = \sum_{i=1}^{N_1} (A_i - \hat{A}_i)^2$ indicates the sum of square errors of the former time series, and $S_3 = \sum_{i=1}^{N_2} (A_i - \hat{A}_i)^2$ indicates the sum of square errors of the latter N_2 number of the time series. c is the number of the estimated parameter in the whole time series.

3.3. Interannual Variation Rate Calculation

A simple linear regression model was used to calculate the interannual variation rate. Using the albedo time series as an example, the interannual variation rate of each pixel is equal to the slope of the trend line via the least-squares regression of the multiyear value in each pixel. The calculation for the slope is as follows:

$$K_A = (n \times \sum_{i=1}^{n} i \times A_i - (\sum_{i=1}^{n} i)(\sum_{i=1}^{n} A_i))/(n \times \sum_{i=1}^{n} i^2 - (\sum_{i=1}^{n} i)^2) \tag{3}$$

where K_A represents the interannual variation rate of albedo, n represents the number of years, i denotes the ith year, and A_i denotes the albedo value in the ith year. A positive slope value indicates an increasing trend, while a negative slope indicates a decreasing trend.

The significance of the calculated tendency is determined by an F test. The calculation formula is expressed as follows:

$$F = R \times (n - 2)/Q \tag{4}$$

where $Q = \sum_{i=1}^{n} (A_i - \hat{A}_i)^2$ indicates the sum of square errors, and $R = \sum_{i=1}^{n} (\hat{A}_i - \overline{A})^2$ represents the regressed square sum. A_i denotes the albedo value in the ith year, $\hat{A}_i$ denotes the albedo regression value in the ith year, $\overline{A}$ denotes the mean albedo value across all years, and n denotes the number of years.

This method has also been applied to the calculation of interannual variation rates for vegetation and urbanization, where K_V and K_U represent the interannual variation rate of vegetation and urbanization, respectively.

3.4. Contribution Analysis

Urbanization and vegetation are two main factors that affect variations in albedo. In our study, the annual variation rate of albedo in each pixel is expressed by the contributions of vegetation (V), urbanization (U), and other factors (Δ) (formula (5)).

$$K_A = C(V) + C(U) + C(\Delta) \tag{5}$$

where K_A represents the interannual variation rate of albedo. C(V), C(U), and C(Δ) represent the contributions of vegetation, urbanization, and other factors to the interannual variation rate of albedo, respectively. The vegetation contribution calculation method [74] is as follows:

$$C(V) = \frac{\partial A}{\partial V} \times K_V \tag{6}$$

K_V represents the slope of the linear regression line for the multiyear EVI time series. A denotes albedo, V denotes EVI, and $\frac{\partial A}{\partial V}$ (S(V)) represents the sensitivity of albedo to EVI. This sensitivity term was derived as a partial derivative via the multiple regression of albedo on EVI and DMSP/OLS. Positive and negative values of this sensitivity term reflect the positive and negative correlations between the analyzed factors and albedo, respectively. The magnitude of the absolute value of the sensitivity coefficient indicates whether the relationship between the factors and albedo is strong or weak (the greater the value, the stronger the relationship). The contribution of urbanization (C(U)) can also be calculated via the sensitivity of albedo to urbanization(S(U)) and K_U in the same way.

Due to the spatial differences in the contribution intensity from vegetation and urbanization in different regions, the relative contribution percentage of vegetation, urbanization, and other factors to changes in albedo can be expressed with the following equation [44]:

$$P(V) = \frac{|C(V)|}{|C(V)| + |C(U)| + |C(\Delta)|} \times 100\% \tag{7}$$

$$P(U) = \frac{|C(U)|}{|C(V)| + |C(U)| + |C(\Delta)|} \times 100\% \tag{8}$$

$$P(\Delta) = 100 - P(V) - P(U) \tag{9}$$

where, P(V), P(U) and P(Δ) represent the relative contribution percentage of vegetation, urbanization, and other factors, respectively.

4. Results

4.1. Albedo Variations and Spatial Patterns

The average albedo of the Jing-Jin-Ji region from 2001 to 2011 was 0.12 ± 0.02 and the spatial distribution of the multiyear mean albedo is shown in Figure 2. According to this spatial pattern, the albedo increased from the Central Business District (CBD) to the suburbs, and a majority of the Core Area had the lowest mean value of albedo compared to that in other areas. Based on the shift linear regression method [42], the results of the breakpoint detection showed that 2005 was the breakpoint year for the ~10 years of albedo data. With the Chow test, the *p*-value of the breakpoint detection was 0.002, which was substantially less than 0.05 and significant. The trend in albedo from 2001–2005 (T1 period) was the highest (6.5×10^{-3} year^{-1}, $p < 0.05$), with a lower trend of 1.2×10^{-3} year^{-1} ($p > 0.05$) from 2006–2011 (T2 period), whereas the lowest trend occurred from 2001–2011 (T3 period; 0.78×10^{-3} year^{-1} at $p > 0.05$). These trends show big differences in the T1 and T2 periods. The albedo trend during T2 was approximately 1/5 the trend during the T1 period, which led to a reduction in albedo of approximately 0.05. This indicates that the long-term growth trend of albedo was suppressed after 2005 due to several influential factors.

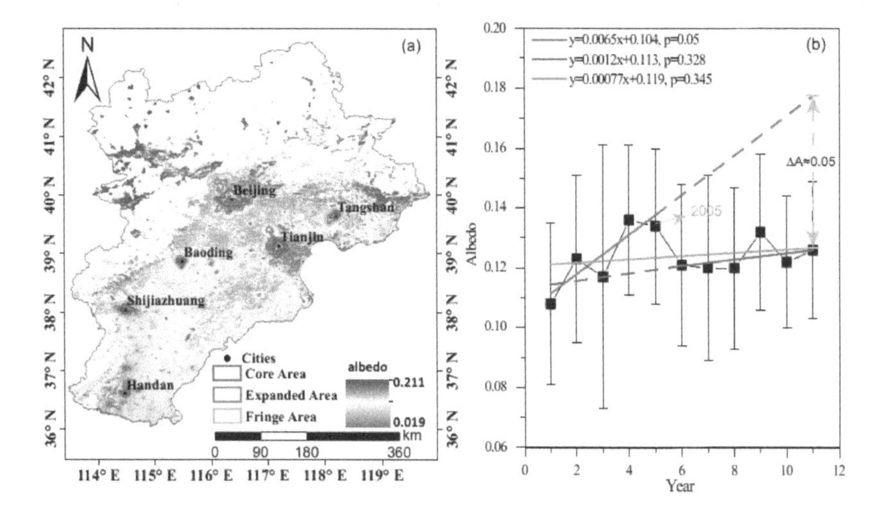

Figure 2. Spatialtemporal variation of albedo in the Jing-Jin-Ji region from 2001 to 2011. (**a**) Spatial distribution of the multiyear average albedo, and (**b**) the yearly average albedo and its trends in 2001–2005 (red line), in 2006–2011 (blue line), and in 2001–2011 (orange line) since 2005 is the breakpoint year. *p* stands for p-value, which is gained from an F test in a simple linear regression model.

To identify the factors contributing to this difference, we explored the spatial patterns of albedo before and after the breakpoint (Figure 3).

From 2001–2005 (Figure 3a), the albedo showed an increasing trend ($k > 0$) in over 99.5% of the whole region, while the number of pixels with a decreasing trend ($k < 0$) in albedo was small and accounted for only 0.5% of the total area. The percentage of the area where the albedo had significant trends ($p < 0.05$) is 13.2%, in which the percentage of the significant increasing trend ($p < 0.05$) was 13.13% and the significant decreasing trend ($p < 0.05$) was 0.07%. Albedo showed a general increasing trend in the study area. Spatially, from southwest to northeast, the interannual variation rate of albedo gradually decreased. Pixels of decreasing trends were mainly distributed across the northern fringe areas and along east coast areas near Bohai Bay.

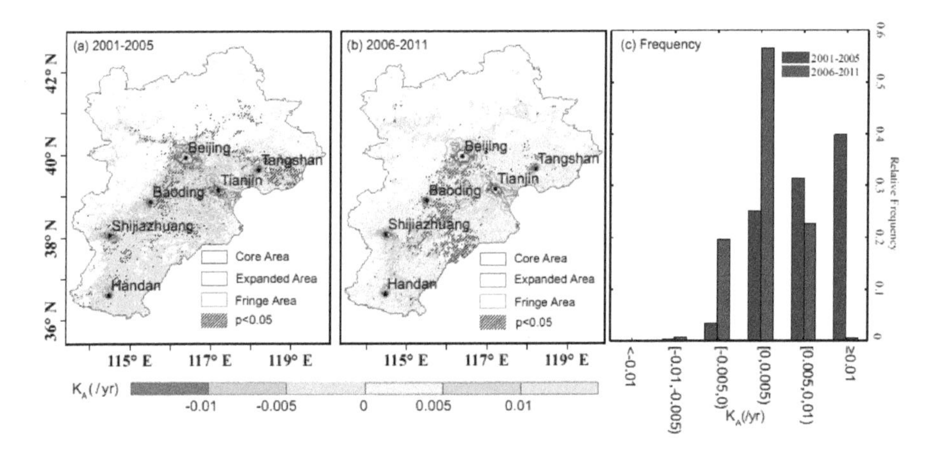

Figure 3. Spatial distribution of the annual change rate of albedo in (**a**) 2001–2005, (**b**) 2006–2011, (**c**) and their relative frequency in different slopes range of the study area.

From 2006–2011 (Figure 3b), the albedo generally had a lower increasing trend than that in 2001–2005. The percentage of the area of increasing albedo trends decreased to 94%. In addition, the area of decreasing albedo trends increased, which accounted for 6% of the total area, distributed in the northern and eastern coastal areas. The area of significant albedo trends ($p < 0.05$) was 11.11%, in which the area of increasing trends ($p < 0.05$) took up 10.44% of the whole area and the area of decreasing trends took up 0.67%.

4.2. Urbanization Spatial Patterns

In this study, we use the interannual variation rate of the DN values from corrected DMSP/OLS nighttime light data to define the urbanization rate (Figure 4) and study the developmental characteristics in the Jing-Jin-Ji region during T1 and T2 periods.

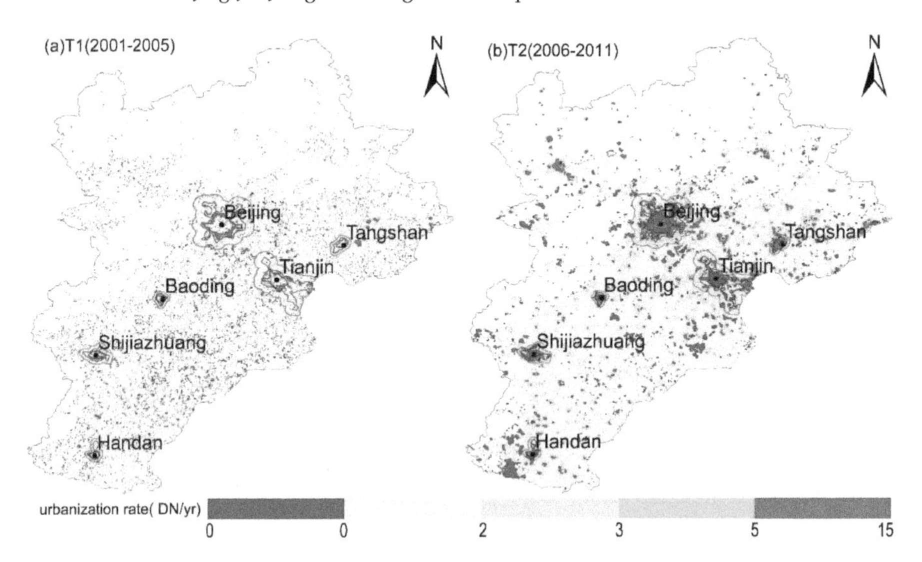

Figure 4. The spatial pattern of urbanization rate. (**a**) Represents the urbanization rate in T1 (2001–2005), and (**b**) represents the T2 (2006–2011).

From 2001 to 2005 (Figure 4a), the urbanization rate in most parts of the Jing-Jin-Ji region was less than 2 year^{-1}, and the urbanization rate in only a few areas of the Expanded Area surrounding Beijing, Tianjin, and Tangshan was greater than 5 year^{-1}. After 2005 (Figure 4b), although the rate of urbanization in the urban Core Area almost decreased to 0, the rate of urbanization in other areas increased significantly, and the urbanization area increased significantly, indicating that there is rapid urbanization occurred after 3 April 2005.

4.3. Sensitivity of Urbanization and Vegetation to Albedo

Numerous studies have shown that vegetation and urbanization are two major factors that cause changes in surface albedo [21,75]. For vegetation, different types of vegetation have different levels of albedo. For example, forests usually have a lower albedo (0.05–0.2) while grasslands have a higher albedo (0.16–0.26) [76]. In addition, changes in surface roughness caused by vegetation growth are also responsible for changes in albedo. In a similar way, surface roughness also changed with the process of urbanization. Land cover changes in the process of urbanization play a decisive role in the properties of three-dimensional surfaces in urban areas, which equally has a decisive influence on albedo in urban areas. The three-dimensional surfaces formed by buildings and roads etc. create large inner spaces and cracks for lights to transfer, which result in the multiple reflection of lights, trapping lights, and leading to a decrease in albedo. Considering that the units of vegetation data and nighttime light data are not uniform, and the range of values for these data is quite different, the corrected DMSP/OLS data in this study has been normalized. The sensitivity of albedo to vegetation and urbanization intensity is calculated by multiple linear regression. The spatial distribution patterns of each factor's sensitivity term during different periods, as well as their variations, are shown in Figure 5. The sensitivity of albedo to vegetation and urbanization displays differences in period T1 and period T2.

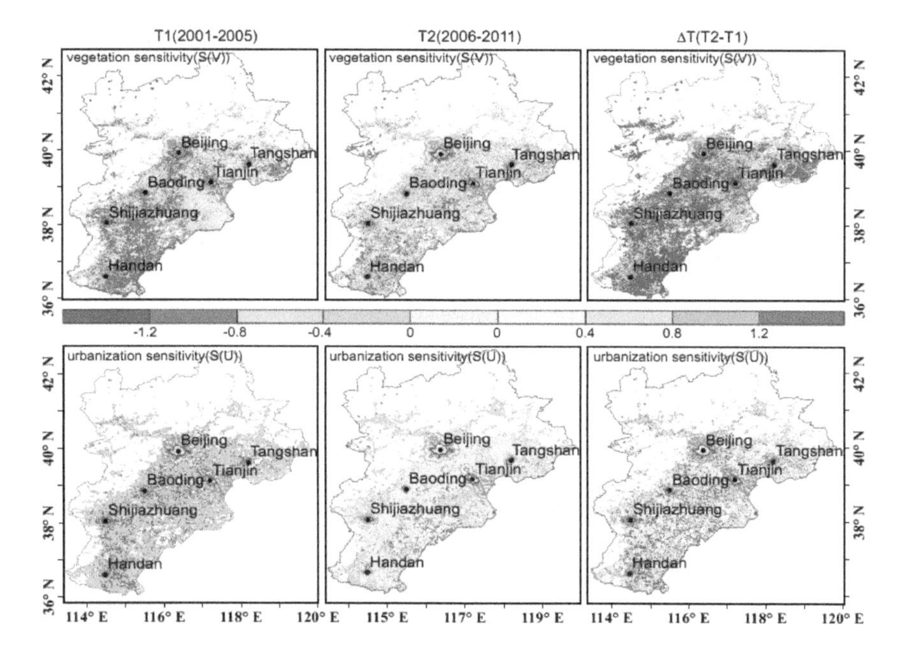

Figure 5. Spatial pattern of albedo sensitivity to vegetation and urbanization. T1 means the period from 2001 to 2005, T2 means 2006 to 2011, ΔT represents difference between T2 and T1, S(V) represents sensitivity of albedo to vegetation ($\frac{\partial A}{\partial V}$) in which A denotes albedo and V denotes EVI, S(U) represents the sensitivity of albedo to urbanization ($\frac{\partial A}{\partial U}$) in which A denotes albedo and U denotes DMSP/OLS, and S(V), S(U) were calculated via multiple regression of albedo to EVI and DMSP/OLS.

In period T1 (2001–2005), the sensitivity of albedo to urbanization showed a significant spatial distribution difference, of which the fifth percentile was −0.006 and the 95th percentile was 1.95. Spatially, the relatively high positive S(U) is mainly concentrated in the surrounding areas of major cities, such as Beijing, Tianjin, and Tangshan, positive-correlated to albedo. In contrast, S(U) in other regions is much smaller and generally negative, which has a weak negative correlation with albedo. In these regions, the sensitivity of albedo to vegetation(S(V)) shows high positive sensitivity, especially in the southeastern plains. In summary, urbanization has stronger promoting effects on albedo around large cities and has much weaker suppressing effects on albedo in other regions, where vegetation plays a promoting role in these areas.

In period T2 (2006–2011), instead of concentrating in areas surrounding large cities, the sensitivity of albedo to urbanization shows a regional diffusion feature compared to that in T1. In total, 72% of the S(U) is positive, and the fifth percentile of S(U) is −0.24 and the 95th percentile is 0.73. As the sensitivity of urbanization increases over a large area, the sensitivity of vegetation decreases significantly and extensively. In total, 53% of the vegetation sensitivity(S(V)) shows negative effects and mainly ranges from −0.4 to 0 (i.e., vegetation tends to inhibit the increase in albedo during this period).

From T1 to T2, the sensitivity of albedo to vegetation generally decreases and changes from positive to negative. The effect that this change causes is that the strong increased effect of vegetation on albedo turns into a weak decreased effect. However, the sensitivity of albedo to urbanization has extensively increased. The increased effects of urbanization on albedo exist not only in areas surrounding large cities during T1, but also in other large areas during T2, although the sensitivity intensity is much larger in T1 than that in T2.

4.4. Effects of the Influential Factors on Changes in Albedo

The sensitivity of albedo to vegetation and urbanization indicates a correlation between albedo and various factors. Because it is dimensionless, this correlation does not quantify the effects of various factors on albedo. Therefore, our study also quantifies the effects of vegetation and urbanization on albedo based on sensitivities. The effects of each factor on the interannual variation rate of albedo are shown in Figure 6.

From 2001–2005, the area with positive effects of vegetation on the interannual variation rate of albedo comprised more than 80% of the entire region (Figure 6). The areas with negative effects of vegetation mostly existed in the surrounding areas of cities and partially in the northern area of the study region. Urbanization had extensive positive effects on albedo, which were distributed around large cities. Shijiazhuang, Handan, and their surrounding areas had greater positive effects on variations in albedo compared to those from Beijing, Tianjin, and Tangshan, with the greatest effects exceeding 0.01 year^{-1}. The effects in other regions were almost 0. Other factors had both positive and negative effects on variations in albedo, but most of these effects were positive and located in Shijiazhuang, Handan, and their surrounding areas, with effects greater than 0.01 year^{-1}. In the northern part of the region, the effects of other factors on albedo were almost in the range of −0.005 to 0.005 year^{-1}. The statistics of the relative contributions of each factor (Figure 7) show that from 2001–2005, the relative contribution percentage of vegetation, urbanization, and other factors was 44%, 15%, and 41%, respectively. Urbanization had the lowest contribution to regional albedo, whereas vegetation and other factors were the two main controlling factors in variations in albedo, which were both greater than two times the amount of contribution from urbanization. One thing that must be explained is that due to the type and distribution differences of each factor, the spatial heterogeneity was relatively obvious. As a result, some pixels were under the absolute control of vegetation, and some were under the absolute control of urbanization, which led to an expected large standard deviation (STD) value for each factor's contribution.

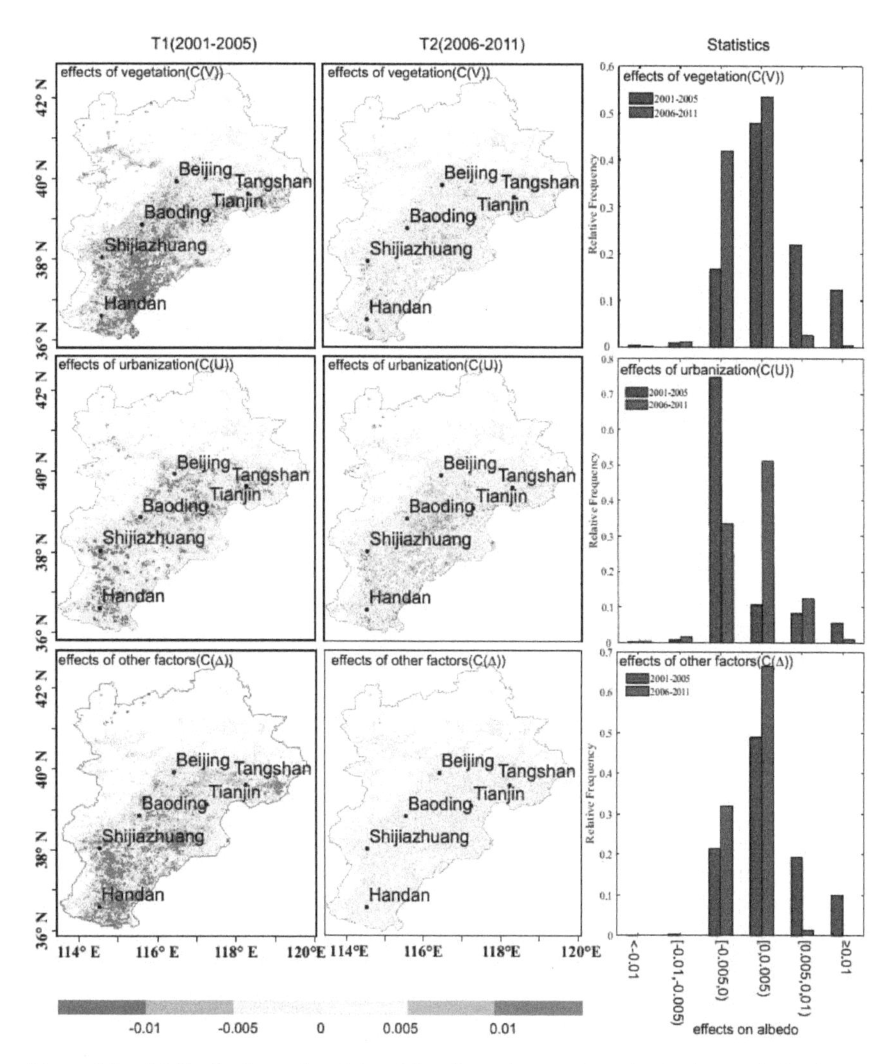

Figure 6. Spatial distribution and statistics of the effects of vegetation, urbanization, and other factors on interannual variation in albedo in T1 (2001–2005) and T2 (2006–2011). C(V) represents effects of vegetation on albedo, C(U) represents the effects of urbanization on albedo, C(Δ) represents effects of other factors on albedo, and the statistics of the relative percentage of vegetation, urbanization, and other factors in T1 and T2 are calculated based on the study area.

From 2006–2011, the effect of vegetation on variations in albedo mainly ranged from -0.005 year^{-1} to 0.005 year^{-1}, which was generally lower than that from 2001–2005 (Figure 6). Over 99% of the urbanization effects on albedo were positive. The areas effected by urbanization expanded although the value of this effect decreased compared to the urbanization effect from 2001–2005. The effect of other factors was distributed uniformly across the study area, ranging from -0.005 to 0.005. From 2006–2011, the relative contribution percentages of vegetation, urbanization, and other factors to albedo (Figure 7) were 24%, 48.5%, and 27.5%, respectively. Urbanization became the highest contribution factor, which increased by more than 200%. In contrast, the contributions from vegetation and other factors decreased by 20% and 13.5%, respectively.

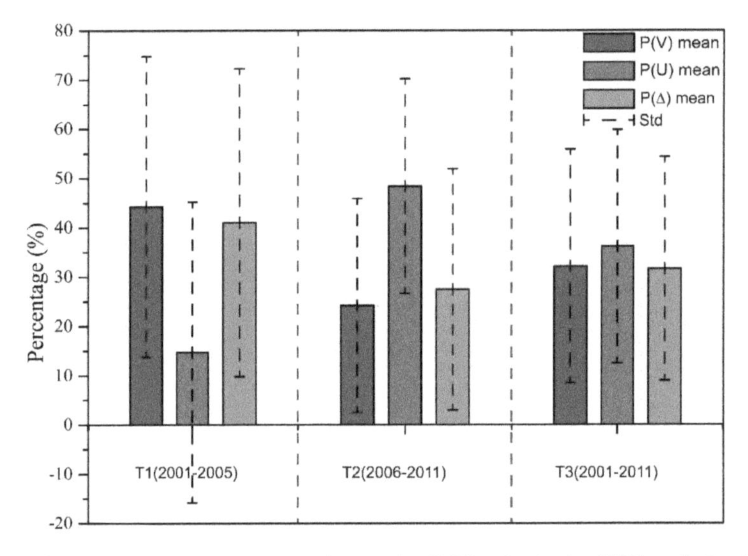

Figure 7. Average contribution percentages of vegetation (P(V)), urbanization (P(U)), and other factors (P(Δ)) to the interannual variation of albedo in the study area in T1 (2001–2005), in T2 (2006–2011), and in T3 (2001–2011). Std represents the standard deviation.

The relative contribution percentages of vegetation, urbanization, and other factors to albedo are different not only in size, but also in spatial distribution pattern (Figure 8).

Although the relative contribution percentage of each factor for the whole region is approximately 30% from 2001 to 2011 (Figure 7), the dominant controlling factors are different in different regions (Figure 8). From 2001–2011, the variations in albedo were mainly controlled by vegetation in the southeast and parts of the eastern region. Other regions were mainly controlled by urbanization, especially in regions surrounding cities, except for the Core Area, which was largely affected by both vegetation and other factors. From 2001–2005, the locations where the urbanization contribution was greater than 60% were the Expanded Area and the Fringe Area, while other regions were mostly controlled by vegetation. Other factors playing dominate roles were distributed in the triangular region formed by Beijing, Tianjin, and Tangshan, as well as near the connecting line between Shijiazhuang and Handan. From 2006 to 2011, the contributions of vegetation and other factors showed a significant reduction. The contribution percentage of vegetation was generally lower than 20%, whereas the contribution percentage of urbanization increased substantially in other regions (generally greater than 60%), except for the Core Area, where the urbanization contribution percentage was equal to zero. Other factors mainly affected the changes in albedo in the Core Area and a partial region near the connecting line between Shijiazhuang and Handan.

4.5. Urbanization in Representative Cities

According to the above analysis, urbanization transformed from a secondary influential factor from 2001–2005 into a major influential factor from 2006–2011, indicating that the effect of urbanization on regional albedo has increased since the breakpoint year (2005). However, the urbanization intensity of each individual city is substantially different. The Core Are, Expanded Area, and Fringe Area in our study represent the initial stage, middle acceleration stage, and final stage of the urbanization process, respectively; why is albedo different during these different urbanization stages? What are the main controlling factors for these areas? Are there any regional differences among the impact factors? We still do not know much about these issues. Therefore, in this paper, we also calculate the relative contribution percentages of vegetation, urbanization, and other factors in different functional areas (Figure 9).

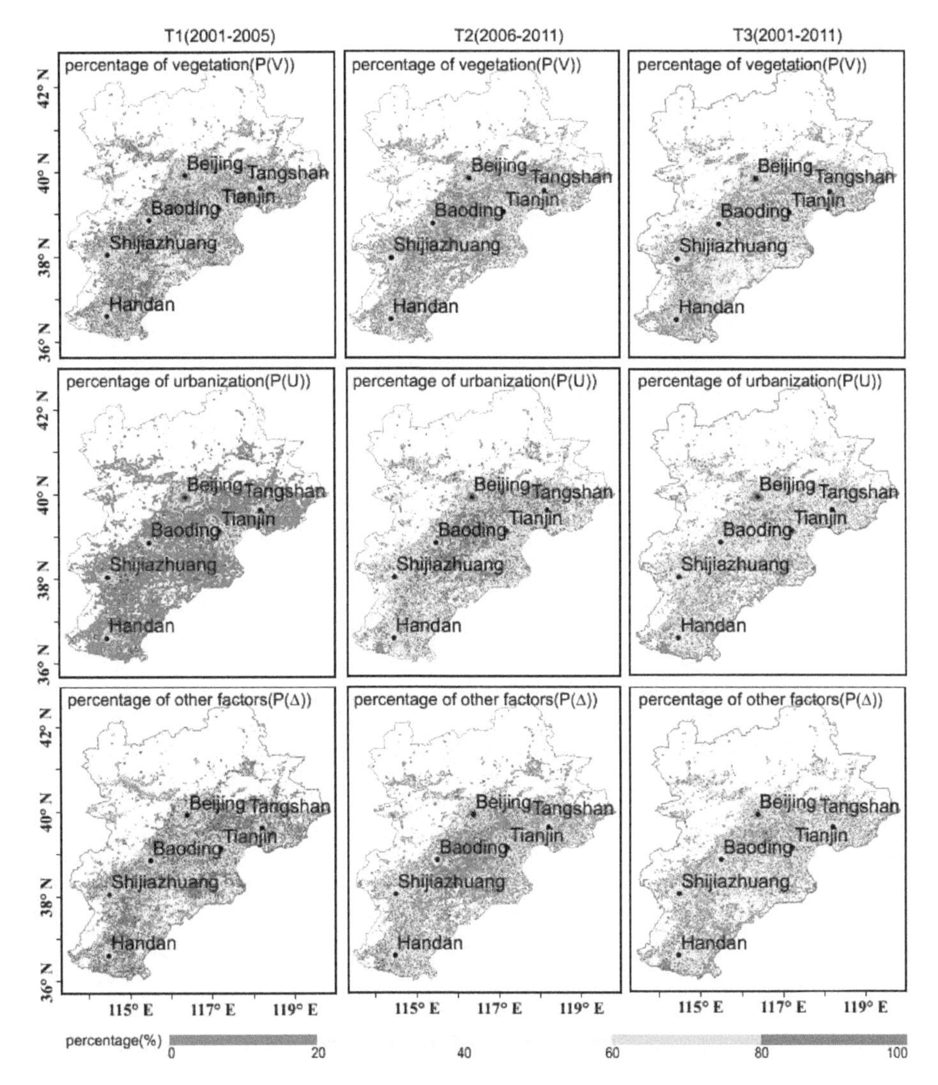

Figure 8. Spatial patterns of relative contribution percentages for vegetation (P(V)), urbanization (P(U)), and other factors (P(Δ)) in the T1, T2, and T3 period. T1 means the period from 2001 to 2005, T2 means 2006 to 2011, and T3 means 2001 to 2011.

For each individual city, the variations in albedo in Core Areas are mostly affected by other factors (Δ), followed by vegetation, and the contribution of urbanization is minimal. The average contribution percentages of vegetation, urbanization, and other factors are 34.5%, 8.4%, and 57.1%, respectively, in the Core Area. This result indicates that because the Core Area is generally composed of old cities that have generally completed urbanization, the influence of human activities on the variations in albedo is basically at a stable level; therefore, the urbanization contribution to albedo (8.4%) is much smaller than the contributions from other factors and vegetation. In the Core Area of the six major cities, other factors contribute more during T2 than those during T1 in 84% of our cities. Vegetation contributes more during T2 than that during T1 in 67% of our cities. The urbanization contribution decreases during T2 compared to that during T1 in 100% of our cities. It is safe to say that the contributions from vegetation and other factors in the Core Area will increase with time, while the contribution from urbanization will decrease.

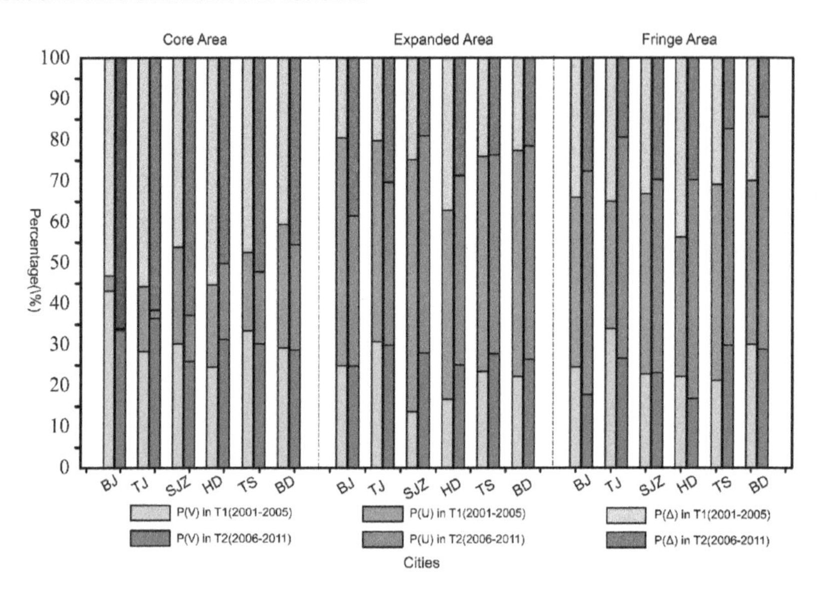

Figure 9. Relative contribution percentages of vegetation (P(V)), urbanization (P(U)), and other factors (P(Δ)) in different functional areas (Core Area, Expanded Area, and Fringe Area) of cities in T1 (2001–2005), and T2 (2006–2011). BJ, TJ, SJZ, HD, TS, and BD represent Beijing, Tianjin, Shijiazhuang, Handan, Tangshan, and Baoding, respectively.

In the Expanded Area, the average contribution percentages of vegetation, urbanization, and other factors were 26.5%, 46.7%, and 26.8%, respectively, and urbanization was the most important contribution factor in the Expanded Area. Spatially, compared to the Core Area, urbanization had the greatest change in the Expanded Area, with an increase of 456%; the contributions of other factors and vegetation decreased by 53% and 23%, respectively. Temporally, the urbanization contribution percentage decreased to various degrees during T2 compared with that during the T1 period. Meanwhile, the vegetation contribution percentage increased in 67% of the cities, while the contribution percentage of other factors decreased. However, Beijing and Tianjin, whose populations are greater than five million, showed the opposite effect; that is, the vegetation contribution percentage decreased slightly in the Expanded Area, while the contribution of other factors increased significantly. For small cities, urbanization accompanying land cover change might have been finished in the Expanded Area and there would be no land cover changes for several years at least in the future, due to the limited population and limited population increasing ability. So, the number of the residents living in this area is relatively stable, which could lead to the stable growth of the vegetation and green space for a comfortable living environment. Thus, the contribution of vegetation will increase. On the contrary, for Beijing and Tianjin, which are the two biggest cities in the Jing-Jin-Ji area, they have been maintaining a fast-speed of urbanization for a long time. With reference to the Globeland30 land cover maps, we know that land cover changes are still happening, and they happened in the Expanded Area. That is to say, the urbanization process is going on in Expanded Areas in Beijing and Tianjin. So, due to the limited urban space, the rate of green space to urban space in Beijing and Tianjin would be smaller than small cities. Thus, the contribution of vegetation in Expanded Area in these big cities is relatively small compare to those in small cities.

In the Fringe Area, the average contribution percentages of vegetation, urbanization, and other factors were 24.9%, 45.8%, and 29.3%, respectively. Compared to the Expanded Area, the urbanization contribution percentage in the Fringe Area decreased, but it was still the dominant contribution factor. However, the vegetation contribution percentage decreased, and the contribution rate of other factors increased. Temporally, the contribution percentage of urbanization increased significantly from period T1 to period T2, and the contribution of other factors decreased significantly. The vegetation contribution percentage also slightly decreased in most of the cities.

In summary, from the Core Area to the Fringe Area, the spatial urbanization contribution percentage rapidly increases at first, followed by a slow increase. The vegetation contribution percentage rapidly declines during period T1, while it slowly declines during period T2. The contribution of other factors decreases quickly during period T1, but slowly increases during period T2. Temporally, the urbanization contribution in the Core Area and the Expanded Area decreases over time, whereas the contribution of vegetation increases. In the Fringe Area, the urbanization contribution increases rapidly with a rapid decrease in vegetation contribution. The other factors only have a significant increase in the Core Area, with decreasing trends in the Expanded Area and Fringe Area.

5. Discussion

In our study, nighttime light data were corrected to characterize the changes in urbanization intensity by the DN value and various functional regions of the city were extracted to express the various stages of urbanization. Combined with the vegetation index data, we quantitatively analyzed the contribution of urbanization and vegetation to variations in albedo. The contribution analysis method used in this study is a partial derivative method, which is widely used in studies on the effects of climate on hydrological dynamics [77,78] and studies on climate response [74,79]. Based on the results of this quantitative analysis, we concluded that the significant increase in the urbanization contribution and the decrease in the vegetation contribution after 2005 were the main reasons for the significant decreasing growth rate of albedo after 2005. In terms of mechanisms, this was consistent with previous conclusions on urban albedo; that is, urbanization could cause a decrease in albedo, which is generally correlated to the surface roughness. For example, albedo observations based on model experiments from Aida [16] showed that multiple reflections of solar radiation in urban canyons increased the absorption of solar radiation in cities, resulting in a reduction in urban albedo. Kondo [80] used the Monte Carlo ray tracing method to show that building height affects albedo, and low-rise buildings have a high albedo. The impact of urban areas, which are one of the most densely populated areas, on urban albedo is multi-fold [81–84]. On one hand, urbanization is accompanied by changes in land cover. In general, the process of transition from a village to a city involves replacing natural surfaces (e.g., farmlands and forests) with impervious surfaces (e.g., cement and asphalt). Due to the changes in the thermal conductivity of the Earth's surface, albedo changes, and the water and heat exchange between the Earth's surface and the atmosphere also changes. The 3D solid surface formed during urbanization has resulted in an increase in surface roughness [16,76,80,85,86] and solar radiation absorption [77], which share the same mechanism with soil roughness and soil albedo. Inner spaces enable the multiple reflection of lights, which increases the absorption of radiation. For this reason, urban areas usually have low albedo [8]. The multiyear average of reflectivity calculated by MODIS albedo products also showed this rule; that is, that albedo in urban areas is generally low (Figure 2). On the other hand, the ability of cities to attract people is also obvious. Urban areas account for approximately 0.5% of the total land area in the world, but they accommodate more than half of the world's population [87]. Due to the complexity and uncertainty of the human activities during the urbanization process, it is very difficult to identify the urbanization effects. The DN value of the night-time lights data is used as the index of urbanization intensity, which could show comprehensive impacts of human activities. This enables us to simplify the impact of urbanization on albedo changes and helps us to quantify the contribution of urbanization and vegetation to the changes in albedo from a macro perspective.

As the parameter indicating the surface's ability to reflect solar radiation, albedo plays a key role in the energy balance at the surface, and the effects of albedo on climate change have raised substantial attention from many scholars. Akbari et al. [88] simulated the long-term effects of urban albedo growth using a mesoscale complex global climate model (UVic Earth System Climate Model), and it was believed that an increase in surface albedo by 0.01 over a square meter could reduce the long-term global temperature by 3×10^{-5} K, which was equivalent to reducing CO_2 emissions by 7 kg [89]. Sailor [8] analyzed the surface albedo in Los Angeles based on a three-dimensional meteorological model and found that albedo increased by 0.14 in urban regions and 0.08 in basin areas, which could reduce the maximum heat by 1.5 °C in summer. Based on a mesoscale atmospheric model, Humdi [40] analyzed the intensity of UHIs and found that the increase in albedo over three types of urban surfaces (walls, roofs, and roads) could reduce the UHI both during the day and at night. Wang et al. [89] analyzed the impact of land use change in urban regions on extreme heat events with the Weather Research and Forecasting model (WRF) in Jing-Jin-Ji and found that an increase of the albedo on urban roofs from 0.12 to 0.85 could reduce the urban mean temperature by 0.51 °C, which was equivalent to 80% of the heat caused by urban expansion in the last 20 years. Menon [7] increased the albedo of roofs and roads using the GEOS-5 basin surface pattern, and the result showed that increasing the albedo of roofs and roads by 0.25 and 0.15, respectively, could result in approximately 57 Gt of CO_2 from global urban regions. Compared with the aforementioned study, it is clear that the decrease of albedo (approximately 0.05) in the Jing-Jin-Ji region, caused by the increasing urbanization contribution and the decreasing contribution of vegetation over 2001–2011, is in a relatively reasonable numerical range, and the effects of this variation in albedo on urban temperature are not negligible. Urbanization could change both the urban morphology and the urban environment. Energy-budget parameters are also varied during this process. The heat-trapping morphology of the 3-D surface and the reduced areas of vegetation and water bodies both contribute to albedo variation and could result in urban climate change [82,90]. The decrease of the albedo (~0.05) in our result showed that there might be a high possibility of temperature changes due to urbanization, and it could also affect the UHI in this area. In this way, the difference between the vegetation and urbanization contribution to albedo might be useful in urban planning to mitigate the intensity of the UHI, and helps to offer better comfort conditions to residents [91–93].

Because of the significant influence of albedo on temperature in urban areas, increasing the albedo in urban areas to mitigate the urban heat island intensity has become an important aspect of urban energy conservation research. Taha [94] found that the changes in urban albedo via whitewashing could save 35% of the cooling peak power and 62% of the cooling energy. The research of Akbari [95] also showed that urban trees and high albedo could potentially reduce air conditioning energy by 20%, which would save about $10 billion a year in energy costs and help improve urban air conditions. Therefore, it is necessary and meaningful to understand the reason why urban albedo changes during the process of urbanization, which would be helpful for future studies on urban climate change and for the development of urban energy conservation strategies.

There have been many detailed studies on the factors that may influence albedo, such as the solar zenith angle [16], the underlying surface regime [21], soil moisture [96,97], and meteorological conditions [98–100]. Based on a mathematical statistics method, our study quantitatively calculated the contributions of multiple factors to urban albedo in each pixel and identified the main contribution factors, which is one of the highlights of our study. However, our study is still insufficient. Because the basis of the calculation method in our study was a comprehensive differential equation, we cannot evaluate the uncertainty of the results. Instead, we can only compare the results with other studies or use other auxiliary data to validate the reliability of our conclusions. Second, one of our study conclusions was that the contribution of albedo in the Expanded Area that also includes the new urban areas and in the suburb-located Fringe Area was mainly affected by urbanization. Since our study simplified the urbanization process, more specific reasoning, such as why urbanization is the dominant factor, still requires more detailed scientific research in the future. Third, we used the DN value of the DMSP/OLS nighttime light data to represent the urbanization intensity, but whether or not this index can assess urban development levels accurately has not been evaluated. In addition, there are many types of nighttime light correction methods, but the methods used in Cao [63] were applied in this study due to the adequate correction effect and more applicable study area (China). However, whether or not this method can be applied at a global scale remains to be explored. Finally, as we have only taken the fastest population growth period (~10 years) into consideration, the length of the time series is relatively short. Therefore, the impact of data length still needs to be evaluated.

6. Conclusions

Based on remote sensing data, we explored the spatiotemporal distribution patterns and variation characteristics of albedo in the Jing-Jin-Ji region. In addition, a quantitative approach based on partial derivatives was applied to calculate the contribution of urbanization and vegetation to albedo variability.

The results showed that albedo changed greatly before and after 2005. Albedo variation was mainly contributed by vegetation (44%) and other factors (41%) before 2005. However, after 2005, large-scale urbanization became the main controlling factor that affected the change in albedo, with a contribution proportion of 48.5%, and the vegetation and other factors contributions were 24% and 27.5%, respectively. Spatially, the contribution percentage of urbanization gradually increased from the Core Area to the Fringe Area, and the contribution percentage of vegetation gradually decreased. Temporally, the contribution of urbanization in the Core Area and Expanded Area decreased with time, and the contribution of vegetation increased. In contrast, the impact of urbanization in the Fringe Area increased, while that of vegetation decreased. Other factors (e.g., extreme weather, natural disasters et al.) contributed more to albedo variation in Core Areas than other areas, and exhibited an increasing trend over time.

In summary, our paper provides a method via mathematical statistics to quantitatively estimate the contribution of urbanization and vegetation to albedo changes in urban areas. Understanding the spatiotemporal differences in albedo via dominant factors will help us perform additional research on urban climate change, especially regarding temperature changes in urban areas. At the same time, it can also provide a data foundation for developing urban energy conservation policies.

Author Contributions: R.T. and X.Z. conceived of and designed the experiments. R.T. processed and analyzed the data. All authors contributed to the ideas, writing, and discussion.

Funding: This study was supported by the National Key Research and Development Program of China (No. 2016YFA0600103) and the State Key Laboratory of Remote Sensing Science (No. 16ZY-06).

Acknowledgments: The authors thank Jia Cheng, Yifeng Peng, Haoyu Wang, and Xiaozheng Du for helpful comments that improved this manuscript.

Conflicts of Interest: The authors declare no conflict of interest.

References

1. Stroeve, J.; Box, J.E.; Gao, F.; Liang, S.; Nolin, A.; Schaaf, C. Accuracy assessment of the modis 16-day albedo product for snow: Comparisons with greenland in situ measurements. *Remote Sens. Environ.* **2005**, *94*, 46–60. [CrossRef]
2. Cess, R.D. Biosphere-albedo feedback and climate modeling. *J. Atmos. Sci.* **1978**, *35*, 1765–1768. [CrossRef]
3. Dickinson, R.E. Land surface processes and climate—Surface albedos and energy balance. *Adv. Geophys.* **1983**, *25*, 305–353.
4. Lofgren, B.M. Surface albedo–climate feedback simulated using two-way coupling. *J. Clim.* **1995**, *8*, 2543–2562. [CrossRef]

5. Liang, S. Narrowband to broadband conversions of land surface albedo I: Algorithms. *Remote Sens. Environ.* **2001**, *76*, 213–238. [CrossRef]
6. Akbari, H.; Menon, S.; Rosenfeld, A. Global cooling: Increasing world-wide urban albedos to offset CO_2. *Clim. Chang.* **2009**, *94*, 275–286. [CrossRef]
7. Menon, S.; Akbari, H.; Mahanama, S.; Sednev, I.; Levinson, R. Radiative forcing and temperature response to changes in urban albedos and associated CO_2 offsets. *Environ. Res. Lett.* **2010**, *5*. [CrossRef]
8. Sailor, D.J. Simulated urban climate response to modifications in surface albedo and vegetative cover. *J. Appl. Meteorol.* **1995**, *34*, 1694–1704. [CrossRef]
9. Daan, B.; Gabriela, S.-S.; Harm, B.; Monique, M.P.D.H.; Trofim, C.M.; Frank, B. The response of arctic vegetation to the summer climate: Relation between shrub cover, ndvi, surface albedo and temperature. *Environ. Res. Lett.* **2011**, *6*. [CrossRef]
10. Flanner, M.G.; Shell, K.M.; Barlage, M.; Perovich, D.K.; Tschudi, M.A. Radiative forcing and albedo feedback from the northern hemisphere cryosphere between 1979 and 2008. *Nat. Geosci.* **2011**, *4*, 151–155. [CrossRef]
11. Hannesp, S.; Davidneil, B. Integration of albedo effects caused by land use change into the climate balance: Should we still account in greenhouse gas units. *For. Ecol. Manag.* **2010**, *260*, 278–286.
12. Meng, X.H.; Evans, J.P.; Mccabe, M.F. The influence of inter-annually varying albedo on regional climate and drought. *Clim. Dyn.* **2014**, *42*, 787–803. [CrossRef]
13. Tedesco, M.; Fettweis, X.; Van den Rroeke, M.R.; Van den Wal, R.S.W.; Smeets, C.J.P.P.; Van den Berg, W.J.; Serreze, M.C.; Box, J.E. The role of albedo and accumulation in the 2010 melting record in greenland. *Environ. Res. Lett.* **2011**, *6*. [CrossRef]
14. Caiazzo, F.; Malina, R.; Staples, M.D.; Wolfe, P.J.; Yim, S.H.L.; Barrett, S.R.H. Quantifying the climate impacts of albedo changes due to biofuel production: A comparison with biogeochemical effects. *Environ. Res. Lett.* **2014**, *9*, 69–75. [CrossRef]
15. Stull, E.; Sun, X.; Zaelke, D. Enhancing urban albedo to fight climate change and save energy. *Sustain. Dev. Law Policy* **2010**, *11*, 5–6.
16. Aida, M. Urban albedo as a function of the urban structure—A model experiment. *Bound. Layer Meteorol.* **1982**, *23*, 405–413. [CrossRef]
17. Guan, X.; Huang, J.; Guo, N.; Jianrong, B.I.; Wang, G. Variability of soil moisture and its relationship with surface albedo and soil thermal parameters over the loess plateau. *Adv. Atmos. Sci.* **2009**, *26*, 692–700. [CrossRef]
18. Chen, X.H. Relationship between surface albedo and some meteorological factors. *J. Chengdu Inst. Meteorol.* **1999**, *14*, 233–238.
19. Ialongo, I.; Buchard, V.; Brogniez, C.; Casale, G.R. Aerosol single scattering albedo retrieval in the uv range: An application to omi satellite validation. *Atmos. Chem. Phys.* **2010**, *10*, 331–340. [CrossRef]
20. Li, G.; Xiao, J. Diurnal variation of surface albedo and relationship between surface albedo and meteorological factors on the western qinghai-tibet plateau. *Sci. Geogr. Sin.* **2007**, *27*, 63–67. [CrossRef]
21. Dana, K.J.; Nayar, S.K.; Ginneken, B.V.; Koenderink, J.J. Reflectance and texture of real-world surfaces. In Proceedings of the IEEE Computer Society Conference on Computer Vision and Pattern Recognition, San Juan, Puerto Rico, 17–19 June 1997.
22. Baret, F. Reflection of radiant energy from soils. *Adv. Space Res.* **1993**, *13*, 130–138.
23. Bowers, S.A.; Smith, S.J. Spectrophotometric determination of soil water content. *Soilence Soc. Am. J.* **1972**, *36*, 978–980. [CrossRef]
24. Cierniewski, J.; Ceglarek, J.; Karnieli, A.; Ben-Dor, E.; Królewicz, S. Shortwave radiation affected by agricultural practices. *Remote Sens.* **2018**, *10*, 419. [CrossRef]
25. Cierniewski, J.; Karnieli, A.; Kaźmierowski, C.; Królewicz, S.; Piekarczyk, J.; Lewińska, K.; Goldberg, A.; Wesołowski, R.; Orzechowski, M. Effects of soil surface irregularities on the diurnal variation of soil broadband blue-sky albedo. *IEEE J. Sel. Top. Appl. Earth Obs. Remote Sens.* **2015**, *8*, 493–502. [CrossRef]
26. Piech, K.R.; Walker, J.E. Interpretation of soils. *Photogramm. Eng.* **1974**, *40*, 87–94.
27. Mikhaĭlova, N.A.; Orlov, D.S.; Rozanov, B.G. *Opticheskie Svoĭstva Pochv I Pochvennykh Komponentov*; Nauka: Moskva, Russia, 1986.
28. Li, X.; Zhou, Y.; Asrar, G.R.; Mao, J.; Li, X.; Li, W. Response of vegetation phenology to urbanization in the conterminous United States. *Glob. Chang. Biol.* **2017**, *23*. [CrossRef] [PubMed]

29. Kalnay, E.; Cai, M. Impact of urbanization and land-use change on climate. *Nature.* **2003**, *423*, 528–531. [CrossRef] [PubMed]
30. Carlson, T.N.; Arthur, S.T. The impact of land use—Land cover changes due to urbanization on surface microclimate and hydrology: A satellite perspective. *Glob. Planet. Chang.* **2000**, *25*, 49–65. [CrossRef]
31. Liu, Y.; Huang, X.; Yang, H.; Zhong, T. Environmental effects of land-use/cover change caused by urbanization and policies in southwest china karst area—A case study of guiyang. *Habitat Int.* **2014**, *44*, 339–348. [CrossRef]
32. Oke, T.R. City size and the urban heat island. *Atmos. Environ.* **1973**, *7*, 769–779. [CrossRef]
33. Oke, T.R. The energetic basis of the urban heat island. *Q. J. R. Meteorol. Soc.* **1982**, *108*, 1–24. [CrossRef]
34. Matson, M.; Mcclain, E.P.; McGinnis, D.F., Jr.; Pritchard, J.A. Satellite detection of urban heat islands. *Mon. Weather Rev.* **1978**, *106*, 1725–1734. [CrossRef]
35. Price, J.C. Assessment of the urban heat island effect through the use of satellite data. *Mon. Weather Rev.* **1979**, *107*, 1554–1557. [CrossRef]
36. Gallo, K.P.; Owen, T.W. Satellite-based adjustments for the urban heat island temperature bias. *J. Appl. Meteorol.* **1999**, *38*, 806–813. [CrossRef]
37. Xu, H.; Chen, B. Remote sensing of the urban heat island and its changes in xiamen city of se china. *J. Environ. Sci.* **2004**, *16*, 276–281.
38. Streutker, D.R. Satellite-measured growth of the urban heat island of Houston, Texas. *Remote Sens. Environ.* **2003**, *85*, 282–289. [CrossRef]
39. Spångmyr, M. *Global Effects of Albedo Change Due to Urbanization*; Lund University: Lund, Sweden, 2010.
40. Hamdi, R.; Schayes, G. Sensitivity study of the urban heat island intensity to urban characteristics. *Int. J. Climatol.* **2008**, *28*, 973–982. [CrossRef]
41. Schaaf, C.B.; Liu, J.; Gao, F.; Strahler, A.H. Aqua and terra modis albedo and reflectance anisotropy products. In *Land Remote Sensing and Global Environmental Change*; Ramachandran, B., Justice, C.O., Abrams, M.J., Eds.; Springer: New York, NY, USA, 2010; pp. 549–561.
42. Chen, B.; Xu, G.; Coops, N.C.; Ciais, P.; Innes, J.L.; Wang, G.; Myneni, R.B.; Wang, T.; Krzyzanowski, J.; Li, Q. Changes in vegetation photosynthetic activity trends across the asia–pacific region over the last three decades. *Remote Sens. Environ.* **2014**, *144*, 28–41. [CrossRef]
43. Huang, Q.; Yang, X.; Gao, B.; Yang, Y.; Zhao, Y. Application of dmsp/ols nighttime light images: A meta-analysis and a systematic literature review. *Remote Sens.* **2014**, *6*, 6844–6866. [CrossRef]
44. Wu, S.; Zhou, S.; Chen, D.; Wei, Z.; Dai, L.; Li, X. Determining the contributions of urbanisation and climate change to npp variations over the last decade in the yangtze river delta, China. *Sci. Total Environ.* **2014**, *472*, 397–406. [CrossRef] [PubMed]
45. Huete, A.; Didan, K.; Leeuwen, W.V.; Miura, T.; Glenn, E. *Modis Vegetation Indices*; Springer: New York, NY, USA, 2010.
46. LI, Y.; Shi, B. Analysis of population distribution in beijing-tianjin-hebei region in 2000–2013 (In Chinese). *Youth Times* **2016**, *13*, 85–86.
47. Jiang, D.; Zhuang, D.; Xu, X.; Lei, Y. Integrated evaluation of urban development suitability based on remote sensing and gis techniques—A case study in jingjinji area, china. *Sensors* **2008**, *8*, 5975–5986.
48. Tan, M.; Li, X.; Xie, H.; Lu, C. Urban land expansion and arable land loss in China—A case study of Beijing–Tianjin–Hebei region. *Land Use Policy* **2005**, *22*, 187–196. [CrossRef]
49. Benas, N.; Chrysoulakis, N. Estimation of the land surface albedo changes in the broader mediterranean area, based on 12 years of satellite observations. *Remote Sens.* **2015**, *7*, 16150–16163. [CrossRef]
50. Qu, Y.; Liu, Q.; Liang, S.; Wang, L.; Liu, N.; Liu, S. Direct-estimation algorithm for mapping daily land-surface broadband albedo from modis data. *IEEE Trans. Geosci. Remote Sens.* **2014**, *52*, 907–919. [CrossRef]
51. Li, Y.; Zhao, M.; Motesharrei, S.; Mu, Q.; Kalnay, E.; Li, S. Local cooling and warming effects of forests based on satellite observations. *Nat. Commun.* **2015**, *6*, 6603. [CrossRef] [PubMed]
52. Peng, S.; Piao, S.; Philippe, C.; Pierre, F.; Catherine, O.; François-Marie, B.; Huijuan, N.; Liming, Z.; Myneni, R.B. Surface urban heat island across 419 global big cities. *Environ. Sci. Technol.* **2012**, *46*, 696–703. [CrossRef] [PubMed]
53. Zhou, D.; Zhao, S.; Liu, S.; Zhang, L.; Zhu, C. Surface urban heat island in China's 32 major cities: Spatial patterns and drivers. *Remote Sens. Environ.* **2014**, *152*, 51–61. [CrossRef]

54. Liang, S.; Shuey, C.J.; Russ, A.L.; Fang, H.; Chen, M.; Walthall, C.L.; Daughtry, C.S.T.; Hunt, R. Narrowband to broadband conversions of land surface albedo: II. Validation. *Remote Sens. Environ.* **2003**, *84*, 25–41. [CrossRef]
55. Wang, K.; Liu, J.; Zhou, X.; Sparrow, M.; Ma, M.; Sun, Z.; Jiang, W. Validation of the modis global land surface albedo product using ground measurements in a semidesert region on the tibetan plateau. *J. Geophys. Res.* **2004**, *109*. [CrossRef]
56. Huete, A.R.; Didan, K.; Van Leeuwen, W. Modis Vegetation Index (mod 13) Algorithm Theoretical Basis Document. Universities of Arizona and Virginia, USA. 1999. Available online: https://modis.gsfc.nasa.gov/data/atbd/atbd_mod13.pdf (accessed on 6 July 2018).
57. Solano, R.; Didan, K.; Jacobson, A.; Huete, A. *Modis Vegetation Index User's Guide (Mod13 Series)*; Vegetation Index and Phenology Lab, The University of Arizona: Tucson, AZ, USA, 2010.
58. Wardlow, B.D.; Egbert, S.L.; Kastens, J.H. Analysis of time-series modis 250 m vegetation index data for crop classification in the us central great plains. *Remote Sens. Environ.* **2007**, *108*, 290–310. [CrossRef]
59. Wu, D.; Zhao, X.; Liang, S.; Zhou, T.; Huang, K.; Tang, B.; Zhao, W. Time-lag effects of global vegetation responses to climate change. *Glob. Chang. Biol.* **2015**, *21*, 3520–3531. [CrossRef] [PubMed]
60. Yu, B.; Shu, S.; Liu, H.; Song, W.; Wu, J.; Wang, L.; Chen, Z. Object-based spatial cluster analysis of urban landscape pattern using nighttime light satellite images: A case study of china. *Int. J. Geogr. Inf. Sci.* **2014**, *28*, 2328–2355. [CrossRef]
61. Zhou, Y.; Smith, S.J.; Elvidge, C.D.; Zhao, K.; Thomson, A.; Imhoff, M. A cluster-based method to map urban area from dmsp/ols nightlights. *Remote Sens. Environ.* **2014**, *147*, 173–185. [CrossRef]
62. Liu, Z.; He, C.; Zhang, Q.; Huang, Q.; Yang, Y. Extracting the dynamics of urban expansion in china using dmsp-ols nighttime light data from 1992 to 2008. *Landsc. Urban Plan.* **2012**, *106*, 62–72. [CrossRef]
63. Cao, Z.; Zhifeng, W.U.; Kuang, Y.; Huang, N. Correction of dmsp/ols night-time light images and its application in china. *J. Geo-Inf. Sci.* **2015**, *3498*, 1010–1016.
64. Elvidge, C.D.; Baugh, K.E.; Dietz, J.B.; Bland, T.; Sutton, P.C.; Kroehl, H.W. Radiance calibration of dmsp-ols low-light imaging data of human settlements. *Remote Sens. Environ.* **1999**, *68*, 77–88. [CrossRef]
65. Wu, J.; He, S.; Peng, J.; Li, W.; Zhong, X. Intercalibration of dmsp-ols night-time light data by the invariant region method. *Int. J. Remote Sens.* **2013**, *34*, 7356–7368. [CrossRef]
66. Chen, J.; Cao, X.; Peng, S.; Ren, H. Analysis and applications of globeland30: A review. *ISPRS Int. J. Geo-Inf.* **2017**, *6*, 230. [CrossRef]
67. Chen, J.; Jin, C.; Liao, A.; Xing, C.; Chen, L.; Chen, X.; Shu, P.; Gang, H.; Zhang, H.; Chaoying, H.E. Concepts and key techniques for 30 m global land cover mapping. *Acta Geod. Cartogr. Sin.* **2014**, *43*, 551–557.
68. Arsanjani, J.J.; Tayyebi, A.; Vaz, E. Globeland30 as an alternative fine-scale global land cover map: Challenges, possibilities, and implications for developing countries. *Habitat Int.* **2016**, *55*, 25–31. [CrossRef]
69. Brovelli, M.A.; Molinari, M.E.; Hussein, E.; Chen, J.; Li, R. The first comprehensive accuracy assessment of globeland30 at a national level&58; methodology and results. *Remote Sens.* **2015**, *7*, 4191–4212.
70. Antrop, M. Landscape change and the urbanization process in europe. *Landsc. Urban Plan.* **2004**, *67*, 9–26. [CrossRef]
71. Gong, P.; Howarth, P.J. The use of structural information for improving land-cover classification accuracies at the rural-urban fringe. *Photogramm. Eng. Remote Sens.* **1990**, *56*, 67–73.
72. Wehrwein, G.S. The rural-urban fringe. *Econ. Geogr.* **1942**, *18*, 217–228. [CrossRef]
73. Wang, X.; Piao, S.; Xu, X.; Ciais, P.; Macbean, N.; Myneni, R.B.; Li, L. Has the advancing onset of spring vegetation green-up slowed down or changed abruptly over the last three decades? *Glob. Ecol. Biogeogr.* **2015**, *24*, 621–631. [CrossRef]
74. Forzieri, G.; Alkama, R.; Miralles, D.G.; Cescatti, A. Satellites reveal contrasting responses of regional climate to the widespread greening of earth. *Science* **2017**, *356*, 1180–1184. [CrossRef] [PubMed]
75. Small, C. High spatial resolution spectral mixture analysis of urban reflectance. *Remote Sens. Environ.* **2003**, *88*, 170–186. [CrossRef]
76. Oke, T.R. *Boundary Layer Climates*, 2nd ed.; Methuen: New York, NY, USA, 1987.
77. Meng, D.; Mo, X. Assessing the effect of climate change on mean annual runoff in the songhua river basin, China. *Hydrol. Process.* **2012**, *26*, 1050–1061. [CrossRef]

78. Rana, G.; Katerji, N. A measurement based sensitivity analysis of the penman-monteith actual evapotranspiration model for crops of different height and in contrasting water status. *Theor. Appl. Climatol.* **1998**, *60*, 141–149. [CrossRef]
79. Zhou, S.; Yu, B.; Schwalm, C.R.; Ciais, P.; Zhang, Y.; Fisher, J.B.; Michalak, A.M.; Wang, W.; Poulter, B.; Huntzinger, D.N. Response of water use efficiency to global environmental change based on output from terrestrial biosphere models. *Glob. Biogeochem. Cycles* **2017**, *31*, 1639–1655. [CrossRef]
80. Kondo, A.; Ueno, M.; Kaga, A.; Yamaguchi, K. The influence of urban canopy configuration on urban albedo. *Bound. Layer Meteorol.* **2001**, *100*, 225–242. [CrossRef]
81. Oke, T.R. *Review of Urban Climatology, 1973–1976*; Secretariat of the World Meteorological Organization: Geneva, Switzerland, 1979.
82. Oke, T. Climatic impacts of urbanization. In *Interactions of Energy and Climate*; Springer: New York, NY, USA, 1980.
83. Wilmers, F. Effects of vegetation on urban climate and buildings. *Energy Build.* **1991**, *15*, 507–514. [CrossRef]
84. Djen, C.S. The urban climate of Shanghai. *Atmos. Environ. Part B* **1992**, *26*, 9–15. [CrossRef]
85. Aida, M.; Gotoh, K. Urban albedo as a function of the urban structure—A two-dimensional numerical simulation. *Bound. Layer Meteorol.* **1982**, *23*, 415–424. [CrossRef]
86. Kanda, M.; Kawai, T.; Nakagawa, K. A simple theoretical radiation scheme for regular building arrays. *Bound. Layer Meteorol.* **2005**, *114*, 71–90. [CrossRef]
87. MacLachlan, A.; Biggs, E.; Roberts, G.; Boruff, B. Urbanisation-induced land cover temperature dynamics for sustainable future urban heat island mitigation. *Urban Sci.* **2017**, *1*, 38. [CrossRef]
88. Akbari, H.; Matthews, H.D.; Seto, D. The long-term effect of increasing the albedo of urban areas. *Environ. Res. Lett.* **2012**, *7*. [CrossRef]
89. Wang, M.; Yan, X.; Liu, J.; Zhang, X. The contribution of urbanization to recent extreme heat events and a potential mitigation strategy in the Beijing–Tianjin–Hebei metropolitan area. *Theor. Appl. Climatol.* **2013**, *114*, 407–416. [CrossRef]
90. Oke, T. The urban energy balance. *Prog. Phys. Geogr.* **1988**, *12*, 471–508. [CrossRef]
91. Morini, E.; Touchaei, A.; Castellani, B.; Rossi, F.; Cotana, F. The impact of albedo increase to mitigate the urban heat island in terni (Italy) using the wrf model. *Sustainability* **2016**, *8*, 999. [CrossRef]
92. Morini, E.; Castellani, B.; Ciantis, S.D.; Anderini, E.; Rossi, F. Planning for cooler urban canyons: Comparative analysis of the influence of façades reflective properties on urban canyon thermal behavior. *Sol. Energy* **2018**, *162*, 14–27. [CrossRef]
93. Rossi, F.; Anderini, E.; Castellani, B.; Nicolini, A.; Morini, E. Integrated improvement of occupants' comfort in urban areas during outdoor events. *Build. Environ.* **2015**, *93*, 285–292. [CrossRef]
94. Taha, H.; Akbari, H.; Rosenfeld, A.; Huang, J. Residential cooling loads and the urban heat island—The effects of albedo. *Build. Environ.* **1988**, *23*, 271–283. [CrossRef]
95. Akbari, H.; Pomerantz, M.; Taha, H. Cool surfaces and shade trees to reduce energy use and improve air quality in urban areas. *Sol. Energy* **2001**, *70*, 295–310. [CrossRef]
96. Liu, H.; Wang, B.; Fu, C. Relationships between surface albedo, soil thermal parameters and soil moisture in the semi-arid area of Tongyu, Northeastern China. *Adv. Atmos. Sci.* **2008**, *25*, 757–764. [CrossRef]
97. Roxy, M.S.; Sumithranand, V.B.; Renuka, G. Variability of soil moisture and its relationship with surface albedo and soil thermal diffusivity at astronomical observatory, Thiruvananthapuram, South Kerala. *J. Earth Syst. Sci.* **2010**, *119*, 507–517. [CrossRef]
98. Twomey, S. Pollution and the planetary albedo. *Atmos. Environ.* **1974**, *8*, 1251–1256. [CrossRef]
99. Twomey, S. The influence of pollution on the shortwave albedo of clouds. *J. Atmos. Sci.* **1977**, *34*, 1149–1152. [CrossRef]
100. Xu, X.; Gregory, J.; Kirchain, R. Climate impacts of surface albedo: Review and comparative analysis. In Proceedings of the Transportation Research Board 95th Annual Meeting, Washington, DC, USA, 10–14 January 2016.